Frontiers of Development in the Amazon

Frontiers of Development in the Amazon

Riches, Risks, and Resistances

Edited by
Antonio Augusto Rossotto Ioris,
Rafael R. Ioris, and Sergei V. Shubin

LEXINGTON BOOKS
Lanham • Boulder • New York • London

Published by Lexington Books
An imprint of The Rowman & Littlefield Publishing Group, Inc.
4501 Forbes Boulevard, Suite 200, Lanham, Maryland 20706
www.rowman.com

6 Tinworth Street, London SE11 5AL, United Kingdom

British Library Cataloguing in Publication Information Available

Library of Congress Cataloging-in-Publication Data Available

ISBN 978-1-4985-9471-4 (cloth)
ISBN 978-1-4985-9472-1 (electronic)

This book is dedicated to those who have been severely impacted and continue to experience the socio-ecological devastation caused in Amazonia in the name of development. Their voices remain largely unheard.

Contents

Acknowledgments

This book is the main outcome of the workshop "The Historical and Cultural Trajectory of Agricultural Frontiers and the Subjectivities of Regional Development in the Amazon," held at Cardiff University in May 2018, which was part of the activities of the Agro-cultures International Network funded by the Arts and Humanities Research Council (AHRC) of the United Kingdom, RC Grant reference: AH/R003645/1. AHRC's support is gratefully acknowledged and appreciated by all academic and nonacademic members of the network. We would like to acknowledge also the sponsorship from the Newton Fund and the British Council under the Researcher Links Workshop Grant Agreement 204503171 on the theme of "Supporting Sustainable Ecosystems for Poverty Alleviation in the Amazon," which enabled the organization of an international interdisciplinary workshop in Manaus in June 2016, attended by early career and researchers from Brazil, Colombia, United States, and the United Kingdom, which likewise laid foundations to this edited volume.

Sergei Shubin would like to acknowledge the support from the UK-ESPA programme (Ecosystems Services and Poverty Alleviation—grant code: NE/I004467/1). I am grateful to the colleagues from the Universidade Federal do Pará, EMBRAPA Amazônia (all Brazil), Instituto Boliviano de Investigación Forestal, Universidad Autónoma Gabriel Rene Moreno (all Bolivia), and especially Dr. Antonio Ioris and Prof. Michelle Pinard (then at the University of Aberdeen) for their help with the project design, collection, and analysis of data. I would also like to thank all of the local officials, practitioners, and local residents in the Amazonian communities who gave their time to work with us and support this research.

Chapter 1

What Is New in the Amazon and What Is Amazonian in the New?

The Twenty-First-Century Brand-Old Frontiers of Exploitation and Development

Antonio Augusto Rossotto Ioris and Rafael R. Ioris

THE RECURRENT OLD FRONTIERS

The Amazon represents one of the world's most complex and consequential living spaces, with a wide range of unique socioeconomic activities, cultural realities, and politico-ecological challenges. Far from homogenous, the region comprises breathtaking animal and vegetal diversity, along with a multitude of landscapes, peoples, languages, cultures, and social compositions. The highly diversified Amazon biome occupies close to half of the vast South American continent, spreading across nine countries and encompassing around 30 million inhabitants. For at least 10,000 years, the interactions of different human populations over such a large territory (around 7 million km^2) have helped shape biological and socio-spatial features of the region, fostering the propagation of new species of fauna and flora that still help to define the uniqueness of this complex, yet fragile, environment.

The complexity of the Amazon is indeed legendary, though not properly understood by disciplinary scientific approaches, which are firmly based on binary notions of Western rationality that tend to dissociate society from the rest of nature, scientific findings from experience-based knowledge, and development from community and subsistence practices. For a long time, the region has been paradoxically and simultaneously celebrated and ignored, praised while also feared and even attacked by scientists and intellectuals who struggled to make sense of its intricate socio-ecological configuration. Hence, the urgency of considering lived and contested geopolitical realities of the Amazon as perennial spatial, material, and ontological frontiers. Specifically, frontiers that are there, signaled and apparently within reach, but in which there are always more terrains and resources beyond what had been seized and where the (metaphorical and empirical) explorer is constantly overtaken by dramatic and challenging events. An exploration into some aspects of the enigmatic features of the Amazon frontiers is precisely our quest and the intellectual journey proposed and presented in the pages of this book.

To begin, it should be clear that far from an isolated, pristine, lush green terrain around an ocean of freshwater, the Amazon has been a dynamic stage where intense human activity has taken place, with consequences not only for the region itself, but also for the world as a whole. From Orellana's first expedition down what was called by Europeans as the Amazon River in the first century of colonization, to today's accelerated capitalist encroachments via global commodity and consumerism chains, the region has persistently elicited degrees of fascination and wonderment, counterpointed by equally expansive transformations and challenges. In effect, from awe-inspired map-making efforts, in the mid-1500s, to the imperial plans for occupation and political oversight, in the mid-1700s,[1] the Amazon River basin was steadily incorporated by intruding foreign powers into transatlantic commercial activities. After emancipation from formal colonial rule by the kingdoms of Portugal and Spain (note also the sustained territorial aspirations of France, Holland, and Britain in the region) early in the 1800s, the political regional context expanded into constant struggles over territorial control, wherein increased human occupation and multiple economic undertakings were displayed.

If negotiations and conflicts between different social groups, as well as fluctuating civilizational dynamics, have marked the region since time immemorial, it is equally clear that Western, colonial, and more recent globalized capitalist projects have restructured the region in unprecedented ways, especially over the last half century. These historical dynamics involved different moments of great historical, anthropological, and geographical transformations, which could be described as the materialization of intense frontier expansion. This continuous process of frontier-making, defined by the confluence of multiple patterns of socio-ecological change, the production of novel spatial configurations, and expanding the boundaries of cultural and ecological activities, has been an integral driving-force throughout the historical trajectory of the Amazon region.

The perpetual, uneven development of capitalism and the recurring production of "old-new" frontiers in the region have been objects of debates that have occupied critical scholars working in several different disciplines for more than a century. The spaces of capitalist frontiers—*frontspaces*—have been repeatedly studied by multiple sets of researchers, whose interpretations range from the neoclassical enthusiasm for the economic outcomes of new frontiers to critical denunciation about growing proletarianization and acts of resistance on the part of various segments of local populations. Although often used interchangeably, Watts (2018) differentiates between "frontier" as a zone of socioeconomic advance and "border" as a line of demarcation between national territories or administrative units. Lund and Rachman (2018) similarly distinguish dynamic frontiers (free-resources zones, where social order is eliminated, property is disrupted, and social contracts are dissolved) from territorialization (spaces with new systems of authority and regulation).

Taking into account this bourgeoning yet still insufficient academic literature, and especially aimed at presenting the varied and multidisciplinary reflections advanced in the different chapters of this book, this introductory chapter reviews the politics, broadly conceived, of frontier-making in the Amazon. First, it is important to note that we conceive frontiers as biophysical and human spaces increasingly subjected to cumulative transformations due to rapid and intense migrations of peoples and the opening of new economic opportunities taking place in contexts where authority and governance are constantly disputed and renegotiated. This is the analytical frame we propose in order to critically examine the manifold locales and landscapes undergoing changes due to the continuous encroachment of new socioeconomic processes dialectically connected to wider trends and broader socioeconomic, political, societal, and cultural dynamics unfolding in various ranges and multiple scales throughout the Amazon.

Central to our analyses is the always complex interfacing between relatively consolidated areas and the so-called frontiers. This involves multiple mechanisms of innovation occurring amid continuation of old patterns of occupation of territory, socioeconomic structures, and political rationalities, constantly at work and consistently manifested along multidimensional and contradictory geographical dynamics. What is more, spatial frontiers are constantly redefined within processes of production, extraction, and politico-ideological containment, not only because of favorable economic opportunities but mainly due to the need to stabilize and reinvigorate core economic and political trends. Especially relevant, this interconnected process of frontier-making has not been interrupted by market-based globalizing trends, since the materialization of new frontiers continues to be predicated on rising tensions in central economic areas and their unstoppable demand for goods and services. In effect, at the frontiers, capitalism can reassert its hegemony with much lower costs and fewer restrictions, since order, authority, and convention are suspended, time and space are reconfigured, and excesses are committed—as Martins (2009: 9) appropriately describes, frontier expanding into the Amazon is the "scenery of intolerance, ambition and death."

The inexorable interdependency between center and frontier is evidently a process with major ideological and political significance (Ioris, 2018). Whereas, in purely conceptual terms, the evolution of the frontier could take many directions and lead to multiple social arrangements, what is witnessed in practice is a great deal of constraint due to the continued hegemonic presence of global capitalist relations along the historical lines of a self-fulfilling prophecy consistently defined by the winners. The Amazonian experience is particularly paradigmatic for this line of reflection as it is increasingly clear that the region has become one of the most disputed frontier tied to the encroachment of Western type of modernity over global territories in the last few decades. This is a phenomenon marked by structural socio-spatial asymmetries as the exploitation of resources in the Amazon, which are transformed and consumed elsewhere, producing a constant loss of value as gains are accumulated outside the region itself (Bunker, 1985). Ironically, this process is regularly carried out under the guise and compelling symbolism of notions such as abundance, anticipated enrichment, and a bright future for all involved; all of which work as powerful narrative devices acting as effective legitimization tools for deeply destructive and socially and environmentally unbalanced outcomes.

Development policies have consistently demanded intensified levels of exploitation of minerals, timber, water, biodiversity, etc., as well as the expansion of plantation farms and the construction of gigantic dams (such as the controversial Belo Monte dam in the lower Xingu River, in the State of Pará, Brazil, the largest of a series of planned schemes along many Amazonian rivers). Thus, new navigation routes tied to intense extractivist activities and farming projects have been devised by policy-makers in order to promote and support higher levels of human presence and economic activity in the region. And although the Amazon has often been perceived by outside intervening forces, such as central agencies of government and international investors, to be persistently mired in unforeseen difficulties, over time such efforts have consolidated awareness and interest in the area's supposed untapped riches.

The extension of the environmental degradation expanded together with the devaluation of the traditional knowledge held by local communities (D'Incão and Silveira, 1994). Peoples and ecosystems previously largely situated beyond the easy reach of mass markets have been increasingly involved in global socioeconomic, cultural, and political dynamics. Likewise, powerful agribusiness activities have, since the mid-1970s, converted vast areas of the Brazilian Amazon (which contains around 80 percent of the vast South American river basin) into cattle ranches, soybean farms, and sprawling towns (Acker, 2017). The new context of globalization, structural adjustment, regional integration, and rapid technological change contributed to accelerate deforestation, beyond the capacity of environmental policies developed in the previous decades designed to address the problems associated with intensive production in tropical agro-industrial frontiers, as in the Amazonia of Bolivia and Brazil (Hecht, 2005).

It is the case that, since the 1990s, there has been a more accurate awareness of the need to change hegemonic trends and encourage more sustainable policies and practices (Hall, 2000). However, because of the balance of power behind the narrative of sustainable development, conservation efforts have normally led to the expansion of state authority and the exacerbation of regional imbalances (Bratman, 2019; de Sartre and Taravella, 2009). Because the majority of the regional population has benefited much less from the belated insertion of the Amazon into national development and integration agendas, we are left with many intriguing questions.

For one, given that the promises of development have been largely proven false, how can forest and regional ecosystems be removed from a spiral of socio-ecological impacts and persistent poverty-making? Similarly, how can we alter the path-consolidated processes of social and environmental justice that have consistently benefited local and national politico-economic elites, most recently through the language of sustainability and progress? Finally, what sort of economic and social relations should be pursued as an alternative so that respect for regional human and biological peculiarities can be attained and long-term preservation of ecosystems guaranteed?

In the pages and chapters that follow, we critically engage with these and related questions on the basis of concrete examinations of multiple cases and experiences taking place in the rich, complex, multidimensional, and challenging reality of the Amazon.

Developing the Amazon: Elusive Goals and Reiterated Justifications

Development, as conceptualized in the post–World War II period, is normally associated with the intensification of the production of goods and services according to the patterns of consumption and waste production, as well as cultural conceptions grounded in and constitutive of Western societies (Arndt, 1987; Escobar, 2012). Along these conceptual lines, the process of mainstream development in the Amazon has historically relied on the construction of large-scale infrastructure and the operation of business enterprises, supported and/or executed by national governments.

Yet, contrary to the critique often raised by NGOs and environmental groups, the Amazonian process of development does not have to be inherently anti-ecological (i.e., in the sense that it demands the destruction of forest ecosystems and their replacement with radically different landscapes), nor should it necessarily be rejected (i.e., following romantic anti-development calls that ignore the material and social needs of deprived populations). In effect, we alternatively believe that the complex realities of the Amazon cannot be apprehended, let alone improved with simplistic and homogenizing frames of analysis but rather demand closer scrutiny and lucid reflections such as those advanced by the different authors here.

It should be noted that it is possible to identify significant shifts and new social experiences, such as the supposed transition to a "green frontier." But although relevant and insightful, these trends will continue to be partial and vulnerable without concrete evidence of real changes in government action and the mentality of decision-makers (Schmink et al., 2019). It is thus, on the grounds of the many urgent needs of the region that we have sought to understand better and critically review the kinds of development plans that have been pursued in the Amazon, which groups, agencies, and companies have mostly benefited from the policies consistently imposed on its peoples and cultures, and which socio-ecological presuppositions have informed how these plans and initiatives have been devised and implemented over time.

In concrete terms, we scrutinized the purported universalistic basis underpinning mainstream development platforms that tend to ignore local circumstances and consistently seek to produce homogenous human and ecological spaces according to foreign values, aesthetics, and needs. Ultimately, we engaged with a complex reality defined by the contradictory dynamic in which the more the Amazon has been inserted into the conventional path of development and the more developed it is said to have become, the more restricted and exclusionary its politico-economic opportunities actually become, thus undermining the prospects for more autonomous and sustainable paths of development.

As alluded earlier, the Amazon region is becoming increasingly, and often violently, inserted into global markets by means of the appropriation of territorial resources and exclusionary policies imposed over its indigenous inhabitants. It goes without saying, that these actions unfold at the expense of environmental conservation and are deeply tied to the expansion of inequalities and social stratification in the region. What is more, the continuous encroachment of Western-inspired conceptions of modernity into the Amazon has been combined with recurrent cultural dynamics shaped by both profound estrangement and sheer fascination in Western eyes. This included legends such as that the Garden of Eden was located in the middle of the forest, that the much sought-after fountain of youth could be found in the region, and that the bounties of *Eldorado* were therein to be found. Finally, and particularly relevant, the alleged existence of uniquely beautiful and powerful women living in an exclusively female world that helped create the tale of the revived society of the Greek Amazons (Slater, 2002). Over time, these mesmerizing narratives became the generic and distorted reference for the region as a whole, obscuring its diverse and complex internal dynamics (Holanda, 2000).

Thus constantly reiterated and/or recurrently pursued, this pattern of territorial conquest and socio-ecological exploitation in search of countless riches and limitless gains sustained a long history of outside ventures into the region. Largely thanks to new technological capabilities, political arrangements, and global economic fluxes, this path of continuous incorporation of the Amazon into ever more capitalist developmental lines has significantly accelerated and become immensely more intricate in the last fifty years. This mainstream and ever more hegemonic model of development greatly depends on access to ever newer territories and ecosystems. And it is precisely the availability of land and resources that fuels the expectations of rapid capital accumulation by encroaching cattle ranchers, agribusiness, miners, loggers, construction companies, dam operators, etc.

Given the novelty of many of these historical dynamics, in both their empirical and analytical dimensions, it is important to point out that far more than profit extracted from production, the process of development in the Amazon relies to a rapidly accelerating extent on today's path of integrating the region into global economic chains on the basis of the appropriation of differential rent (in the Ricardian sense) from forested land and river systems. Indeed, in contrast to colonial and early postcolonial times, when the riches of the Amazon were captured and removed, conventional contemporary developmental initiatives are based on the primacy of private ownership of land in the region and the associated reorganization of social relations in function of the political power of private owners of farming, mining, and industrial enterprises, or similarly in function of state-owned property serving economic growth, such as dams, roads, and ports.

The operation of these manifold economic activities has become an integral force of sociocultural change, and their negative impacts have been the object of biased, market-based forms of environmental regulation that in practice only marginally mitigate the fundamental socio-ecological problems of development. Some of the most harmful attributes of the process of development are its anti-nature and anti-commons imperatives, that is, the dominant model of development depends on the conversion of collectively owned and jointly managed tracts of the Amazon into the sphere of private property relations for short-term gain at any socio-ecological cost. The central contradictions of regional development are directly and primarily related to this simultaneous control and exploitation of the human and nonhuman dimensions of the Amazon region (Ioris, 2016).

In effect, articulating the logic of efficiency by economists working on anti-commons theorizations, capitalist modalities of development have consistently codified inequality and perpetuated poverty, regardless of the level of environmental conservation and restoration of forests. This path of change under the aegis of development policies unfolded through many different scales of eco-class-race struggles that propagate and reinforce poverty and inequality. And due to these growing powerful anti-socio-natural pressures, the forest is being brutally altered (in symbolic and material terms) from its long-established, dynamic condition, giving rise to both continuous and persistent environmental disruption and social exploitation.

Along these historical lines, far from being a remote land of exuberant biological formations that fascinated explorers for many centuries, as indicated above, the Amazon region has been brought, always in a subordinate position, to the center of national development policies aimed at replicating similar technological, institutional, and political approaches. This implies that wealth is increasingly accumulated not only at the expense of the appropriation of land and territorial resources, but more importantly through the dissolution of collectively shared opportunities and the reduction of socio-ecological complexity to the sphere of market transactions (which results in socioecological degradation and the reinforcement of poverty). In the end, private land concentration has also intensified sociopolitical conflicts and violence against the poor.

And though often still defined by the condition existing at the margins of the so-called *civilization*, centers of economic activity, and national territories, the very internal dynamics of the Amazon has always involved in ongoing liminal negotiations and conflicts between different groups of people. What is more, in the last few decades, the region has maintained a subservient stance, both in terms of economic and political interests, especially in the Brazilian case—as richly demonstrated by several of the upcoming chapters. As a result, instead of being the permanent and ever-elusive final frontier, in the sense of an isolated, marginal place, the Amazon needs to be understood as the newest location for the confluence of global economic, political, and cultural conflictive dynamics that are intensified according to a logic typical of frontier regions. These are social and spatial realities experiencing the exacerbation of conflicts existing elsewhere but that are dramatically accentuated in the often-violent collisions of worlds that constitute developmental, social, and cultural frontier dynamics.

Much in the same way, along ever more intrusive, Western-defined developmental lines, poverty and affluence are constant outcomes of the process of appropriation and transformation of forest ecosystems; a path that greatly enhances patterns of social stratification. Governmental economic incentives (such as credit, subsidies, and the granting of private property) and infrastructural investments (in the form of roads, ports, and warehouses) have played central roles in these processes. These inducements have sought to attract different contingents of people to the Amazon, who, once there, have only marginally benefited from the ongoing development initiatives deployed therein, especially during recent decades, themes that are also examined in the several different reflections included in this book.

The end result of the acceleration of these complex, multidimensional, uneven, contradictory and unsustainable dynamics is that Amazonian landscapes today are shaped by massive inequalities and significant asymmetries spreading from the local to the regional, to the national and international scales. So much for the so-called development and integration of the Amazon!

Together, these are some of the many dynamic and challenging socioeconomic, political, cultural, and environmental realities that the different chapters here will try to assess. In effect, considering the above observations, it should be noted that the following case studies are all grounded on extensive empirical research and reflect the disciplinary frames of each of the different contributors. This collective effort also aims to update and expand previous approaches to frontier and frontier-making in the Amazon, such as the highly influential work of Schmink and Wood (1992), particularly considering the preeminence of agribusiness and the associated flows of capital in the region.

We have tried to provide some general analytical lines, mostly along the notion of what is meant by the, broadly conceived, expanding frontiers of the Amazon, with the aim of fostering a coherent effort of reflecting about the manifold complexities of the Amazonian human and natural realities. Yet, we accepted the idiosyncratic features of each reflection as an asset of a collective project conceived as having a multidisciplinary, as well as interdisciplinary repercussions.

Rationale, Aims, and Scope of the Book

Although the Amazon region has always been the subject of greed and controversy due to the advance of capitalist relations of production and reproduction, this reality has never been as clearly manifested and as challenging as it is today. In the final years of the second decade of the twenty-first century, the Brazilian Amazon in particular finds itself in a political, economic, and civilizational crossroads. Mapping out this context by revisiting long-term and recent historical developments, as well as making sense of today's national, regional, and local realities by means of an examination of the region's dramatic expanding economic, political, social, and cultural frontiers are much needed scholarly initiatives, and this was a central goal guiding the authors combined in this book.

The paradoxical reality of rich ecosystems rife with poverty provided us with several questions dealing with the purpose and the actual beneficiaries of the process of development and the premises and interventions of local, regional, and ever more global state and private actors. The overarching notion of expanding frontiers proved useful in structuring the following works. In this sense, we are confident the present book represents an innovative attempt at advancing new perspectives on the huge challenge of assessing the developmental trajectory of the Amazonian region in a critical, multidimensional, and interdisciplinary fashion.

The present book embodies many outcomes of a new international collaboration involving academics and nonacademics based in South America, the United Kingdom and other European countries, and the United States. The authors have been engaged in several research projects and taken part in various events and joint activities. An introductory seminar took place at the Federal University of Amazonas (UFAM) in Manaus in June 2016, supported by the British Council, the Newton Fund, and the Amazonas State Research Support Foundation (FAPEAM), which involved senior and early career academics working on the interface between poverty and ecosystem services.

An additional preliminary workshop on Amazon frontiers brought a small number of colleagues to Cardiff University in July 2017 (supported by the university), which paved the way for the formation of the Agrocultures network, funded by the UK-Arts and Humanities Research Council (AHRC), which has allowed the organization a series of workshops in South America and the United Kingdom attended by social and socio-ecological scientists. It has stimulated and facilitated the involvement of numerous colleagues, researchers, and graduate students in shared critical reflection and investigation of the Amazon's complexity.

In the first event of the Agrocultures series, which took place at Cardiff University's School of Geography and Planning in May 2018, the proposal to organize this edited book was agreed among the main participants, as well as the decision to extend the invitation to colleagues involved in the network (more at www.agrocultures.org). The invaluable assistance of those grant schemes and the institutional support from the various universities where the authors are based are strongly and warmly acknowledged. The convergence of scholars trained in different disciplinary traditions and based in multiple parts of the globe was intended to address the unevenness and risky patterns of regional development, the centrality of environmental conservation and sociocultural recognition, and the need to promote more inclusive, just, and sustainable development strategies throughout the Amazon.

We indeed believe that despite growing scholarly work among social scientists on socioeconomic, agrarian, and demographic aspects of the advance of the Amazon's agricultural frontiers (particularly in Brazil), we are still in need of better analyses and reflections about how global influences and past legacies are incorporated and retransformed in the production of alterities and cultural expressions. We are likewise certain that by framing the analyses contained herein within the broad notion of dynamic and perpetually unfinished frontiers, the reflections advanced in this book will shed new light on different elements involved in the recent larger-scale and accelerated path of the encroachment of mainstream development into the Amazon region. Much in the same way, the authors here are confident that their contributions are original and relevant to any sound assessment of the ever-more complex and challenging realities of the Amazon.

What is more, we share the conviction that social scientists, broadly defined, fully and meaningfully engaged with the diversity of civil society, government, and activist communities across multiple scales, can greatly contribute toward resolving environmental and developmental dilemmas in the Amazon region, particularly considering the aggressive advance of agricultural modernization and the movement of people to new areas, which have resulted in complex cultural exchanges, rapid erosion of old practices, and the formation of new, hybrid identities (Ioris, 2019). In the case of the growing number of cattle ranches, Hoelle (2015) has demonstrated that cattle are about much more than beef production and deforestation, but it is deeply associated with the spread of a vibrant "cattle culture," a patriarchal and cattle-centered vision of rural life and regional development that exacerbates tensions with conservation and undermines the prospects of social inclusion.

Finally, we see that governments at national, regional, and local levels, as well as the international community represented in multilateral bodies of governance, all share a clear responsibility to swiftly respond to the emerging challenges presented by impacts on local traditions, customs, and knowledges, mounting socioeconomic inequalities and ecosystem degradation, and the lack of proper recognition of the agency, distinctiveness, and subjectivities of subaltern groups of old and new residents (such as extractive communities, squatters, family farmers, and indigenous groups) in areas of agricultural frontier.

Having taken these multiple challenges into account, we aim to continue to strengthen international collaboration around the study of the history and geography of development and sociocultural diversity in areas of agricultural expansion and economic intensification in the Amazon region, paying special attention to local interactions, socio-spatial inequalities, legacies from the past, and the lived experience of urban and rural communities. We expect that the activities of this group of academics and activists will continue to contribute toward much-needed academic and political debates on sustainable development alternatives and more inclusive economic options for the peoples of the Amazon region and beyond.

Main Contributions of the Book and Chapter Structure

The prevailing nonacademic narrative of the economic and sociocultural frontiers in the Amazon is derived from a social imaginary dominated by the supposed heroism, entrepreneurialism, and economic success of large-scale farmers. It is true that the vast scholarly literature on the topic contains a near-universal criticism of the attempts at heroic insertions of grant planning and capital influxes in the region; however, the perennial unfolding of development frontiers, the range of environmental injustices and the interpersonal connections between different social groups, their subjectivities and cultural baggage, emerging identities and household practices are still not adequately considered.

A significant anthropological literature on indigenous tribes, covering elements of their oral histories and cultures has emerged. But little is known about their participation and reaction to the advance of agricultural frontiers. Likewise, the Amazon still seems to be occupying an ambiguous place between the ancient and modern worlds, not ancient enough yet not modern enough, which has resulted in the material culture of the modern region being largely absent from anthropological (and in fact historical) studies. In fact, the Amazon has thus far been mostly defined by its natural rather than its cultural histories, something that can be observed in recurrent representations of the region mostly as a physical, largely natural setting, thus only marginally taking into consideration its rich and vast cultural dynamics.

This being the case, one of our departure points here was the conceptualization of Amazonian frontiers as cultural melting pots where national and global influences have clashed and generated increased social heterogeneity. There is, to be sure, a burgeoning interdisciplinary agenda of global agriculture and food systems, which calls for ground-breaking work on the study of cultural, interpersonal, and historical dimensions. Thus, by conceiving culture from its margins, from liminal sociabilities, from hybridities, and in-betweens (or, as it were, through frontier-thinking), we seek to contribute to the reflexive project of enlightening ourselves about—and thus relativizing—its basic, often tacit, assumptions.

The result of these complex historical events is a social and cultural hybridity, which is an increasingly relevant topic for social and interdisciplinary scholars, both in theoretical and empirical terms. In effect, we see our work in general, and this book in particular, as a gateway to new understandings of historical, cultural, linguistic, and religious components of agricultural frontiers, as spaces of dramatic socio-ecological transformation, and where old traditions combine and clash with hyper-modern trends. Culture is conceived here neither as a homogeneous entity, nor as a container comprehending either meanings or peoples (depending on one's theoretical preference), but rather as an internally coherent collection of communicatory processes and resources that enable, constitute, and organize the sociality and socialization of different groups of stakeholders.

The activities leading to the writing of this book were significantly inspired by the goal of fostering three main lines of reflection and engagement:

1. Emerging cultures and interpersonal subjectivities of frontier-based interactions: In seeking to promote new ways of understanding the complexities involved in the so-called frontier-based personal and collective dynamics, we recognize the dynamic formation of identities, values, and otherness in agricultural frontiers, as the result of population migration, multiple struggles, and intensive interaction with socio-ecological change. This involves complex politics of belonging formed by institutional forces, imaginaries and negotiated relatedness to one another, and individuals' standing in the emerging social tissue. Regrettably, much of the existing academic literature has almost no examples of studies that investigate and contrast the subjectivity of different stakeholders, especially the values, aspirations, and frustrations of large-scale farmers and landowners. Interpreting the new and ever-dynamic agricultural frontier requires a proper consideration of the genesis of personal and social subjectivities, and the extent to which interactions between individual subjects, organizations, and the environment produce shared meanings, consciousness, and personal identities. How different socioeconomic and political frameworks and concrete dynamics unfold in the light of these cultural, personal, and collective subjectivities is a central component of the different reflections pursued in the chapters that follow.
2. Frontiers as socio-ecological territories: The second intellectual pillar of our work deals with the politico-ecological significance of environmental change, agricultural expansion, and the mobilization of natural resources, which has come into sharp focus in recent years as major uncertainties exist around the sustainability of production and distribution systems, as well as a focus on issues of justice and equity in conventional economic systems. The political ecology of development is a growing field of investigation, as it integrates different approaches to study the combined historico-geographical accounts with political and socio-cultural factors. Place-based interactions and local contexts have also increasingly been recognized by political ecologists, as it matters to all where and how nature is defined, transformed, and disputed. Research conducted in the Amazon demonstrates that regional development and economic intensification trends are translated into values and practices at the local level, which affect political mobilization and the ability of farmers to adopt technologies and respond to pressures.
3. Food sovereignty and environmental justice in frontier regions: Our work here was also inspired by the rapidly growing literature, based on transdisciplinary theoretical and empirical work, on food sovereignty and environmental justice (starting with interrogation of the actual contribution of agribusiness expansion and the search for socio-technical alternatives). Emerging from radical social movements and giving direct

> voice to the concerns of land-based communities and civil society, this literature emphasizes the autonomy of local communities and a critique of techno-centric approaches that neglect the socioeconomic, political, and cultural dimensions of agricultural lifestyles. In particular, the recognition of food sovereignty (the right of peoples to healthy and culturally appropriate food) as a socio-spatial relation, whose conceptualization includes the rights of nations, peoples, regions, and states to craft agrarian policy according to their culture and multiple values.

Informed by the academic and activist discourse of food sovereignty and environmental justice, the texts herein make an essential contribution to the emerging role of multidisciplinary studies on agri-food systems and rural/regional development. That is achieved explicitly in the text of several chapters dealing with food and agriculture and their associated land-rights struggles, threats to indigenous cultures, violence perpetrated against environmental activists and the uneven distribution of the environmental impacts of development, as discussed in other parts of the book. What is more, illuminated by these three important contributions, the discussions and ensuing analyses in these pages have sought to address the intensive cultural exchanges and historical transformation of agricultural frontiers in the Amazon.

The project has represented a space for discussion and North-South interaction between academics and nonacademics focused on local knowledges, habits, languages, and subjectivities in the context of a rapidly changing spatial order under the influence of Western-like modernization. We recognize that neither governments nor frontier communities have so far formulated stable routines for dealing with sociocultural estrangement, and that the multiple and constantly evolving new alterities and emerging identities in the region call for better understandings of the sociocultural and intersubjective uniqueness of agricultural frontiers so as to deal with economic, ecological, and social dilemmas.

Addressing these issues requires moving away from the reductionist approaches that have thus far characterized the work of policy-makers and development agencies working on the unfolding development and agricultural frontiers in the Amazon. We believe that any valid new interpretation about the region needs to take into account the growing importance of the continually redefined yet persistent collective bonds among local communities, their access to lands, and the many social issues tied to making agriculture production more sustainable, just, and resilient. Equally, social and environmental justice in these complex frontier regions of the Amazon call more than ever for a proper consideration of sensibilities, identities, and sensations, which are not simply epiphenomenal to the economy, but rather reveal the continuing complexities of uneven and exploitative sociopolitical arrangements and cultural dynamics. All in all, guided by these manifold notions and goals, this book is structured in three main areas, as follows.

In the part I, "Thinking and Making of the Amazon Frontier," three chapters revisit the conceptualization and political justification behind the promotion of regional development, particularly in the last half-century. First, A. Ioris reflects on assumptions and practices during the occupation and exploitation of the Amazon region and uses those controversies as background to propose a theoretical contribution to frontier theory. The spatiality of frontier-making goes beyond the more immediate relocation from "core" to "periphery," entailing a deep relational interaction between old and new areas. The chapter discusses the interdependence between center and frontier, and suggests that this happens through the "law of scarcity-abundance" (*Political Geography*, 65, 98–106, 2018). This "law" synthesizes the general tendency to deal with mounting scarcity in central areas through the pledge of abundance at the frontier, although in practice new rounds of scarcity emerge in both areas due to the internal dynamics of capitalism (notably, the exploitation of society and the rest of nature). This means that the evolution of capitalist relations of production and reproduction is also, and fundamentally, based on accumulation through frontier-making. This conceptual framework is then applied to Brazil, a country largely shaped by territorial conquest and the expansion of internal economic frontiers. The State of Mato Grosso, in the southern tract of the Amazon, has been at the forefront of frontier-making for many centuries, recently accelerated by the spiraling growth of neoliberalized agribusiness. Mato Grosso may now have reached the center of the national political and economic landscape because of the crucial importance of agribusiness exports, but nonetheless it remains a frontier space where abundance and scarcity continue to jointly materialize. Frontier-making never ended in Mato Grosso, but remains a persistent necessity, much more than a simple contingency.

In the following chapter, Freitas examines the controversial designation of conservation areas in the Amazon, as an emblematic example of the compromise between zones where development was a priority and other areas considered less important or less suitable for the plans of the military government ruling Brazil in the 1970s. The history of the creation of parks and reserves—carefully examined in the text with reference to a wealth of plans and legislation—reveals how processes of state-led frontier expansion can paradoxically create conditions for the introduction of ambitious conservation policies. The introduction of new objectives and strategies for Amazonian development not only brought technical and logistical competences, but also fostered new spatial representations of the region, which allowed park proponents, within and outside the government, to identify areas to be conserved in the spaces between proposed development zones. It is widely accepted today that

Amazonia constitutes a fragile socio-ecological frontier that requires protection, and this view frames any conversation, planning, and policy related to the regional ecosystems. However, this is still an unfinished process, with a good number of protected areas still to be implemented, existing only on paper, which tells us a great deal about the overall controversies and ideologies of development and national integration.

This is followed by a discussion by Castro on community-based management systems (CBMSs), which have often been grounded on monolithic institutional, social, and ecological perspectives focusing on the commons as the only local actor, collective territorial rights as the only local tenure system, and the managed resource unit or ecosystem as the only contested resource driving collective action (a previous version appeared in 2016 in the *International Journal of the Commons*, 10(1), 1–20). However, CBMSs are embedded in local social-ecological systems usually characterized by multiple ruling systems, different local groups, and heterogeneous ecological systems. The chapter examines how the floodplain tenure system is negotiated and rearranged between two local groups—community residents and large landholders. This complex and dynamic arrangement comprises three layers of property rights, which are combined according to changing ecological and social contexts in the Amazon. Based on longitudinal empirical data spanning twenty years of research, the chapter describes the history of contemporary human occupation, and the most recent regional socioeconomic and institutional changes, in order to unpack the dynamics of the floodplain tenure in the region. The author concludes that it is overly simplistic to assume that integrating local management systems into a formal legal framework will be sufficient to achieve an efficient co-management system. Despite major structural changes in the formal tenure framework, power relations between different local users may remain unchanged unless local perceptions and everyday practices of power relations are challenged. Unpacking the multiple ruling systems and everyday practices that mediate interactions between different local actors is fundamental to understand how the commons are appropriated at the local level. Therefore, a local contextualization of the social and ecological structure is crucial to reveal potential barriers to the development of an inclusive and sustainable production system.

The part II, "Questioning the Basis of Progress, Power, and Poverty," approaches frontier dynamics with a focus on claims that economic growth is a surrogate for development, and also on the multidimensionality of poverty amid extremely rich ecosystems. R. Ioris and Schneider examine how, over the last quarter-century, Brazil has reinserted itself in the global economy with a booming agribusiness sector, but with a subsequent bust revealing the distortions and vulnerabilities of combining technologically advanced agriculture with oligarchic-style political economy. As Brazil lost its hard-fought status as a manufactured goods exporter and stepped back into its traditional role as a major commodity exporter, Brazilian agricultural elites 1) modernized their activities by means of technological innovations and more complex global trade and international agribusiness alliances, and 2) reasserted their prominence as central economic and political players. This process included a huge expansion into new agricultural frontiers, this time more than ever around the Amazon region. These developments present several ironies. First, oligarchic emergence occurred amid the strengthening of the country's formal democratic institutions. In effect, rather than being forced to concede to a more sustainable path of territorial occupation and a more labor-inclusive mode of production, latifundia and traditional land-based oligarchic elites exploited their importance as generators of export revenues and control of political office. When economic fortunes and the dominance of a coalition of left-wing parties dipped, these oligarchic elites were the first to abandon the government, pursue undemocratic control of power, and seek to reverse the few social gains they had allowed during the boom. This chapter addresses these processes from a critical perspective, paying particular attention to both the economic transformation taking place in agribusiness in the Amazon frontier and the political dimensions of recent agriculture-led development.

The following chapter complements and expands the analysis of the two previous texts. Schor and Azenha consider the contradictions of agri-food policies and the evolution of food acquisition and eating habits. Changes in eating habits are a strong indicator of changes in perceptions, uses, and engagements with nature, providing a useful vehicle for examining the gap between the myths of sustainability and the reality of rapid urbanization and changing livelihoods in the contemporary Amazon. The authors argue that a political ecology of food regimes in the Brazilian Amazonian floodplain—drawing on anthropological and geographic approaches and insights—provides a privileged vantage from which to illuminate the contradictions of current development trajectories and the socioenvironmental disparities they engender, potentially contributing to the articulation of more effective sustainable development and social inclusion policies. The conclusion is that more inclusive and sustainable development approaches require genuine democratic participation in environmental governance and development planning in terms of alleviation of poverty, addressing the issue of land concentration, addressing educational and infrastructural inequalities, promoting diversification of agricultural and extractive enterprises, and the continuation of efforts to promote an increased valuation of agricultural production of food crops.

In the subsequent chapter, Shubin explores the relationships between deforestation and poverty in the context of "contested frontiers" in the Brazilian Amazon. The author makes use of post-structuralist conceptual approaches and Actor-Network Theory to understand the ways in which collective forces generate socio-natural systems in the Amazon and produce poverty. First, the chapter considers different temporalities and spatialities of poverty emerging as a result of historical and circumstantial combinations of people, natural entities, and forces. It explores the role of institutional power in setting up boundaries and determining where and when certain activities may or may not take place. Building on memories, histories, and expectations, the chapter considers the effects of conflicting and contradictory land claims on natural resource use and the creation of knowledge about ecosystems, sustainability, and poverty. Second, it studies the process of gathering and coming into being of socio-natural systems and uneven topographies of poverty developing in various parts of the world. Drawing on fieldwork examples form a year-long study in the Brazilian Amazon, the chapter reveals different scales, energies, and countercultures of poverty that exceed existing categorizations of disadvantage. The chapter raises theoretically informed observations about the possibilities of creative and collective reworking of ecosystem management and alternative approaches to poverty alleviation.

That is followed by an analysis by Holland of the controversial expansion of illegal gold mining in the Peruvian Amazon and the range of impacts and reactions involved. The chapter discusses the national context under the influence of global trends, as well as the circumstances of specific mining sites. The main focus is on Madre de Dios, a hotspot of mining activity and growing policy-making and political problems. It is relevant for the distinction made between informal and illegal mining practices, which are exacerbated by the hegemonic influence of neoliberalizing tendencies in the country. The results demonstrate that the future of the Amazon and the response to intricate environment-development questions greatly depend on the livelihood of miners and the needs of local social groups, such as indigenous peoples and riparian, rural communities. The thorny question of prior and informed consent remains largely ignored by decision-makers and the permission to extract natural resources is regularly justified by the (questionable) need to promote economic growth. The author calls for a more comprehensive and inclusive approach, combining different labor relations with the cross-scale attention to environmental dilemmas.

Part III, "Identities, Cultures, and Subjectivities," includes works on the interactions and clashes between indigenous peoples and a range of other groups with contrasting cosmovisions and attitudes to socio-ecological processes of change. Oliveira begins with a reflection based on more than four decades of anthropological work with indigenous groups in the Western tracts of the Amazon. The author claims that the Amazon is an important presence in Brazilian social thought, described in polarized ways, either as a land of promise or as a huge green hell. What is common to both portrayals is that the data and interpretations they use are framed in a process of occidental and capitalist expansion, conceived almost as spontaneous and natural, in which the Amazon and its populations appear as minor and isolated presences, performing only secondary roles and functions. The objective of this chapter, based on recent research by Brazilian anthropologists and historians, is to make explicit and to reflect on the narrative tropes and ideological biases that feed a very limited view of the frontier, and to incorporate the agency, strategies and protagonism of the indigenous populations into analysis. By emphasizing the conflicting and polyphonic dimension of the frontier situation, we hope to contribute to a critical and dialogic understanding of existing sources and interpretations, stimulating the analytical explorations required for a new use and conceptualization of the social construction of territories, borders, and frontiers.

Gebara then considers the Amazon as the largest area of tropical forest in the world, which hosts a complexity of ecosystems and fulfills vital functions in global climate regulation. The chapter elaborates on Stenger's and Latour's studies on the emerging movement toward post-natural politics, moving beyond the human-nature dichotomy to reimagine political regimes outside capitalism. The text builds on the concept of cosmopolitics and on Amazonian indigenous cosmologies to provide insights into the phenomenological path toward thinking of human and other-than-human beings as earthed and embedded with—not in—their environment (i.e., "earth beings"). It asks how this provocative movement might inform understandings of both the opportunities for capitalism that have been created by contemporary environmental crises and the possibilities for critical political reframing and responses. The current political narratives transform nature into new commodity fractions called "ecosystem services," which are a main pillar of contemporary structures for forest governance (i.e., payments for ecosystems services, REDD+, natural capital). It then looks at the framing of cosmopolitics to discuss the antagonism between such contemporary structures and indigenous peoples' cosmologies and views of nature. With a focus on Amazonian cosmologies, it is argued that to move toward long-term solutions for the complex challenges that forest governance entails, policy-makers need to step outside of Western rationalities to conceive and experience our relationship with nature in ways that allow us to creatively respond to the challenges of forest conservation.

In the next chapter, Little presents a framework for research on territorial conflicts in Amazonian frontiers using an engaged ethnographic approach. This chapter seeks not only to understand the dynamics of specific conflicts but also to track the many ongoing efforts to resolve them and, as such, has both empirical and normative dimensions. A brief review of Amazonian frontiers reveals a common tendency toward rapid land-use change driven by developmentalist ideologies and global demand for commodities accompanied by the construction of large-scale infrastructure projects and high rates of deforestation and environmental degradation. The concept of territory provides a key anthropological entry point into the study of frontier dynamics and is used to identify the principal social groups along with their respective, and often conflicting, claims to territory. Analyses of the various mechanisms for conflict resolution undergird the presentation of possibilities for new, more equitable systems of post-frontier resource governance regarding territorial rights, food security, environmental justice, ethnogenesis, and sustainable development.

The final chapter is a very personal recounting of the experience of the director of the 2017 movie *Beyond Forldlândia*, which deals with the disastrous attempt by Henry Ford to implement large-scale rubber farms in the Brazilian State of Pará in the interwar years and until the end of World War II. Colón's movie is particularly influenced by Nixon's concept of "slow violence," defined as a long-term, gradual process of degradation and oppression occurring across time and space, but that is not normally described as violence at all. There is an interesting association between slow violence and the relevance of viewing the Amazon slowly. During the preparation and shooting of the movie, the author revisits his personal perception of Amazonian issues and the role of art in the search for alternatives to the failures of regional development. It is quite disturbing to realize that the bitter lessons from Fords's folly have not been learned and the regional economy is again largely dominated by soybean plantation farms. The chapter makes reference to national and international scholars, as in the case of Euclides da Cunha, Mário de Andrade, Leandro Tocantins Bruno Latour, and Walter Benjamin, who are all connected to the critical assessment of Amazonian troubles, from the personal to the global scale. The movie has not only artistic value, but also helps to reflect on continuities between past and present, colonial legacies and political manipulation, the territorialization of power and also vitality of culture and diversity of socio-ecological practices, among other complex elements of frontier-making in the Amazon.

As can be seen from the above, the sequence of chapters was conceived as a cumulative and interconnected critique of political, representational, and material trends, considering particularly the growing tensions of frontier-making and the unjust, exclusionary basis of regional development in the Amazon. Much in the same way, in organizing this book, we aimed to uncover and make sense of present-day human and environmental realities rapidly and dramatically unfolding in the Amazon at the dawn of the new millennium. One main inspiration guided the project throughout: that frontier developments are not only closely dependent on and connected to events taking place at the so-called centers of power, but are indeed tied together within an interdependent and mutually reinforcing dynamics.

Inspired by the great Brazilian sociologist Florestan Fernandes who stated, already in the 1970s, that the history of capitalism reveals itself more clearly and intensely in the periphery (Fernandes, 1979), we take frontier dynamics as constitutive expressions of center-based contradictory historical developments produced by conflicting social dynamics between materially and politically unevenly defined social actors. We believe that the complex reality of frontiers needs to be understood within an analysis capable of integrating, or at least take into consideration, the equally complex case of so-called centers, as well as the interplay between the two. We hope the following chapters will live up to this challenging and much-needed goal.

NOTE

1. Such as the notorious Directory of the Marquis of Pombal of 1757, which forced the reorganization of indigenous life in settlements aimed at promoting higher levels of miscegenation and acculturation.

BIBLIOGRAPHY

Acker, A. 2017. *Volkswagen in the Amazon: The Tragedy of Global Development in Modern Brazil* (Global and International History). Cambridge University Press: Cambridge.

Arndt, H. W. 1987. *Economic Development: The History of an Idea*. University of Chicago Press: Chicago.

Bratman, E. Z. 2019. *Governing the Rainforest: Sustainable Development Politics in the Brazilian Amazon*. Oxford University Press: Oxford.

Bunker, S. 1985. *Underdeveloping the Amazon*. University of Illinois Press: Urbana.

De Sartre, X. A. and Taravella, R. 2009. National Sovereignty vs. Sustainable Development: Lessons from the Narrative on the Internationalization of the Brazilian Amazon. *Political Geography*, 28(7), 406–15.

D'Incão, M. A. and Silveira, I. M. (eds.). 1994. *A Amazônia e a Crise da Modernização*. Museu Paraense Emílio Goeldi: Belém.

Escobar, A. 2012. *Encountering Development: The Making and Unmaking of the Third World*. 2nd ed. Princeton University Press: Princeton.

Fernandes, F. 1979. *Foreword of book Dependent Development: The Alliance of Multinational, State and Local Capital in Brazil*, by Peter Evans. Princeton University Press: Princeton.

Hall, A. (ed.). 2000. *Amazonia at the Crossroads: The Challenge of Sustainable Development*. Brookings Institution Press: Washington, DC.

Hecht, S. B. 2005. Soybeans, Development and Conservation on the Amazon Frontier. *Development and Change*, 36(2), 375–404.

Hoelle, J. 2015. *Rainforest Cowboys: The Rise of Ranching and Cattle Culture in Western Amazon*. University of Texas Press: Austin.

Holanda, S. B. 2000. *Visão do Paraíso*. Brasiliense: São Paulo.

Ioris, A. A. R. 2016. The Paradox of Poverty in Rich Ecosystems: Impoverishment and Development in the Amazon of Brazil and Bolivia. *The Geographical Journal*, 182(2), 178–89.

Ioris, A. A. R. 2018. Amazon's Dead Ends: Frontier-Making the Centre. *Political Geography*, 65, 98–106.

Ioris, A. A. R. 2019. Peasant Farming in the Southern Tracts of the Amazon: The Reluctant Alterity of Agribusiness. *Perspectives on Global Development and Technology*, 18, 375–400.

Lund, C. and Rachman, N. F. 2018. Indirect Recognition. Frontiers and Territorialization around Mount Halimun-Salak National Park, Indonesia. *World Development*, 101, 417–28.

Martins, J. S. 2009. *Fronteira: A Degradação do Outro nos Confins do Humano*. Contexto: São Paulo.

Schmink, M. and Wood, C. (eds.). 1992. *Contested Frontiers in Amazonia*. Columbia University Press: New York.

Schmink, M., Hoelle, J., Gomes, C. V. A. and Thaler, G. M. 2019. From Contested to "Green" Frontiers in the Amazon? A Long-term Analysis of São Félix do Xingu, Brazil. *Journal of Peasant Studies*, 46(2), 377–99.

Slater, C. 2002. *Entangled Edens: Visions of the Amazon*. University of California Press: Berkley.

Watts, M. J. 2018. Frontiers: Authority, Precarity, and Insurgency at the Edge of the State. *World Development*, 101, 477–88.

Part I

THINKING AND MAKING OF THE AMAZON FRONTIER

Chapter 2

Revisiting Frontier Theory and the Experience of Frontier-Making

Antonio Augusto Rossotto Ioris

INTRODUCTION

There is little doubt that we live a world characterized by mounting dilemmas and fierce controversies. New problems accumulate and old ones deepen, while there is limited room for comprehensive, long-term solutions. Instead of addressing the distortions and inequalities associated with mainstream social, political, and economic institutions, public policies and private initiatives are often diverted toward the production of new spatial configurations. This movement, away from where the trouble originated, seems to suggest that central areas have become saturated with themselves, prompting the dislocation and reterritorialization of people and enterprises. Attempts to evade and transfer national or location-specific dilemmas to other socioeconomic settings seem to be a hallmark of capitalism's unsettling powers and inherent contradictions. Since the European Renaissance, frontier-making has been a favorite response, to poverty, unemployment, land and housing deficits, resource exhaustion, environmental degradation, market saturation, and political persecution. The mitigation of socioeconomic tensions and the search for novel money-making opportunities in newly opened spatial frontiers have been crucial for the affirmation of capitalist modernity and the functioning of the modern world. Watts (1992: 116–17) considers frontiers as "particular sorts of spaces" that represent "the first wave of modernity to break on the shores of an uncharted heartland" with "their own territorial form of law and (dis)order."

All this means that capitalism is also, and fundamentally, based on "accumulation by frontier-making," a concept that needs to be properly theorized and adequately investigated beyond simplistic geographical conceptualizations. The spatiality of frontier-making goes beyond the more immediate relocation from "core" to "periphery," entailing a relational interaction and joint processes of exploitation, realignment, and reinforcement in both old and new areas. At the frontier, enclosure, extraction, and production are recreated and integrated into wider politico-economic arrangements, which are themselves transformed through the emergence of new spatial frontiers. In the end, the incorporation of new territories only temporarily alleviates tensions, and without challenging existing relations of production and reproduction. Problems are naturalized, fragmented, and depoliticized, new accumulation mechanisms are activated and reinforced, while those who have been most seriously affected by socioeconomic developments are compelled to move or risk being blamed for their own difficulties. The lack of opportunity for some groups and individuals triggers the imagination (of a different reality and a possible better life) but without any guarantee of success (there are none).

Our goal here is to theorize frontier-making and associate it with the politics of scale, that is, the shared experience of capitalist relations of production and reproduction occurring at distinct, but interconnected, scales of socio-ecological interaction. "Scale is not necessarily a preordained hierarchical framework for ordering the world—local, regional, national and global. It is instead a contingent outcome of the tensions that exist between structural forces and the practices of human agents" (Marston, 2000: 220). The politics of scale is directly implicated in the production of space given that scale is constituted and reconstituted through relations of production, reproduction, and consumption interwoven with space. The uneven development of the socioeconomic forces of capitalism is essentially multiscale and unfolds through the dynamic political economy of old and new regions (Agnew, 2000). Consequently, daily life and class-based differences at the local scale are permeated, and help to shape or transgress wider economic and politico-ideological constructs. Likewise, the "reproduction of the household enterprise is dependent simultaneously on the economic relations of production and on the political relations necessary to protect those relations" (Smith, 1989: 24).

In this chapter, we will also revisit what has happened in the Brazilian Amazon, a region notoriously associated with the expansion of internal and external economic frontiers. The symbolism of frontier-making in the Amazon has pervaded the social and scientific imaginary. More recently, the expression "agricultural frontier" has also become prevalent and is extensively used by academics and scientists. Yet, the ongoing encroachment of export-driven agribusiness into the Amazon region is only the most recent chapter in a long history of the pursuit of new economic frontiers in Brazil. In order to understand the long-term drivers of frontier-making it is essential to consider that the opening of new production areas is not the leftover or the excess of national development, but that it has been central to cultural, political, and social change throughout the country. Socioeconomic inequalities and socio-ecological exploitation have also been managed through the proliferation of frontiers and the prospect "of something better elsewhere" (when "here" is no longer enough). In that way, the responsibility for problems is shifted back to those exploited and marginalized in the core areas, implying that it would be their own fault if they refused to embark on the journey to a more promising reality at the frontier. Contemporary agricultural frontiers in the Amazon re-enact, once again, the dreams of modernity and prosperity that for generations attracted migrants to the Brazilian west (Ioris, 2017a).

To demonstrate the contested politics of scale behind frontier-making in the Amazon we will particularly focus on agribusiness activity in the State of Mato Grosso, in the Brazilian Amazon. Mato Grosso has been at the forefront of economic and political frontier-making for many centuries, but the process has accelerated in the last few decades due to the spiraling growth of soybean-based agribusiness. This experience vividly illustrates the paradoxes and extravagances associated with frontier-making; Mato Grosso now accounts for around 10 percent of global soybean production, but the state is an enormous food desert that, like most economic frontiers, still depends for its supply of food on the core, "consolidated" economic areas in the south and southeast of the country. In addition, the intense economic activity and commodity exports in the Amazon have further moved Brazil toward the periphery of market-based globalization and reinforced the old pattern of extractivism and socio-ecological waste (often disguised by calls for efficiency and narratives of sustainable development along the lines of ecological modernization).

Our interpretation is informed by empirical work conducted between 2013 and 2016 that involved various data collection campaigns and regular visits to different locations in the north of Mato Grosso at the Upper Teles Pires River basin, which is the main soybean production area in Brazil today. Even more revealing, in terms of frontier-making, is that the Upper Teles Pires is situated exactly at the transition between forest and savannah ecosystems. The study takes on board the recommendation of Pred and Watts (1992: 2) to consider the "various historical configurations and reconfigurations of capitalism in an effort to understand how difference, connectedness and structure are produced and reproduced within some sort of contradictory global system, within a totality of fragments." Furthermore, the challenging complexity of the Mato Grosso agribusiness frontier is considered as not only a socio-spatial construction, but also a true analytical tool and a basis for proposing new investigations (Pacheco de Oliveira, 2016).

The most innovative contribution of our investigation is in offering a meta-theoretical framework and an associated reflection on the specific frontier-making experience in order to reconceptualize the wider

Amazonian politico-economic trajectory. To achieve that goal, the text is divided into two main parts. In the first, after this brief introduction, a theoretical and interpretative perspective is presented, which goes beyond traditional accounts of frontier-making in order to emphasize socio-spatial interconnections and interlocked scarcity and abundance (consolidated under the "law of scarcity-abundance"). In the second part, the evolution of the agribusiness frontier in Mato Grosso and, indirectly, in the rest of the Amazon is critically examined, making reference to the roots of the sustained processes of violence, exclusion, and hierarchization that have characterized the long history and contested geography of frontier-making in the region. The case study is followed by overall conclusions and implications for future studies.

CENTERED FRONTIERS: ABUNDANCES AND SCARCITIES

The emergence of new spatial frontiers remains a ubiquitous process in the contemporary world, considering that the decline of frontier-making is still an unfulfilled aspect of globalization and a postmodernist fantasy (i.e., the proclamation of a borderless world, instantly connected and horizontally networked). Novel spatial settings, distinguished by their own patterns of economic production and socio-ecological organization, continue to appear, with significant repercussions for national and global societies. One main consequence is that the ontological complexity of frontiers persists as a real challenge for social scientists. Imamura (2015) appropriately recommends that the analysis should begin with an inquiry outside academia to observe how the word "frontier" is used in ordinary speech before it is scrutinized by academics. In the United States in particular, the public imaginary is influenced by Frederick J. Turner's persuasive argument about settlement frontiers, basically the claim that spatial frontiers provided the elemental conditions for freedom and social opportunities in North America (Billington, 1963). However, as in the case of Turner's, most interpretations seem to miss the multiscale political, social, and economic ramifications of frontier-making. Scholars have typically described various types of frontier—political, agricultural, resource, commodity, etc.—but have failed to properly take into consideration the range of interests, social differences, and political disputes that help to shape frontier spatiality. For example, Demangeon (1932: 636) bluntly considers it "an exceptional fortune" [*une fortune exceptionnelle*] for a country to have pioneering frontiers and Webb (1952) argues that Western European civilization was the fortunate result of the opening up of world economic frontiers, which started with Columbus and continued until the twentieth century, but it is rare to find studies that effectively connect local, lived activities with wider politico-economic scales.

On that regard, Hennessy (1978: 12) rightly observes that frontiers "have encouraged dichotomies, as they invite Manichean schemes of thought" about developed and underdeveloped regions around the world and the associated division between advanced and primitive societies. For instance, development economists, such as Di Tella (1982), identify frontiers as stocks of untouched resources and sources of wealth waiting to become economically viable (as in the case of "virgin" land incorporated by the expansion of Germany, Spain, Russia, and Portugal). Findlay and Lundahl (2016) likewise explain frontier-making in relation to capital investment in land use and in the mobilization of labor to satisfy the demands of manufacturing in core areas; according to this conceptual model, the frontier is extended and agricultural production increases following initial investments in manufacturing, thus creating a virtuous circle that can lead to the general expansion of the economy. Into the bargain, Alston et al. (1999) emphasize the importance of defining rights to private land to promote market transactions and investments, which will gradually replace the first adventurers with more entrepreneurial farmers. Rationalizations of economic frontiers such as these have been criticized for being naïve, proselytizing expressions of environmental determinism and because they ignore the actual actors, historical differences and circumstances involved in frontier-making (Mikesell, 1960). That is, although each frontier location is different, their unstable and mismatched patterns are taken for granted and seem to justify common disdain for their internal configuration. The main consequence of such economic and institutionalist reductionisms—which broadly follow Locke's philosophy of private property—is the naturalization of the exploitative and hierarchical basis of frontier-making. They ignore the fact that the "empty spaces" of the American continent, as well as those in Africa, Russia, Oceania, etc., were targeted for the opening of new frontiers—in areas considered *terra nullius*—on the basis of deliberate disregard for the native inhabitants and their socio-ecological conditions (Richards, 2006).

Contrary to most accounts, frontier-making does not happen merely because of the opportunities available in remote new areas, nor does the center depend on the frontier for its continued existence. The relevance of frontier-making derives primarily from the fact that capitalism is not only crisis ridden but also crisis dependent—considering that recurrent crises serve to induce changes designed to defend or restore profits—and new spatial frontiers play a fundamental role in the mitigation and reorganization of politico-economic structures. Frontiers not only have location advantages and constitute favorable investment areas but they are also where old and new socio-spatial features coalesce, resulting in the revitalization of social and economic patterns of the (relatively consolidated) center. Regarding the internal tensions of capitalism and its crisis-dependency, O'Connor (1998) suggests that the basic antagonism between the forces and relations of production (originally described by Marx) needs to be complemented with a "second contradiction" between relations and conditions of production. The second contradiction thesis is one of the main reference points of political ecology and ecosocialism in the critical literature available today. However, it is possible to append a third layer to the twofold contradictions of capitalism described by O'Connor, which is the clash between, on the one hand, the homogenization of lifestyles, consumption and social attitudes and, on the other, mounting inequalities, politico-economic hierarchies and context-specific problems. In other words, there is a widening gap between the appearance of equality and the crude experience of inequality. This "third contradiction of capitalism" is profoundly geographical in nature, insofar as exploitation and capital accumulation depend on the imposition of socio-spatial and ideological homogeneity.

It is this third contradiction in particular that frontier-making helps to mitigate through the restitution of geographical differences that invigorate capitalist relations (without ever addressing the central irrationality and wastefulness of capitalism). Capitalist development is not only uneven, but also essentially expansionist and expands in different directions, into different places and practices, to avoid having to confront their own contradictions. In the case of the Brazilian Amazon, Galaz et al. (2018) show that, between October 2000 and August 2011, 68 percent of all investigated foreign capital to nine focal companies in the soy and beef sectors was transferred through one, or several, known tax havens, what represents as much as 90 to 100 percent of foreign capital for some companies investigated by these authors. The expansionist drive of capital is nothing other than the manifestation of its capacity to creatively reconstruct the world in its own image while avoiding serious challenges to its internal logic (basically, the extraction of surplus value and the commodification of labor and nature). Frontier-making is an integral component of capital's intrinsic struggle to suspend socio-spatial dilemmas (including financial cycles, market crashes, and socio-ecological exhaustion—see Solimano, 2017), while privatization, exploitation, and accumulation can be expanded to new areas. This stabilizing basis of frontier-making goes beyond what Luxemburg (1951: 362) considers the insufficiencies of core capitalist areas and the need to spread out to new regions where excessive surplus value can be transformed into productive capital (described as the exploitation of "territories where the white man cannot work"). In actual fact, frontiers are primarily required because the capitalist center has become saturated with itself, fraught with uncertainties about its actual legitimacy vis-à-vis its widespread impacts, hesitant about how to cope with past legacies and constantly emerging challenges (migration, urban chaos, food insecurity, social breakdown, to name a few), and consequently needs escape mechanisms to dislocate contradictions to other areas. On the other hand, Luxemburg rightly questioned the idea of rigid distinctions (suggested at the time by people like Kautsky to dismiss the idea that the Soviet revolution could be replicated in Germany) between Western Europe and "backward" Russia by comparing the complex capitalist regions with the exploitation of Russian agrarian society through a form of capitalism emanating from Western Europe.

To fully grasp the peculiar geographical features of frontier-making, it is necessary to "subvert" any artificial separation between (comparatively consolidated) central areas and (relatively undeveloped) frontiers. It is particularly mistaken to describe frontiers as young areas destined to replicate the politico-economic system of the center, which, as warned by Massey (2006), constitutes a reduction of geography to history or space to time. Instead of such linear, sequential conceptualization of the association between center and frontier, Massey argues that spatial connectedness needs to be considered more seriously, in particular the understanding that space is the outcome of relations and practices, that it is basically a multiplicity and a constellation of ongoing trajectories. The unique features of center and frontier must be correctly appreciated, as spatial identities and sociocultural subjectivities are relational and co-constituted through engagement. Based on this, it is possible to infer that frontier activities are distinctive but also predicted in the contradictions and achievements of the center, just as the center needs to expand into new frontiers in order to maintain its particularism. Both the frontier and the center are singular spatial settings but belong to wider processes of space production and its related contestation. Frontier-making is a transient suspension of mounting trends and circumstances consolidated in central areas, while also reinstating (in the frontier) similar relations and patterns to those responsible for mounting socio-ecological problems, economic collapse, and inequality in the center.

The main claim so far is that the frontier is not merely an expansion or projection of the center, but that center and frontier are permanently connected and interdependent. It means that new frontiers may be seen as a departure from the center, but in effect they are reproducing it. Frontier-making is a systemic process that entails a spatial separation but only through organic interdependencies between the so-called "center" and "periphery" (in the Spinozian sense, the frontier is *in* the center, as much as the center is *in* the frontier). The frontier ultimately results from the center being saturated with mounting tensions and multiple contradictions, thus frontier-making plays a role in the reduction of socio-ecological problems and the revitalization of economic growth and development. We could, therefore, redefine *frontier-making as a range of interconnected processes of socio-ecological disruption and spatial reorganization affecting regions increasingly under the sphere of influence of, and primarily established to attend, national and international capitalist centers.* This definition allows us to go beyond a focus merely on the newly incorporated areas and systematically consider interdependencies and continuities. The basic feature of frontier-making is the movement of occupation, transformation, and landscape discipline in the new zones of appropriation, but tightly connected with the process of internal restructuring in central areas (Moore, 2015). Frontier-making may have an appearance of chance and chaos, but in practice it is highly instrumental for the reaffirmation of the "old" center through the production of "new" relations (that are, in effect, projections of the relations established at the center). Frontier-making therefore has important parallels with the ambiguous manner in which Thomas More described *Utopia*, a new world that at first sight seems to be the opposite of England but which in practice replicates, in subtle and distorted ways, its many problems.

This first observation leads us to a second crucial postulation: the interconnections between center and frontier are primarily manifested through the promise of abundance and, paradoxically, the multiplication of scarcity. The interlocking of abundance and scarcity takes place in both central and frontier spaces. Capitalist scarcities, as well as abundances, are human-made and produced through historically determined relations, pressures, demands, and technologies. Landed property in particular is necessarily relational, violent, and political, because land is an economic asset and a source of revenue held against others. At the frontier, the private property of land and resources is established through the imposition, via multiple social and political mediations, of a rational, legitimate institution upon what was considered an anomic, savage, and unlawful pre-frontier reality. "The frontier, which appears as a neutral boundary, serves as a condition of possibility for property's violence" (Bromley, 2003: 135), considering that "violence, law and bureaucracy work in complementary fashion to mediate the struggle for land on the frontier" (Foweraker, 1981: 25). Campbell (2015) even describes the genesis of private property in the Amazon as something that is "conjured" and that both reconstructs the past and anticipates the future. Nothing is really settled, but land tenure can evolve according to political disputes and the emergence of opportunities. Crucially, because these interconnected spatial settings are gradually inserted into the capitalist logic of accumulation, there are fundamental synergies between selective abundance and widespread scarcity in both the center and the frontier. No capitalist frontier is ontologically given or pre-determined from the outset, but it is possible to anticipate that its unique multifaceted socio-spatial developments will be influenced by abundance and scarcity in the homeland and that it will be marked by the production of new scarcities and abundances at the frontier. These trends are evidently consequences of class-based relations and labor exploitation, which cannot be dissociated from the exploitation of the rest of nature. As examined by Marx, the subordination of nature to capital requires the control and exploitation of labor, and this double process of exploitation through capitalist relations connects everyday life with wider, more long-term socio-economic structures.

The classical example is the colonization of the Americas, which from the sixteenth century involved the disruption of the culture and livelihoods of the Amerindians and the appropriation of land by migrant settlers. Frontier-making in the American continent was essentially a "boom and bust" economy, where, after intense exploitation, resources were depleted (even if only partially) and new frontiers were needed. Particularly the poorest and most marginalized segments of metropolitan society were encouraged, or forced, to migrate to the European colonies under the promise of accessible abundance (land, resources, social opportunities), but the majority of those who migrated would only experience harshness and fresh manifestations of scarcity. Frontier-making primarily benefited the bourgeoisie and the political establishment, but to some extent also helped workers in core economic areas, although at the cost of the exploitation of people and resources in the newly incorporated areas (see Bukharin, 1929). The experience in the West Indies was paradigmatic, with the decimation of the indigenous population by violence and diseases, and the increasing cultivation of land for sugarcane and cattle (Richards, 2006). The removal of "anti-capitalist tendencies" at the frontier was facilitated by governmental control of land prices and limited access to private property, which compelled immigrants to work long hours for low wages, exploited by the wealthier, capitalist farmers. In the nineteenth century, imperialism expanded frontier-making scarcities in the form of land expropriation and displacement in Asia, Africa, and Oceania. After World War II, international development was maximized by the Green Revolution and associated with the commodification of even larger tracts of land. Finally, with the advance of agro-neoliberalism in the last quarter of the twentieth century, the intensification of investment in land and agriculture has led to systematic land grabbing.

Therefore, the tendency toward social and economic scarcity in one location is, to a certain extent, compensated—as a counter-tendency of the capitalist system (Mészáros, 1995)—by the material and symbolic construction of abundance at the frontier. During more than five centuries of European colonial and imperialist history, the seizure and accumulation of land evolved according to the history and geography of capitalism, and agri-food and natural resource systems were likewise restructured in accordance with the need to renew the accumulation of capital following the incorporation of new economic areas (Dixon, 2014). This happened through a powerful politics of scale in which the interaction between local, national, and international dynamics was operationalized according to the privileges of a small elite while the majority of the population lived with scarcity. The exhaustion of primitive accumulation in Western Europe led to the colonization of new frontiers (as in North America), but capitalists strived to contain the relative liberties of economic migrants and had to resort to the "power of the mother country" and "use force to clear out of the way the modes of production and appropriation which rest on the personal labor of the independent producer" (Marx, 1976: 931). It is important to realize that manifestations of scarcity and abundance take place at different moments in the center and the frontier. There is a time gap, and when scarcity starts to increase in the center there is the prospect of abundance at the frontier. It can be seen that scarcity is never a single process caused by the shortage of means or resources; it is a social relation that unevenly and cyclically alternately affects groups and locations (Ioris, 2016a). Certain economic features become scarce, which creates the preconditions for the production of abundance elsewhere. Through the instrumentalization of scarcity and abundance many of the worrying tensions that emerge in consolidated areas can be partly released with the movement of (surplus) labor and overaccumulated capital to the frontier.

It is possible to summarize the interplay between (promised) abundance and (widespread) scarcity that underpins frontier-making as the *law of scarcity-abundance* (LSA). The LSA is a nonreductionist synthesis of the escalation of problems in politico-economic core areas in association with promises of a solution through frontier-making. It is important to note that this is a "law" in the sociological, nonpositivistic sense, that is, the consolidation of social institutions and politico-economic tendencies that result from long-term processes of conflict and cooperation in the history and geography of capitalism. More importantly, the LSA is not an argument in favor of economic determinism or a teleological model of geographical change, but rather a constructivist explanation that takes into account the complex relationship between culture and political economy. The LSA helps to clarify that, while frontier-making has the appearance of a centrifugal process (a movement away from the center), it is in effect mainly a centripetal force, that is, the frontier is instrumental in reinforcing and reinstating the center. Frontiers happen because of the deferment of scarcity in core areas through the imagined space of abundance projected elsewhere; nonetheless, this promise is normally frustrated and only materialized selectively and at great social and ecological cost. Moore (2015: 87) points out that the "great secret and the great accomplishment of capitalist civilization has been to not pay its bills. Frontiers made that possible." Frontier abundances—as pronounced by Turner (1920), these were at the edge of "free land" in the American west—derive from the fact that these are considered zones beyond the customary rule of law, therefore lawless and up for grabs. At the frontier, authority is still open to challenge, there are combinations of order and chaos, with the state unable or unwilling to exercise more effective control (Watts, 2018).

The various forms of violence that characterize frontier-making are rationalized by religious, civilizational, or economic discourses that characterize the frontier as a space of inclusion and shared gain (Elliot, 2016). "In the interest of the so-called wealth of the nation, he [the political economist, "sycophant of capital"] seeks for artificial means to ensure the poverty of the people. Here his apologetic armour crumbles off, piece by piece, like rotten touchwood" (Marx, 1976: 932). On the other hand, the perverse and conservative dynamics of the frontier can trigger novel forms of reaction that challenge institutions as soon as they become slightly consolidated. Frontiers are functional for capitalism, but also expose its micro and macro internal fissures, which can lead to smaller or larger forms of resistance. Frontiers are not only places of scarcity, but also "sites of potential" (Li, 2014: 13). An ontological inspection of the frontier reveals a mosaic of disruption and reconstruction, cross-scale interconnections and multiple repercussions of the transposition of values and practices, and collective action based on acute individualism. The frontier is ultimately history through geography and geography with a forward-looking past, as it always brings the old back through the exercise of an uncertain and tentative future. Instead of respecting the role of the subject, the construction of capitalist frontier shatters the possibilities of autonomy both in core and frontier areas. The periphery is the other, the civilizational alter-ego of the center. That makes any frontier inherently complex and politicized, far from egalitarian. The "chaos" of the frontier is, in this sense, the offspring of development and of economic problems in the core areas that cannot be easily controlled, let alone harmonized. In the next section, the contested geography of Mato Grosso will illustrate these controversies.

AMAZON'S MAIN AGRIBUSINESS FRONTIER

The two main claims presented above—the permanent interdependence between center and frontier, as well as the LSA—will help us now to examine the general trends of frontier-making in the Amazon, from colonial times to the ongoing advance of neoliberalized agribusiness. These two conceptual lenses will be particularly useful in addressing the specificities of the agricultural frontier in Mato Grosso, in the south of the Brazilian Amazon, which is shaped by modernization, commodity exports, and global integration demands. The geographical changes in the State of Mato Grosso clearly illustrate the mitigation of socioeconomic problems in "core" national areas and, in addition, the asymmetry of opportunities associated with frontier locations. The "conquest" of Mato Grosso by agribusiness in the last half-century has been a process of accumulation through frontier-making that in the end has served primarily to reinforce a highly centralized and exclusionary national society. The appropriation and transformation of Mato Grosso into a perennial economic frontier—still unfolding and even accelerating today—reveals a great deal about the unfairness of Brazilian capitalist modernity and how this is projected across different spatial scales. As explained in more detail below, Mato Grosso is a triple frontier, as it has historically been situated at the margins of Amazonian development, of the Brazilian economy and of the Portuguese empire. These three interlaced frontiers have essentially evolved through the recurrent production of scarcity and (selective) abundance.

The roots of frontier-making in the Amazon are to be found not only in Brazilian history, but also in the foundations of Portuguese society. Whereas Brazil was initially only a small piece of a much broader commercial and colonial machine, Portugal was itself a nation marked by opposition to the neighboring Spanish kingdom and by the search for new economic frontiers (Bastos, 1998). Santos (1993) even argues that the Portuguese have an "identity of the frontier" that was sociologically translated into a predisposition toward maritime navigation and global trade. Portugal had already pioneered the route to India around Africa when it decided to claim possession of its South American territory granted by the pope in 1494 under the terms of the Tordesillas Treaty. From the time of the first letter written in 1500 by Caminha, secretary of Cabral's conquering fleet, to the Portuguese king, the new colony was characterized by vast resources and an enormous agricultural potential. In practice, nonetheless, the occupation of Brazil was fraught with violence, ecological degradation, the exploitation of slaves and free workers, as well as constant efforts to suppress foreign attacks and local insurgences. To start with, there was a perennial shortage of workers due to the high cost of African slaves and the recurrent difficulty of enslaving the natives (particularly because of their vulnerability to alien diseases such as malaria and smallpox). More significantly, despite the immensity of the colony there were persistent complaints about insufficient land and the struggle of poor migrants for survival. Structural land scarcity was obviously caused by the elitist and exclusionary basis of colonial institutions. In the words of Alencastro (2001: 41), "land and labor [in Brazil] did not come up as independent factors, but as two variables that stem from the driving-forces of commercial capitalism." These were clear manifestations of the LSA, in the sense that impoverished, landless people were led to migrate from Portugal to the colony only to again be deprived of land and to become entangled in the wider processes of colonial exploitation, abject slavery, and a hierarchical society.

The growing scarcity of land along the Atlantic coast, together with the prospect of finding precious stones and other riches, was an incentive to explore the vast interior of South America. However, enlarging the colonial Brazilian economy presupposed the transgression of the political border established by Tordesillas. For the Portuguese, most of the land where treasures could be found was beyond reach and could only be accessed through regular incursions into what was formally Spanish territory. "Indeed, one of the most striking features of South American history has been the success of Brazilian expansion at the expense of its Spanish-speaking neighbours" (Hennessy, 1978: 106–7). One of the most remarkable consequences of such coordinated, though "illegal," expeditions into the continent was the appropriation and conversion of the majority of the Amazon into a single Portuguese province. The Amazon, a separate administrative unit since the seventeenth century, never existed before the Portuguese invasion and its systematic clashes with the other four competing European powers (not only Spain, but also England, France, and Holland). Ultimately, Portugal maintained logistical and military supremacy over the Amazon River basin, given that the only practicable and easy means of access to Amazonia lay on the Portuguese side of the continent (Prado Jr., 1967). Ever since, frontier-making has been a perennial phenomenon in the Amazon, directly associated with cyclical commodity booms. "In Amazonia, frontiers have not only been opened and closed but reopened and reclosed again and again" (Little, 2001: 3). The process has predictably followed the overall logic of perceived abundance that recreates material scarcity due to pillage, wastage, and socioecological disruption (Heckenberger, 2005). It depended on the violent assault and cultural obliteration of the native population and the miserable impoverishment of the survivors (Hemming, 1987). From the perspective of the indigenous population, the advance of the frontier represented absolute scarcity, that is, the denial of the very possibility of social existence and, for the majority, of physical survival.

The history of what is now Mato Grosso started in the beginning of the eighteenth century, with the increasing presence of Brazilians and Portuguese involved in the extraction of diamonds and gold in what was officially Spanish territory. The foundation of the first settlements gradually displaced the activity of Spanish missionaries and weakened the influence of Asunción and Santa Cruz de la Sierra (Lucidio, 2013). New political borders were agreed later in the century by a series of international treaties that granted Portugal the possession of much larger tracts of land, approximately the size of contemporary Brazilian territory. After national independence in 1822, the Brazilian government promoted fierce policies of national unity and containment of the numerous rebellions. The Regency period (1831–1840) was particularly turbulent and there was real prospect of territorial fragmentation because of accumulated socioeconomic tensions and a strong sense of political contempt among regional elites. It was a long and painful road to national unification, in which scarcity and abundance were managed by able political leaders on behalf, primarily, of land and slave owners. One key feature of nation-building, according to Santos (1993), was that Brazil internalized center and periphery in order to keep the territory united. It was an astute plan that involved the consolidation of international political borders and the subordinate integration of national economic frontiers. Although "Mato Grosso never became a center of more than secondary importance" (Prado Jr., 1967: 78), its relative significance increased with the outbreak of a large-scale war (1864–1870) that followed the invasion of Brazilian territory by Paraguayan troops. After the war, Mato Grosso was again reduced to a mineral and agricultural periphery, with some incipient farming activity and only marginally involved in the "rubber boom" that dominated the Amazon economy at the turn of the twentieth century (Wilcox, 2017).

Despite some infrastructure and operational improvements, the economic situation of Mato Grosso remained highly peripheral to the rest of the country during the first decades of the twentieth century. While the south and southeast of Brazil were gradually becoming more urban and industrialized, in Mato Grosso there were only meager signs of agricultural intensification and economic growth. The process of frontier-making had operated more or less unabated for several centuries based on the prospect of abundant minerals, land, and resources, which corresponded to short-lived periods of prosperity quickly succeeded by the more enduring condition of isolation and renewed forms of scarcity. Mato Grosso was at the periphery of a peripheral country that was at the time highly dependent on the export of coffee and a few other primary commodities. The main economic frontiers at that point were located in the west of the state of São Paulo, which offered the best socioeconomic prospects for the poorer segments of society. With a limited domestic market and scant capital investment, one of the only options available for the majority of the population was internal migration, although this necessarily represented the "sacrifice of certain zones in favor of others" (Sodré, 1964: 313). Furtado (2006) demonstrates the concurrence of scarcity and abundance underpinning the national economy of the time. After the stock market crash in 1929, the price of the main Brazilian commodity (coffee) suffered a serious deterioration, although production continued to increase (peaking in 1933); excess coffee production was caused by a stagnant market and the country was forced to adopt radical anticyclical policies, including the physical destruction of stocks. Additional economic strategies started to focus on import substitution industrialization, emphasis on the domestic market, and greater economic integration of the different parts of the country.

In the 1930s, through resolute interventions by the federal government, the Centre-West region was targeted to become a main development frontier. The "March to the West" was a government program launched in 1938 in a context of national mobilization and economic expansion promoted by the authoritarian Vargas administration (1937–1945). The march involved exploratory surveys, road construction, agricultural colonization, and various economic incentives. The intention was to integrate and modernize the Centre-West, but the ultimate long-term goal was to prepare the terrain for the future economic incorporation of the Amazon. Mato Grosso, located both in the Centre-West and in the Amazon, then became the Holy Grail of land speculation, fueled by the easy, almost cost-free, concession of public land to private petitioners (Moreno, 2007). In the post–World War II period alone, the government of Mato Grosso indiscriminately titled around 4.2 million hectares of public land (Garfield, 2001, in Jepson, 2006). In the 1950s and 1960s, while Mato Grosso's vast territory was being partitioned among farmers and speculators, central economic areas in the eastern states were marked by escalating social conflicts and calls for labor reform and agrarian legislation. It was a situation of mounting scarcity felt by the majority of the population, which contrasted with the concentration of wealth resulting from industrialization and other developmentalist policies. Instead of seeking inclusive and democratic solutions to these fundamental challenges, the long Brazilian tradition of political shortcuts prevailed once again. In 1964, a military coup, expected and supported by the United States, instituted a long dictatorship (twenty-one years) that worked in favor of hyper-conservative modernization along the lines of state gigantism and technocratic developmentalism.

To satisfy public opinion, the generals even legislated on agrarian matters—notably, the 1964 Rural Land Statute—although this was a bureaucratic, anti-reform response that did not address structural problems. Their main agrarian strategy was actually the reinforcement of frontier-making in Mato Grosso and the rest of the Amazon. A vast agricultural frontier was established in the 1970s by the military government, which basically reinstated the same ideological orientation adopted in previous decades under the March to the West (Velho, 2009). Instead of food production or rural development goals, what actually happened was the formation of a "frontier of landed property" in state-owned areas. It was a process loaded with ideological and political claims, according to the long-established logic of promising abundance to overcome scarcity. Such massive internal migration closely observed the LSA, given that impoverished groups in the south and northeast were compelled to move to colonization projects in the newly opened frontiers, only to struggle with new rounds of scarcity and, in many cases, were forced to migrate again further inland. The Amazon was particularly targeted for the relocation of "troublemakers," that is, landless groups who were demanding, within the limits of a brutal military dictatorship, land for workers and peasants. Especially in the southernmost state of Rio Grande do Sul, farming areas established by the migration, more than 100 years earlier, of peasant families from Italy, Germany, and other European countries were suffering from acute land scarcity caused by the unequal agrarian structure, aggravated since the 1950s by the advance of agribusiness in the context of the Green Revolution. Many impoverished families were hurriedly transferred to the Amazon and, as a rule, found it very difficult to settle, produce and survive because of various economic and social adversities, which soon led to the abandonment and concentration of land. Relevant here is the observation by Sato (2000) that frontiers are actually in-between spaces that are occupied by different groups of in-between people.

In parallel with colonization schemes involving peasant families, private individuals and companies could acquire land from public authorities in order to secure property rights and then claim public subsidies and related benefits. If land along the Trans-Amazononian Highway in the state of Pará was the priority frontier for settling peasants in public colonization projects, Mato Grosso was the main arena for entrepreneurial resettlement.[1] Application for land titles was massively tainted by corruption and happened without even the most basic attention to circumstances on the ground, topographic features, local settlements, or even other documents issued for the same area (Ioris, 2017b). Land allocation typically involved the mere drawing of lines on a highly imprecise and crude map by government officials working hundreds of kilometers away from the area. The result was that the expansion of private property had very limited agricultural results, but it was nonetheless legitimized by the sheer abundance of land and, more importantly, generous government assistance and rural credit (Schwantes, 1989). As in other countries, the new farmers were more like speculators with no interest in working the land as a personal enterprise or in building up a long-term tenant estate, instead purchasing large sections of unimproved land, intending to sell after land values had risen sufficiently to make their sale remunerative (Bogue and Bogue, 1957). In total, the production of Mato Grosso's agricultural frontier involved more than 100 resettlement projects and incorporated around 4 million hectares between 1970 and 1990 (Jepson, 2006). Particularly in the Upper Teles Pires River basin in the center-north of Mato Grosso, land settlement was encouraged by the construction of a federal road, the BR-163, by the Brazilian army (Bernardes et al., 2016). Later, after intense timber extraction, this region became the main national center for soybean production (see below).[2]

Mato Grosso's land allocation experience deviated somewhat from the classical frontier-making model described in the literature, which hypothesizes the initial arrival of the first pioneers, typically subsistence farmers and adventurers with little to lose, who are then gradually replaced by more commercial farmers and private companies, leading to various degrees of tensions between the different groups. In Mato Grosso, there was no spontaneous transition from "farming" to an "agricultural sector" (as dealt with by Bernstein, 2010); what actually happened was the writing off of the socio-ecological circumstances and the imposition, through top-down state action, of exogenous practices and bureaucratic goals informed by values and priorities decided elsewhere. The aggressive mobilization of land and resources to create a large-scale agricultural region was triggered by the political and ideological vision of the federal state (especially after the 1964 military coup) on behalf of a conservative vision of modernity. The government championed national development and integration plans (financed largely from multilateral loans fueled by petrodollars) that triggered internal migration of farmers and fostered new rounds of primitive accumulation by private companies entitled to receive generous subsidies once they started operating in the Amazon (de Sartre, 2006). This is related to the claim by Hung (2014) that frontiers are based on simultaneous dualities (e.g., inclusion/exclusion, periphery/connection, modernity/primitiveness) that constitute essential dilemmas through which the state can "tailor" different meanings to meet contingent market demands.

Rather than bringing development to an area considered destitute (as hypothesized by Bauer, 1991), frontier-making in Mato Grosso was speculative and aimed to mitigate the scarcity of land and limited social opportunities elsewhere in the country. As result, there were serious doubts about how resilient this agricultural frontier would be without sustained and generous state assistance. Because of the gradual exhaustion of the frontier-making model introduced in the 1970s, which was highly dependent on public funds and direct state intervention, some authors concluded that it had become a decadent frontier and even anticipated the retreat of capitalism from the Amazon (Cleary, 1993). The frontier-making model introduced by the Brazilian military proved to be too costly, scandalously inefficient and obsolete, which became evident after the end of the dictatorial regime in 1985 and the subsequent fiscal crisis. Nonetheless, after a turbulent period of adjustments, and to the surprise of many, the agricultural frontier re-emerged and flourished from the late 1990s onwards, under a favorable convergence of agrarian reorganization (i.e., further land concentration), global market opportunities (the commodity boom in the 2000s), and macroeconomic liberalization policies (centered on monetary stability, global trade, financial speculation, and deindustrialization). This corroborated the claim by Pichón (1997) that land use at the frontier of agricultural development and colonization is always complicated and tends to be adaptive to new circumstances.

In the last two decades, Mato Grosso has transformed into one of the main agribusiness production hotspots in the planet, particularly powered by the strong demand for soybean from Asian markets. While the Brazilian economy has rapidly deindustrialized and the public deficit soared, Mato Grosso's agribusiness is one of the few economic sectors in the country able to demonstrate high levels of efficiency and significant contributions to trade surplus. Soybean exports have been particularly important because they provide foreign currency, which helps to stabilize the national economy and reduce the public deficit. The agricultural frontier, originally opened by Vargas and expanded by the military, has been reinvented as "corporate Amazonia" today under the sphere of influence of land investors, private banks, and transnational corporations (not only the big international names, Cargill, Bungle, ADM, and Dreyfus, but also impacted by new Brazilian TNCs, such as Amaggi, BRF, Marfrig, and JBS). The strategic road BR-163, build by the army during the military dictatorship to connect Mato Grosso to the neighboring Pará and to the important port city of Santarém, has been partly privatized and its operation transferred in 2013 to Odebrecht (one of the companies heavily involved in corruption scandals in recent years in Brazil). Mato Grosso is now, for the first time in its history, praised as an example of economic success and entrepreneurial efficiency. If one listens inattentively to the narrative of production efficiency and economic triumph—repeated daily by political leaders and echoed through the mass media—one would probably come to the conclusion that Mato Grosso now occupies the center of gravity in the Brazilian economy and political landscape. However, the acclaimed victory of Mato Grosso's agribusiness is actually an "interlocked concept" that preserves the perverse basis of frontier-making. Despite its crucial economic role and the growing influence of regional political elites, Mato Grosso never stopped being a frontier. More than that frontier-making never ended in Mato Grosso, but remains a perennial necessity, much more than a simple contingency.

Regardless of discursive and symbolic constructions, the current state of affairs in Mato Grosso organically depends on new and more sophisticated rounds of frontier-making. First of all, the defensive rhetoric repeatedly employed by agribusiness leaders, their academic allies and political representatives (particularly the powerful *ruralista* group present in the national congress)—for instance, the common claim that "Mato Grosso is the Brazil that is doing well" [*o Brazil que dá certo*]—is in itself an indication that relations are not yet settled, but that the region is still undergoing a process of socio-spatial consolidation. GMO soybean, which was for several years banned in Brazil, became the main economic activity in the state, justified and rationalized by the rhetoric of "sustainable tropical agriculture." Second, Mato Grosso's export-oriented agribusiness remains firmly subordinate to financial, technological, and political centers that determine what and how commodities should be produced and commercialized. Third, because of growing market demands and mounting socio-ecological degradation, as well as the rising price of rural land, the agribusiness sector is constantly in search of new production frontiers deeper inside the Amazon region. Fourth, Mato Grosso is an economic frontier in which abundance and scarcity are organically connected and follow each other. Neoliberalized agribusiness, since the late 1990s, has revived the promise of abundance, but even more rapidly restricted socioeconomic opportunities. According to the state and national statistics, 44 percent of the population lives in a situation of social vulnerability and 18 percent below the poverty line (Circuito Mato Grosso, 2016). Agribusiness accounts for around a third of state GDP, but social inequality and the quality of public services have been seriously affected by widespread corruption, which affects almost the whole political system and has led to the imprisonment of many corrupt authority figures, including a former state governor (El País, 2017). Since the 1970s, the opening of new farms and the intensification of agriculture production in Mato Grosso has been often associated with cases of grave violence and even slavery (first denounced by the catholic bishop Dom Pedro Casaldáliga in 1971), which has required the formulation of novel and more proactive forms of control by state agencies (Vieira, 2018). One of the most tragic results is the growing urban periphery, with abject levels of poverty, which is also a problem in the towns and villages in the main agribusiness areas. Increasing land prices over the last two decades have limited the possibility for smaller farmers to acquire their own property, creating incentives for intrastate migration and for further deforestation and associated environmental degradation (Ioris, 2016b).

The specific conditions for agribusiness expansion in Mato Grosso ultimately reproduced the monopoly of land-holding—a dwindling number of increasingly wealthier landowners—that characterized previous rounds of frontier-making. This situation of selective abundance and shared scarcity is further reinforced at the local scale of farms and production units. Frontier-making involves a dialectic between unique socio-spatial experiences and strong homogenization pressures. There is growing homogenization of production, given that almost all farms use the same technological package (GMO soybeans, large machinery, digital technology, agro-chemicals, etc.),[3] which has resulted in biodiversity erosion, water pollution and soil degradation. These socio-ecological impacts are all manifestations of growing scarcity. In addition, technological uniformity corresponds to a politico-ideological homogenization, as it is not easy to find public spaces to question and criticize regional development trends (even in the local universities). The associations that represent the agribusiness sector (e.g., FAMATO and Aprosoja) are centralized organizations with a façade of public participation but subservient to the politicians and large-scale farmers who virtually control the public sector of Mato Grosso. One notable example of the scarcity of alternative thinking is the appropriation of environmental regulation by the agribusiness sector, along the lines of ecological modernization (Baletti, 2014, describes it as the "greening" of soybean production via environmental governance). In this way, the agricultural frontier not only encroaches upon ecosystems, but also gradually appropriates the practice and rationale of environmental law, obviously focusing on market-based solutions (e.g., carbon trade, compensatory measures for past degradation in a location different from the one impacted, deforestation amnesties, etc.).

In the end, agribusiness, particularly the neoliberalized version that now prevails in Mato Grosso, has never undermined or reduced the importance of capitalist frontier-making. On the contrary, frontier-making continues to serve the stronger interests of the Mato Grosso elite (such an elite is itself the product of frontier-making) and the hegemonic demands of core political and economic centers elsewhere in the country (Brasília and São Paulo, above all). A supposed post-frontier stage, as announced by some scholars, has not materialized in the region, as frontier-making remains predicated upon the need to compensate for national and local tensions. Just like diamonds, rubber, and cattle in the past, the soybean-based economy of Mato Grosso persistently faces many challenges (aggravated by serious ecological degradation and global warming), but without fundamental politico-economic transformations, new cycles of frontier-making are likely to continue to ensue. In that context, the national state—as the key neoliberal player (Ioris, 2012)—maintains a leading role in frontier-making, both through the facilitation of processes of land grabbing, utility privatization and agriculture financialization, and through the flexibilization of social and environmental regulation.[4] Neoliberalism, as a class-based attempt to reorganize production and re-enact exploitation, proves to be an institutional and moral frontier of capitalism and, at the same time, encourages the production of new spatial frontiers.

CONCLUSIONS

Frontier-making has been an integral component of the evolution and renovation of capitalism, not merely because of the conquest of territories and resources, but because of the strategic interactions between old and new areas. The advance of capitalist relations of production and reproduction continues to rely today on accumulation by frontier-making, as illustrated by the persistent formation and constant reinforcement of socio-spatial frontiers in the Global South. The main reason is that the process of frontier-making simultaneously connects and transforms both the core and periphery circuits of capital. Marx and Engels (1970: 88) even argued that settlement frontiers, as in the case of North America, served to accelerate the progress of capitalism as the settlers bring "advanced forms of intercourse" that they were not able to establish in the old (European) countries, which is a phenomenon that will eventually impact the original areas from where the settlers departed. There is, therefore, a totality of relations that underpins capitalist frontier-making and that connects unique socio-spatial circumstances with the wider complexity of capitalist processes of production, reproduction, and legitimization. Frontier-making persistently happens because of the overflow of class-based disputes in geographically "consolidated" areas and the specific opportunities to establish novel socioeconomic mechanisms at the frontier. The unique space of each frontier is what Pred and Watts (1992) describe as "multiple modernities" consistent with "multiple capitalisms." In addition, Li (2014) perceptively points out that capitalism is not simply a totalizing system that incorporates everybody, and that the adoption of capitalist forces in frontier regions is not an inevitable phenomenon, but full of conjunctures (i.e., subject to various conditions).

Capitalist frontier-making was historically initiated as part of the European expansion around the world and over time it became a fundamental pillar of Brazilian political, social, and economic organization. Following the law of scarcity-abundance—which encapsulates the overall tendency to deal with mounting scarcity in central areas through the promise of abundance at the frontier—families and groups have, for many generations, departed from areas fraught with (human made) scarcity, attracted by the promises of abundance and a better life elsewhere in the country, as in the case of Mato Grosso (located in the southern section of the Amazon biome, at the transition between savannah and forested ecosystems, and renowned for its abundance of land, water, and biodiversity). According to the intertwined pledges of abundance and the more regular delivery of multiple scarcities, the expansion of Mato Grosso's economic frontier has deliberately been used to help regulate long-term, unresolved socioeconomic contradictions in the national economic centers. In that way, the perverse configuration of Brazilian society, especially and prominently characterized by acute inequalities and abject gulfs between poor and rich groups, was reinforced by the recurrent instrumentalization of frontier-making. Frontier-making in Mato Grosso also contributed to the spatialization of class struggle at different, nested scales of interaction: from local disputes over land and resources, to nation building and global flows of commodities and geopolitical concerns.

Frontier-making was not only important during colonial and early independence periods, but the modernization of Brazil unremittingly depended on the perennial production of new frontiers. It remained a crucial escape mechanism needed to alleviate mounting socioeconomic and socio-ecological tensions. As a consequence, Mato Grosso was converted into a triple frontier—at the periphery of Portugal, Brazil, and the Amazon region—and today it still functions as a technological, ethical, and political frontier, despite the fact that it currently contributes much more significantly to the national economy via the export of agribusiness commodities. The process of frontier-making was never interrupted or weakened, even now when Mato Grosso occupies a prominent position in the national politico-economic arena. Mato Grosso continues to be a zone of experimentation, migration, and reconstitution, where identities and allegiances are still shifting and the democratic rule of law has not yet materialized. The frontier was always promoted by the national state but the same state never had much interest in guaranteeing a socially and spatially inclusive institutional order to resolve land-related conflicts. Formal laws are diluted and largely replaced by other codes marked by pragmatism and etched by the balance of power between different social groups. For all these reasons, frontier-making in Mato Grosso is both a concrete experience and also a metaphor for the uncertainties created by the geography of capitalism in Brazil and the constant need to provide (albeit circumstantial and transient) responses to the problems accumulated in core and newly produced spatial areas.

NOTES

1. The focus on larger properties in Mato Grosso (under the 1974–1979 government of General Geisel) came a few years after the largely unsuccessful resettlement of peasants to small-scale properties in Pará (during the 1969–1974 administration of General Medici).

2. Interestingly, the Upper Teles Pires is a frontier within the wider frontier, as it is situated at the uncertain boundaries between savannah [*cerrado*] and the Amazon rainforest (the former has lower legal protection compared with forested ecosystems). The fuzzy frontier of the forest has been used by the farming community as an important excuse to clear more land and claim their "environmental prerogatives" (some even call it their "green passport"); a typical narrative is exemplified in this interview with a municipal authority: "we protect the forest because we mainly cultivate land previously occupied by the savannah."

3. Mechanization, digital equipment, and other related technologies help to control labor (a perennial problem in agriculture) and facilitate capital investment in farming.

4. The interpretation of frontier-making needs to be firmly situated in the political ecology of the state (Ioris, 2014) in order to understand the multiple ramifications of the power asymmetries that have helped to shape the frontier, including this movement from "nature as resource" to "nature as business opportunity."

BIBLIOGRAPHY

Agnew, J. 2000. From the Political Economy of Regions to Regional Political Economy. *Progress in Human Geography*, 24(1), 101–10.

Alencastro, L. P. 2001. *O Trato dos Viventes: Formação do Brasil no Atlântico Sul*. Companhia das Letras: São Paulo.

Alston, L. J., Libecap, G. D., and Mueller, B. 1999. *Titles, Conflict, and Land Use: The Development of Property Rights and Land Reform on the Brazilian Amazon Frontier*. University of Michigan Press: Ann Arbor.

Baletti, B. 2014. Saving the Amazon? Sustainable Soy and the New Extractivism. *Environment and Planning A*, 46, 5–25.

Bastos, C. 1998. (In)visible Borders: Ideologies of Sameness and Otherness in a Portuguese Context. *Portuguese Literary and Cultural Studies*, 1, 19–32.

Bauer, P. 1991. *The Development Frontier: Essays in Applied Economics*. Harvester Wheatsheaf: London.

Bernardes, J. A., Buhler, E. A., and Costa, M. V. V. (eds.). 2016. *As Novas Fronteiras do Agronegócio: Transformações Territoriais em Mato Grosso*. Lamparina: Rio de Janeiro.

Bernstein, H. 2010. *Class Dynamics of Agrarian Change*. Fernwood Publishing: Halifax and Winnipeg.

Billington, R. A. 1963. *America's Frontier Heritage*. University of New Mexico Press: Albuquerque.

Bogue, A. G. and Bogue, M. B. 1957. Profits' and the Frontier Land Speculator. *Journal of Economic History*, 17(1), 1–24.

Bromley, N. 2003. Law, Property, and the Geography of Violence. *Annals of the Association of American Geographers*, 93(1), 121–41.

Bukharin, N. 1929 [1917]. *Imperialism and World Economy*. Martin Lawrence: London.

Campbell, J. M. 2015. *Conjuring Property*. University of Washington Press: Seattle and London.

Circuito Mato Grosso. 2016. 44% da População de Mato Grosso Vive em Vulnerabilidade Social. Available at: http://circuitomt.com.br/editorias/cidades/91234-44-da-populaaao-de-mt-vive-na-linha-da-pobreza.html.

Cleary, D. 1993. After the Frontier: Problems with Political Economy in the Modern Brazilian Amazon. *Journal of Latin American Studies*, 25(2), 331–49.

Demangeon, A. 1932. Pionniers et Fronts de Colonisation. *Annales de Géographie*, 41, 631–36.

de Sartre, X. A. 2006. *Front Pionniers d'Amazonie: Les Dynamiques Paysannes au Brésil*. CNRS Éditions: Paris.

Di Tella, G. 1982. The Economics of the Frontier. In: *Economics in the Long View*, Kindleberger, C.P. and di Tella, G. (eds.). Palgrave Macmillan: London, pp. 210–27.

Dixon, M. 2014. The Land Grab, Finance Capital, and Food Regime Restructuring: The Case of Egypt. *Review of African Political Economy*, 41(140), 232–48.

Elliot, H. 2016. Planning, Property and Plots at the Gateway to Kenya's 'New Frontier'. *Journal of Eastern African Studies*, 10(3), 511–29.

El País. 2017. Mato Grosso, o Estado Rico Onde os Escândalos Políticos "Monstruosos" se Sucedem. Available at: https://brasil.elpais.com/brasil/2017/10/24/politica/1508811228_204258.html.

Findlay, R. and Lundahl, M. 2016. *The Economics of the Frontier: Conquest and Settlement*. Palgrave Macmillan: London.

Foweraker, J. 1981. *The Struggle for Land: A Political Economy of the Pioneer Frontier in Brazil from the 1930s to the Present Day*. Cambridge University Press: Cambridge.

Furtado, C. 2006. *Formação Econômica do Brasil*. Companhia das Letras: São Paulo.

Galaz, V., Crona, B., Dauriach, A., Jouffray, J.-B., Österblom, H., and Fichtner, J. 2018. Tax Havens and Global Environmental Degradation. *Nature Ecology & Evolution*, 2, 1352–57.

Heckenberger, M. 2005. *The Ecology of Power: Culture, Place, and Personhood in the Southern Amazon, A.D. 1000–2000*. Routledge: Abingdon.

Hemming, J. 1987. *Amazon Frontier: The Defeat of the Brazilian Indians*. Papermac: London.

Hennessy, A. 1978. *The Frontier in Latin American History*. Edward Arnold: London.

Hung, P.-Y. 2014. Frontiers as Dilemma: The Incompatible Desires for Tea Production in Southwest China. *Area*, 46(4), 369–76.

Imamura, M. 2015. Rethinking Frontier and Frontier Studies. *Political Geography*, 45, 96–97.

Ioris, A. A. R. 2012. The Neoliberalization of Water in Lima, Peru. *Political Geography*, 31(5), 266–78.

Ioris, A. A. R. 2014. *The Political Ecology of the State: The Basis and the Evolution of Environmental Statehood*. Routledge: London.

Ioris, A. A. R. 2016a. Water Scarcity and the Exclusionary City: The Struggle for Water Justice in Lima, Peru. *Water International*, 41(1), 125–39.

Ioris, A. A. R. 2016b. The Paradox of Poverty in Rich Ecosystems: Impoverishment and Development in the Amazon of Brazil and Bolivia. *The Geographical Journal*, 182(2), 178–89.

Ioris, A. A. R. 2017a. Places of Agribusiness: Displacement, Replacement, and Misplacement in Mato Grosso, Brazil. *Geographical Review*, 107(3), 452–75.

Ioris, A. A. R. 2017b. *Agribusiness and the Neoliberal Food System in Brazil: Frontiers and Fissures of Agro-neoliberalism*. Routledge: London.

Jepson, W. 2006. Producing a Modern Agricultural Frontier: Firms and Cooperatives in Eastern Mato Grosso, Brazil. *Economic Geography*, 82(3), 289–316.

Li, T. M. 2014. *Land's End: Capitalist Relations on an Indigenous Frontier*. Duke University Press: Durham and London.

Little, P. 2001. *Amazonia*. John Hopkins: Baltimore.

Lucidio, J. A. B. 2013. "A Ocidente do Imenso Brasil": As Conquistas dos Rios Paraguai e Guaporé (1680–1750). Ph.D. Thesis. Universidade Nova de Lisboa: Lisbon.

Luxemburg, R. 1951 [1913]. *The Accumulation of Capital*. Trans. Schwarzschild, A. Routledge and Kegan Paul: London.
Marston, S. A. 2000. The Social Construction of Scale. *Progress in Human Geography*, 24(2), 219–42.
Marx, K. 1976 [1867]. *Capital*. Vol. 1. Trans. B. Fowkes. Penguin: London.
Marx, K. and Engels, F. 1970 [1846]. *The German Ideology*. Lawrence & Wishart: London.
Massey, D. 2006. Space, Time and Political Responsibility in the Midst of Global Inequality. *Erdkunde*, 60(2), 89–95.
Mészáros, I. 1995. *Beyond Capital: Toward a Theory of Transition*. Merlin Press: London.
Mikesell, M. W. 1960. Comparative Studies in Frontier History. *Annals of the Association of American Geographers*, 50(1), 62–74.
Moore, J. W. 2015. *Capitalism in the Web of Life: Ecology and the Accumulation of Capital*. Verso: London and New York.
Moreno, G. 2007. *Terra e Poder em Mato Grosso: Política e Mecanismos de Burla 1892–1992*. Entrelinhas: Cuiabá.
O'Connor, J. 1998. *Natural Causes: Essays in Ecological Marxism*. The Guilford Press: New York and London.
Pacheco de Oliveira, J. 2016. *O Nascimento do Brasil e outros Ensaios: "Pacificação", Regime Tutelar e Formação de Alteridades*. Contra Capa: Rio de Janeiro.
Pichón, F. J. 1997. Colonist Land-allocation Decisions, Land Use, and Deforestation in the Ecuadorian Amazon Frontier. *Economic Development and Cultural Change*, 45(4), 707–44.
Prado Jr., C. 1967. *The Colonial Background of Modern Brazil*. Trans. S. Macedo. University of California Press: Berkley and Los Angeles.
Pred, A. and Watts, M. J. 1992. *Reworking Modernity: Capitalisms and Symbolic Discontent*. Rutgers University Press: New Brunswick.
Richards, J. F. 2006. *The Unending Frontier: An Environmental History of the Modern World*. University of California Press: Berkeley.
Santos, B. S. 1993. Modernidade, Identidade e a Cultura da Fronteira. *Revista Crítica de Ciências Sociais*, 38, 11–39.
Sato, J. 2000. People in Between: Conversion and Conservation of Forest Lands in Thailand. *Development and Change*, 31, 155–77.
Schwantes, N. 1989. *Uma Cruz em Terranova*. Scritta: São Paulo.
Smith, G. 1989. *Livelihood and Resistance: Peasants and the Politics of Land in Peru*. University of California Press: Berkeley.
Sodré, N. W. 1964. *Formação Histórica do Brasil*. 3rd ed. Brasiliense: São Paulo.
Solimano, A. 2017. *Global Capitalism at Disarray: Inequality, Debt and Austerity*. Oxford University Press: Oxford.
Turner, F. J. 1920. *The Frontier in American History*. Holt: New York.
Velho, O. G. 2009 [1976]. *Capitalismo Autoritário e Campesinato: Um Estudo Comparativo a Partir da Fronteira em Movimento*. Centro Edelstein de Pesquisas Sociais: Rio de Janeiro.

Vieira, M. M. R. 2018. Trabalho Escravo Contemporâneo em Mato Grosso: Denúncias e Estratégias de Combate. In: *Nova História do Mato Grosso Contemporâneo*, Rodrigues, C. and Joanoni Neto, V. (eds.). EdUFMT: Cuiabá, pp. 523–50.

Watts, M. J. 1992. Space for Everything (A Commentary). *Cultural Anthropology*, 7(1), 115–29.

Watts, M. J. 2018. Frontiers: Authority, Precarity, and Insurgency at the Edge of the State. *World Development*, 101, 477–88.

Webb, W. P. 1952. *The Great Frontier*. University of Texas Press: Austin.

Wilcox, R. W. 2017. *Cattle in the Backlands: Mato Grosso and the Evolution of Ranching in the Brazilian Tropics*. University of Texas Press: Austin.

Chapter 3

Conservation Frontier

The Creation of Protected Areas in the Brazilian Amazonia

Frederico Freitas

On June 5, 1979, the president of Brazil, General João Batista Figueiredo, visited the headquarters of the Brazilian Institute for Forest Development (IBDF) in Brasília for a fifteen-minute ceremony celebrating the United Nations' World Environment Day. Figueiredo brought along an entourage that included the ministers of Justice, the Interior, Agriculture, and the Air Force, representing some of the offices responsible for frontier development projects in Brazil during the country's military regime (1964–1985). At the event, Figueiredo signed presidential decrees to establish one biological reserve and two national parks. One of the new protected areas, the Pico da Neblina National Park, encompassed 8,670 square miles of mountainous tropical forest at the border with Venezuela, becoming the largest national park in Latin America at the time. Pico da Neblina was part of a group of extensive national parks established between 1974 and 1980 in Amazonia, during which time the Brazilian government set aside 27,220 square miles of rainforest, an area the size of Ireland, for protection.[1]

In the short speeches that preceded the signature of the decrees, the president of the IBDF and the Minister of Agriculture emphasized the government's effort in "promoting policies of territorial occupation that did not harm the environment" (Jornal do Brasil, 1979). Worthy of note in their presentation was the alignment of environmental protection inside the parks with the colonization policies deployed in Amazonia at the time. To be sure, starting in the 1930s, government officials systematically targeted the region as part of frontier development plans, a path that increased in scale and scope in the 1970s when the military regime implemented an array of initiatives to bring roads, settlers, and investment to the country's expansive northwestern territories. The same period witnessed the mobilization of a small group of Brazilian and foreign conservationists working at federal agencies, such as the IBDF, and international bodies, such as the Food and Agriculture Organization of the United Nations (FAO), who managed to convince the Brazilian military to include nature reserves in their grand plans for Amazonian frontier development.

Pico da Neblina was one of the first national parks created in Brazilian Amazonia, though protected areas had already been established in other frontier areas before the 1970s. In 1939, for example, the Brazilian federal government created the Iguaçu National Park on the southern border with Paraguay and Argentina, and between 1959 and 1961 four new parks were established in the Cerrado, Brazil's interior savanna.[2] As offshoots of plans for frontier development, these earlier national parks were paired with infrastructure projects such as roads, airports, and hydroelectric dams, although they lacked the commitment to wildlife conservation that guided the creation of the new Amazonian national parks of the late 1970s.[3] The establishment of the new parks brought about four critical changes in the way the Brazilian state managed nature conservation and territorial control. First, it helped to establish a survey-based, scientific method for proposing national parks that made them more palatable for politicians, technocrats, and the military outside conservationist circles. Second, it cemented the notion of cordoning off crucial wildlife habitats from impending economic growth in the vocabulary of Amazonian territorial planning. Third, it provided the military, and the subsequent civilian administrations, with a convenient tool to respond to the international criticism of the negative effects of Brazilian development. Fourth, it presented the territory of Amazonia through a different lens, as a frontier to be protected, eventually contributing to the vision of the region as the world's natural heritage after decades of being imagined solely as Brazil's development frontier.

As Acker (2017) argues in his study on the backlash against a Volkswagen project in Amazonia, this move from a national frontier (to be developed) to an international one (to be conserved) happened in several spheres—local, national, and international. This chapter focuses on one specific aspect of this process: the establishment of protected areas as a result of a technocratic reimagination of Amazonian geography. Despite being a commonsense notion today, commitment to nature protection in Amazonia was a novelty in the 1970s, the chronic failure of the Brazilian state to live up to its promises notwithstanding. Protected areas have turned into one of the main instruments used by the Brazilian state to promote its post-1970s mandate of controlling deforestation in Amazonia (Barber et al., 2012, 2014; Nepstad et al., 2006). The national parks established by Figueiredo in 1979 and 1980 were important for introducing concrete measures in response to a new conception of

Amazonia as a *conservation frontier*—a territory with endangered environmental features demanding new regimes of land use and nature protection.

As demonstrated by Foresta (1991)—and to a lesser extent by Drummond (2016)—the Amazonian protected areas resulted from an alliance between foreign scientists working for intergovernmental agencies and Brazilian civil servants acting within the government. In the late 1970s, these actors succeeded in conceptualizing protected areas in a way that the generals who controlled Brazil during the dictatorship could understand and accept. Crucial in this process was reimagining the spatial configuration of Amazonia as a frontier of development projects with room for conservation initiatives. This piece investigates how decades of territorial planning in frontier areas led to the establishment of extensive protected areas in Brazilian Amazonia. This is first done by retracing the evolution of federal plans for the occupation and development of the region between the 1930s and the 1970s. This was a period when Brazil created its first eighteen national parks, mostly outside Amazonia, a process that is examined in the subsequent section. In the final portion of the chapter, I demonstrate how the introduction of new plans for Amazonian development created the conditions for the establishment of the first network of federal protected areas in the region. These new development plans brought not only technical and logistical resources, but also a new spatial representation of Amazonia, which allowed park proponents, within and outside the government, to identify areas to be conserved in the space between proposed development zones. The history of the creation, in the 1970s, of the first protected areas in Amazonia reveals how processes of state-led frontier expansion can create the institutional conditions for the introduction of ambitious conservation policies.

ENVISIONING AMAZONIA AS A DEVELOPMENT FRONTIER

Before being designated as a recipient of development projects and conservation programs, Amazonia needed first to be imagined as a developmental frontier at the higher echelons of the Brazilian government. There is a common thread connecting ideas about the Brazilian frontier in the late 1930s to the plans for Amazonian development and conservation in the 1970s. This thread links all those who thought of the frontier as the place where Brazil would develop its full potential as a nation. The need to develop Brazil's northwestern frontier was first placed at the center of the national debate during the *Estado Novo*—the period between 1937 and 1945 when President Getúlio Vargas shut down Congress and assumed dictatorial powers. In 1937, in a radio broadcast on New Year's Eve, Vargas announced that the "true spirit of Brazilianness" rested on the integration of the country's hinterland with its economic core, a campaign baptized as "The March to the West" (Vargas, 2011). Vargas presented the Brazilian people with their mission: the creation of a unique civilization in the New World. This demanded the integration of the entire territory into a harmonious whole, one in which the rational exploitation of natural resources in the western frontier region would take place under his guidance (Vargas, 1938: 163–66; 1941: 23–24, 32–33; Ricardo, 1940). Vargas's March to the West inaugurated the modern era of frontier development in Brazil, a process that continued through the civil and military governments that followed the end of the Estado Novo in 1945. The military who controlled Brazil between 1964 and 1985 gave special attention to Amazonia, articulating geopolitical concerns to justify the need for frontier development in a context of political, economic, and demographic pressure. Military leaders believed their mission included not only fighting external aggression and internal insurgency in the context of the Cold War but also promoting economic development and the physical integration of national territories. The latter meant resuming the March to the West toward the country's final frontier, Amazonia (Child, 1979; Couto e Silva, 1957: 37–39).

General Golbery do Couto e Silva was undoubtedly the central figure influencing the Brazilian military regime in its frontier development plans for Amazonia in the 1960s and 1970s. During his long career in the Brazilian Army he occupied many positions: chief of operations of the General Staff, head of the National Security Council, director of the National Intelligence Service, lecturer and researcher at the Brazilian War College (ESG), and close adviser to three of the five military presidents of Brazil between 1964 and 1985. Couto e Silva formed his ideas on frontier development in Amazonia in the 1950s when he was a lecturer at the ESG. His tenure at the military college provided him with a platform to make his vision of the geopolitical future of Brazil known among military officers (Child, 1979; Hecht and Cockburn, 1989: 102–3; Keck and Hochstetler, 2007: 143–45; Pádua and Quintão, 1982). In essays written between 1952 and 1960, Couto e Silva portrayed the Brazilian territory as a collection of isolated economic and cultural regions. The largest of these regions, Amazonia, was also the most isolated, lacking land connections to the agrarian areas of the Northeast and the industrial core of the Southeast, rather relying on backward river and coastal shipping for trade. Thus, in the eyes of this prominent advocate of top-down development plans, Brazil's primary task was to "integrate" the national territory via modern land transportation routes (Couto e Silva, 1957: 37–44). He saw Amazonia as a "green desert"—internally homogeneous in its lack of population and development (Couto e Silva, 1967: 47–49, 56). It was necessary to develop new overland routes to enable migratory flows from the east and south, consolidating national unity through "an incessant and leveling acculturation," "indiscriminate miscegenation," and economic and social development in the Amazonian frontier (pp. 111–13, 136–38). With his lectures at the ESG, he presented Amazonia to a military audience as a coherent and unvaried region, one to be penetrated, settled, and developed. In this way, he provided the military regime coming to power in 1964 with a mission and a sense of purpose—in his own words, settling Amazonia would represent a "gigantic task that will require careful long-term planning and take years to realize" (Couto e Silva, 1957: 40).

The settlement of Amazonia demanded significant planning. Vargas's *Marcha para o Oeste* (March to the West) dealt primarily with the country's Central-West region, but his regime also focused on Amazonia, launching a program during World War II to revive rubber production via the import of laborers from the drought-ridden Northeast (Garfield, 2013; Lourenço, 2009; Mahar, 1979: 3–6). Interest in developing the region continued in the subsequently elected administrations that followed the end of the Estado Novo in 1945. The writers of the post-Vargas constitution earmarked 3 percent of the federal tax revenue to Amazonia, and Vargas himself, after his return to the presidency in 1951, as an elected president, created the Superintendence for the Economic Valorization of Amazonia (SPVEA) in 1953 to manage the new budget. The agency helped build the first two major roads into the forest: the BR-14, connecting Brasília, the new capital built under President Juscelino Kubitschek (1957–1961), to the Amazonian port city of Belém; and the BR-29, linking Brasília to the Amazonian territory of Acre (Sawyer, 1984). A 1953 law introduced an official definition and delimited boundaries for the region—Brazilian Amazonia now legally encompassed almost 60 percent of the country's territory (Mahar, 1979: 6–12). The new limits created Amazonia as a concrete object of federal policy, introducing a unique land and fiscal regime that set it apart from the rest of the country.[4] This new legislation would also reinforce the Amazon basin as a defined territory, one that could be identified and objectified in the maps and minds of federal planners. The Amazonia conceived by SPVEA stemmed from a technocratic view of territorial intervention, a vision that would not only guide the implementation of development programs in the 1970s, but also inform the establishment of the region's first protected areas in the same decade (Mougey, 2018).

By the 1960s, during the civil government of João Goulart (1961–1964), SPVEA was being accused of corruption and misuse of funds, and in December 1963, government technicians, intellectuals, and scientists met in Rio de Janeiro to devise a plan to overhaul the agency. The plan, named "Operation Amazonia," aimed to steer the agency toward the promotion of import substitution industrialization in the region through a combination of official investment, subsidized credit, and fiscal incentives. A few months later, in April 1964, the military forcefully ousted Goulart before he could seriously consider the new plans for SPVEA (Correio da Manhã, 1964; Diário de Notícias, 1964; O Jornal do Rio de Janeiro, 1963). The new military president, General Humberto Castelo Branco, established a five-member committee to launch his own, and more radical version, of Operation Amazonia. The committee replaced SPVEA with the Superintendency of Development for the Amazon (SUDAM) and created a new regional bank, the Amazonia Bank (BASA), to finance projects in mining and energy in the region.[5] The military version of Operation Amazonia differed from the 1963 proposals by focusing on attracting international investment to the forest. However, it also included some strong geopolitical content, influenced by Couto e Silva's views on the importance of colonizing Amazonia (Acker, 2017: 56; Mahar, 1979: 10–13). In a speech in 1966, introducing the new program to local businesspeople and politicians in Manaus, Castelo Branco reproduced some of Couto e Silva's ideas about Amazonia, arguing that the nation had a "superior and permanent interest" in diverting the "internal migratory flows" of eastern Brazil to encourage the "human occupation" of the region and in carrying out the "gradual conquest, incremental and planned, of its empty spaces" (Castelo Branco, 1966; Correio da Manhã, 1966; Jornal do Comércio, 1966).

Amazonia covered 60 percent of Brazilian territory but was home to only about 5 percent of its population in the early 1960s. The military saw in this demographic void a problem to be solved with the provision of accessible land and jobs in the region, and launched their first significant plan for Amazonian development in 1970 to bring in migrants from other parts of Brazil (Foresta, 1991: 24–25; Wood and Schmink, 1993). The military's plan to integrate Amazonia with the rest of the Brazil involved the creation of self-sufficient settlements and businesses in frontier areas. To achieve this goal, the government established the National Institute for Colonization and Agrarian Reform (INCRA), a new agency with extensive powers to expropriate land and implement colonization projects at the frontier.[6] The military also introduced the Program of National Integration (PIN), which focused on the combined development of Amazonia and the Northeast through the construction of two new highways into the rainforest: the famous 2,500-mile long east-west Trans-Amazonian Highway and the 1,100-mile long north-south Cuiabá-Santarém Highway.[7] The Trans-Amazonian Highway was designed to encourage peasants from the drought-ridden, impoverished Northeast to move to well-watered, sparsely populated Amazonia to the west. The military planned to settle Northeasterners in ten-kilometer swathes along both sides of the planned roads (Brazil, 1971; Andreazza, 1972). These projects would, in the words of Brazilian president Emílio Garrastazu Médici, solve two problems, "that of the men without lands in the Northeast, and that of land without men in Amazonia" (Correio da Manhã, 1970; Mahar, 1979: 17–21; Wozniewicz, 1974; Hecht and Cockburn, 1989: 108–10).

Though actively pursued, PIN was mired in failure and disappointment. After four years, the Brazilian government had attracted to the forest only 6,000 of the 100,000 families initially planned, and, by 1975, PIN had been abandoned by the Brazilian government (Foresta, 1991: 27–28; Guimarães, 1991: 121; Moran, 1990: 78; Wood and Schmink, 1993). In that same year, the military launched a new five-year plan to help the Brazilian economy cope with the oil crisis and to overhaul frontier development in Amazonia (Velloso, 1974). The plan, co-signed by Golbery do Couto e Silva, now acting as Chief of Staff of the Presidency, introduced a new strategy for Amazonia that emphasized the heterogeneity of an area 1.5 times the size of India, identifying fifteen development poles with different profiles to receive investments from the federal government and private companies (Brazil, 1975).[8]

The new plan, called Program for Agricultural, Cattle, and Mineral Poles in Amazonia (POLAMAZONIA), was devised to correct the failures of the PIN and the "disorderly settlement process" that the opening of new roads and the previous federally funded colonization project had created.[9] POLAMAZONIA was aimed at promoting the "rational" occupation of Amazonia through the identification of distinct zones with diverse developmental prospects within the region (SUDAM, 1976). It steered the government's focus from settlement with small farmers to large-scale development projects led by private actors. New remote sensing data helped to define poles of colonization, mechanized agriculture, cattle ranching, mining, and industrialization (see figure 3.1). Where zones with fertile soil would receive settlements and plantation agriculture, other areas, unsuitable for crop production, would accommodate large-scale cattle ranching, which was seen as a method to improve soil fertility (Acker, 2017: 62–63; SUDAM, 1976). Planners forecasted the introduction of 5 million head of cattle to the region by 1980. Fiscal incentives and abundant cheap credit were used to attract national and international capital to Amazonia, creating the conditions for the exploitation of bauxite, aluminum, iron ore, and tree plantations for pulp and paper. Logging would be "rationalized" in the zones not selected for colonization, agriculture, and mining (Acker, 2017; Foresta, 1991: 29–31; Guimarães, 1991: 176–86; Mahar, 1979: 27–28; Panagides and Magalhães, 1974; SUDAM, 1976; Wood and Schmink, 1993).

Several key elements of POLAMAZONIA were already present in the PIN plan: fiscal incentives, development poles, mining projects. The difference between the two plans was one of emphasis. Whereas PIN focused on colonizing the area with peasants, POLAMAZONIA's objective was to attract capital. In this sense, the latter plan emphasized the myriad investment opportunities in Amazonia—mining, energy, agribusiness, livestock, logging, and manufacturing. This variety of opportunities demanded a more sophisticated view of Amazonia as a frontier of national development, which, along with new data

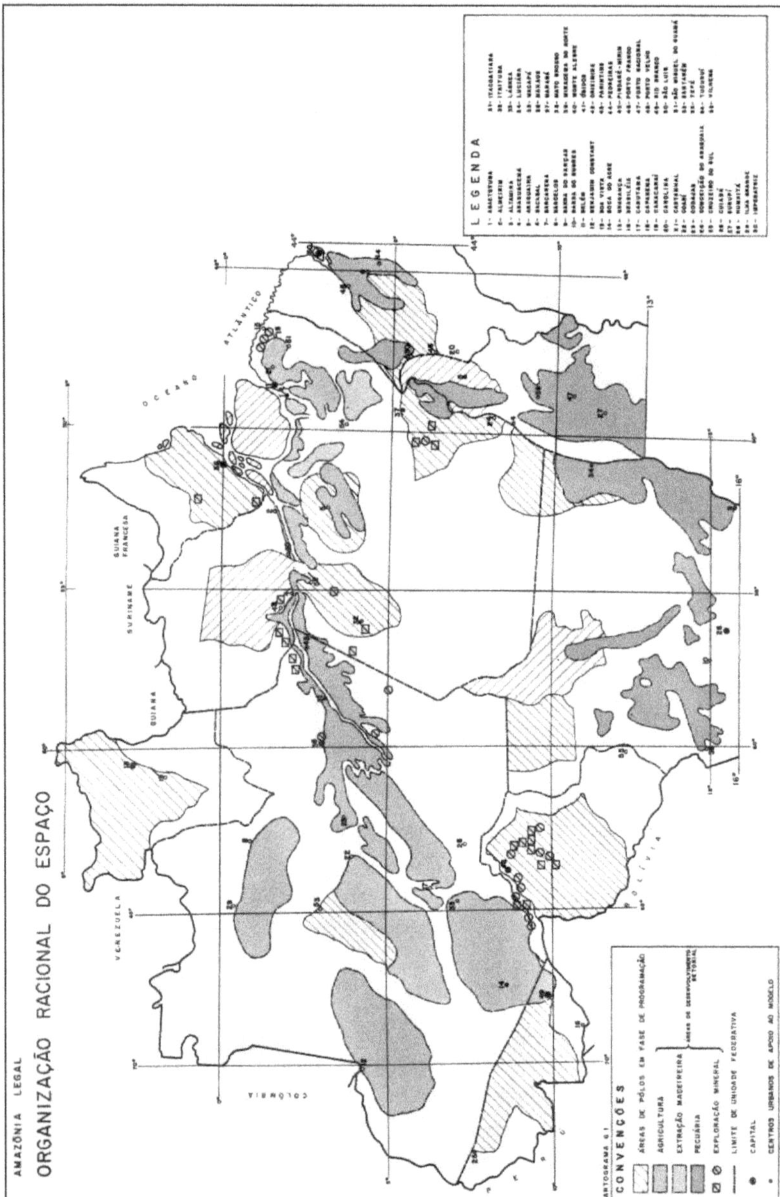

Figure 3.1 Legal Amazonia: Rational Organization of Space. *Source:* SUDAM, 1976.

on ecology, geology, and pedology, led the writers of POLAMAZONIA to plan for a "spatially discontinuous" occupation of the region (SUDAM, 1976: 50). However, if the program brought the land zoning of Amazonia to the forefront of planning, it had little to say about conservation. A single sentence in the 150-page document introducing the program suggested the need to define "national parks and national forests, biological reserves, and hunting parks" as part of the "development of forest resources" (Brazil, 1975: 68). The maps produced by POLAMAZONIA were also silent on conservation, showing no protected areas in the mosaic of agricultural, cattle ranching, and mining zones (see figure 3.1). Still, these maps contained several areas devoid of planned economic activities. In the following years, Brazilian conservationists working in the country's environmental agencies, aligned with foreign FAO technicians, would use these blank spaces to their advantage, proposing protected areas that avoided overlap with regions set for development.

In the three decades that followed the launch of the March to the West in 1937, Brazil's civilian and military governments gradually increased the level of territorial intervention in Amazonia. As shown in the following sections, this growing federal engagement in the northwestern frontier was followed by the first proposals for protected areas made by conservationists working within the government bureaucracy. However, these initial proposals were largely unsuccessful, and Brazil had no protected area created in central Amazonia in the 1950s and 1960s. The creation of the first coherent system of national parks in Amazonia would have to wait for a change in the way policy-makers imagined the region in their development plans, a shift introduced by POLAMAZONIA. Acceptance of national park proposals would also require convincing technical criteria to persuade the generals to set apart land for conservation. The latter was developed by U.S. technicians, who adapted to Amazonia a systematic approach to national parks first developed for other areas of Latin America.

THE FIRST PROPOSALS FOR PROTECTED AREAS IN AMAZONIA

Brazilian conservationists and government technicians had proposed the creation of protected areas in Amazonia before the 1970s, but without success. In 1911, a group of scientists and politicians convinced the Brazilian government to designate a forest reserve in the Territory of Acre, but the protected area was never implemented (Vital, 2018). Half a century later, President Jânio Quadros created nine "forest reserves" in the region in 1961. As explained by Drummond and Franco (2013: 343), these protected areas were the brainchild of Victor Abdennur Farah, executive director of the

Brazilian Foundation for the Conservation of Nature (FBCN—a conservationist group founded in 1953), member of the Federal Forest Council, and personal friend of President Quadros.[10] However, the Brazilian government never acted to demarcate these reserves. Indeed, the category "forest reserve" was eliminated in an overhaul of Brazilian environmental law in 1965, which threw the areas created in 1961 into legal limbo. By 1978, these nine reserves existed only as part of an outdated piece of legislation, and new settlements had already sprung up within their boundaries.[11]

FAO forestry technicians working in Brazil in the 1960s unsuccessfully proposed another group of Amazonian forest reserves. Since 1954, FAO had been working with the SPVEA to survey the forestry potential of Brazilian Amazonia. The agency's forestry department established a local office in Belém and brought in foreign technicians, mostly Europeans, to study different areas of rainforest (Dinius, 2019). In the period between 1954 and 1962, the agency's technicians produced seven different surveys in eastern Amazonia with embedded forest reserve proposals (s and Smit, 1962; IBDF and IBRA, 1969: 28; Heinsdijk, 1961; McGrath, Gachot, and Gallant, 1953). The last survey, produced by Glerum and Smit in 1962, involved FAO technicians attempting to use forest reserves to mitigate the unintended effects on the rainforest of frontier development projects. The survey was carried out at the construction site of the highway BR-14, alongside which FAO technicians "strongly recommended" the creation of a large, 2,400-square-mile forest reserve that would allow limited selective logging, but would prevent the "ruthless destruction" of valuable timber by newcomers following the opening of the road. The 1962 survey provides an early example of international technicians working with their Brazilian counterparts to propose protected areas in combination with development projects in Amazonia; a trend that strengthened from the 1970s onwards.

The Brazilian government never established the forest reserves planned by FAO technicians in 1962, but new proposals were presented in 1966 when the Ministry of Agriculture designated a national park committee to assess the status of all the Brazilian national parks created up to that point (FBCN, 1967: 22; IBDF and IBRA, 1969). It would be the first time government technicians visited most national parks as part of a larger initiative to restructure Brazil's haphazard environmental policy under a single national agency. A year later, in February 1967, the government created the IBDF as the result of the merging of several smaller agencies under the aegis of the Ministry of Agriculture.[12] A committee led by Brazilian botanists Harold E. Strang and Henrique P. Veloso traveled over 18,000 miles by airplane, automobile, boat, and horse, through isolated areas with limited transport infrastructure. Surprisingly, little of what they saw was Amazonia (Strang and Veloso, 1970). The Brazilian national parks established at the time were mostly

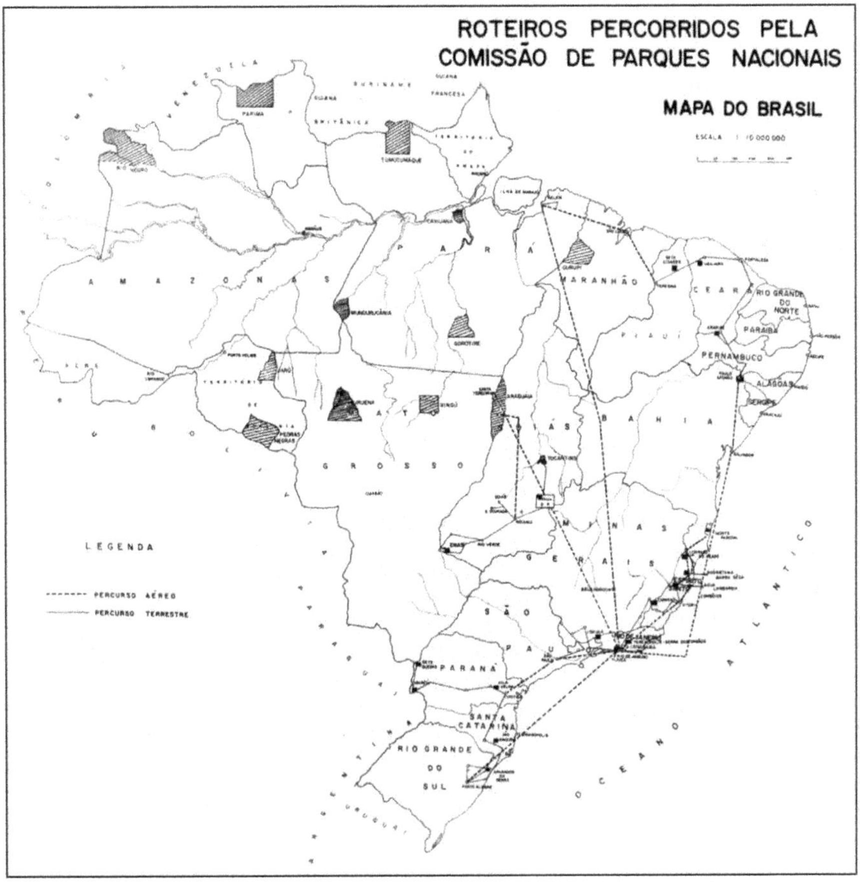

Figure 3.2 Routes Traveled by the National Park Commission. *Source*: IBDF and IBRA, 1969.

located outside the region, in the eastern half of the country (see figure 3.2). The committee did visit the fringes of the rainforest in the Araguaia National Park, situated in the transition zone between Amazonia and Cerrado. They also attended a symposium on the biota of the rainforest in Belém, the main Atlantic port serving the Amazon region. However, they never managed to reach the core of Amazonia and acquired most of their knowledge on the region from scientists and technicians attending the event. Despite their little experience with the rainforest, government officials produced a report strong on Amazonian policy, stressing that a national park "had yet to be established" in the region (IBDF and IBRA, 1969: 13). The document recognized that the "forest reserves" designated in 1961 (depicted in figure 3.2) had lost their legal standing after 1965, and suggested new legislation to change their denomination and adapt them to the new Forest Code (pp. 22–26). In fact, following the lead of local scientists, the national park committee recommended the preservation of at least 10 percent of Amazonia to protect wildlife (e.g., turtles, manatees) and plant species (e.g., mahogany) from overexploitation (pp. 26–28). Two of the protected areas proposed in the 1966 report would later be established in 1979: the aforementioned Pico da Neblina, and a turtle nesting area at the Trombetas River, the latter established as a biological reserve (pp. 73–74).

In the 1950s and 1960s, FBCN conservationists, FAO technicians, and employees at the Ministry of Agriculture introduced several plans for establishing protected areas in Amazonia.[13] Despite all the surveying, planning, and even new legislation, none of these proposals were implemented. Parks were established only when the military regime started carrying out its plans to integrate the Amazonian frontier with the rest of Brazil in the 1970s, in a clear case where development and colonization created the conditions for a state conservation apparatus. The first national park established in Brazilian Amazonia was the Amazonia National Park, in 1974—a direct result of the construction of the Trans-Amazonian Highway. Following its plans to use frontier colonization and the expropriation of idle lands in Amazonia to solve Brazil's agrarian problem, the federal government nationalized in 1971, 13 million acres along the middle section of the planned Trans-Amazonian Highway to establish colonization projects based on family agriculture and cattle ranching. In February 1974, a committee formed by members of INCRA, SUDAM, IBDF, and the Ministry of Agriculture, agreed to set apart a section of this vast expanse of expropriated land as forest reserves. The federal government designated the area as the Amazonia National Park and the Tapajós National Forest and transferred the land—already under INCRA's control—to the IBDF.[14] The new national park alone contained more than 2.4 million acres of protected lands, about the size of Yellowstone, becoming the largest national park in the country, with a territory ten times bigger than the average Brazilian national park at the time. Technicians at IBDF's national park office saw this as an opportunity to easily create the first real national park at the core of Amazonia at a time when the government invested heavily in the development of the region. A few months later, in September 1974, the new administration of general Ernesto Geisel switched from the PIN to POLAMAZONIA. The new program established mining and agricultural projects adjacent to the new national park, which gave the IBDF the chance to request POLAMAZONIA funding for the implementation of the park (Brazil, 1971: 27–31; Drummond, 2016; Ministério da Agricultura, IBDF, and POLAMAZONIA, 1978: 4–6). The first major protected area in the region owed its existence to the federal institutions created to promote economic development and colonization in Amazonia.

Technicians from INCRA, a land and colonization agency, had decided the location and shape of the new Amazonia National Park without scientific input from the IBDF, the country's forestry and environmental agency. The park lacked both notable geological formations suitable to become monuments and endemic endangered species worthy of protection (Foresta, 1991: 33, 49). What it offered instead was a solution to the burgeoning problem of deforestation. The 1970s witnessed the growth of the debate about tropical deforestation, as conservationists in Brazil and abroad began studying the effects of development projects in Amazonia (Lovejoy, 1973; Keck and Hochstetler, 2007: 157–60; Keck and Sikkink, 1998; 134–35; Macekura, 2015).[15] International organizations such as the International Union for Conservation of Nature (IUCN) started pressuring the Brazilian government to include national parks and other protected areas in the country's development plans (IUCN, 1972: 18). The establishment of a national park by INCRA was a response to pressure from conservationists abroad. Establishing a national park in a sparsely inhabited area in Amazonia was an easy way to deflect international criticism—the government had already parceled the land, and a park with no visitation potential would require little infrastructure investment.[16]

The creation of the Amazonia National Park reflected the incomplete knowledge of the region demonstrated in the planning of the PIN in 1975 and the lack of a systematic approach to conservation by the IBDF. Changes followed the introduction of POLAMAZONIA, whose authors abandoned the simplistic vision of Amazonia as a homogeneous expanse that had implicitly informed the PIN. They drew from surveys and scientific studies to understand the forest's heterogeneity, identifying smaller zones suitable for different types of economic development. The new understanding of Amazonia as a mosaic of development zones ushered in the creation of more protected areas, as conservationists started to propose parks in the spaces between these zones. It was only after Amazonia had been reimagined by government planners as a collection of places to be developed that the proponents of national parks could identify in the region places to be protected. Technicians within the IBDF would play an important role here, partnering with FAO to introduce to the country a systematic approach to national parks tested in other Latin American nations. FAO's engagement with national parks started in 1962 when its forestry commission established the Latin American Committee on National Parks and Wildlife to help countries in the region with crafting environmental legislation, conducting studies, proposing protected areas, and training personnel (Wetterberg, 1974; Pizarro, 2012). By the early 1970s, the Committee on National Parks had successfully worked with governments as diverse as Colombia and Cuba in the planning of protected areas, but in Brazil, the FAO limited itself to agriculture and forestry development (Barreto Filho, 1997: 5–7; IBDF, 1975; Ministério da Agricultura, 1972).

The establishment of protected areas in Brazilian Amazonia after 1979 was the combination of a new engagement of the FAO in national park planning in Brazil with a push within IBDF to establish a systematic process for proposing national parks based on field surveys, ecological parameters, and political calculus. Whereas FAO provided the scientific rationale for national parks in the region, IBDF translated it to the language of regional development spoken by the technocrats crafting development plans for Amazonia. Two people were crucial in this effort: on the FAO side, Gary B. Wetterberg, an American conservation scientist sent to Brazil by the Committee on National Park and Wildlife; in the IBDF, Maria Tereza Jorge Pádua, a Brazilian soils scientist working on the small department of nature preserves within the federal agency. Pádua, along with a few allies in the IBDF, was the main force behind the overhaul of the Brazilian national park system, carrying out the technical surveys and producing the political groundwork that convinced the Brazilian military to sign the decrees establishing a slew of new protected areas. The role of the IBDF in designing and negotiating a national park system for Brazil has been discussed at length by different authors, including Pádua herself (Pádua and Quintão 1982; Urban, 1998; Pádua, 2014a; Franco and Drummond, 2013; Drummond 2016). Less known is the connection between the creation of Amazonian national parks in 1970s Brazil and FAO's systematic approach to national parks, which had first been introduced to other Latin American countries. The following section casts light on this connection, focusing on Wetterberg's role in adapting the FAO's approach to Brazilian Amazonia during a time the region was being targeted for massive development projects.

NATIONAL PARKS TRAILING AMAZONIAN DEVELOPMENT

FAO had developed several projects in Brazil in the 1950s and early 1960s, but the 1964 coup brought in a regime wary of the agency for its association with opposition politicians such as Josué de Castro, FAO's former chairman.[17] Beginning in the 1970s, Brazilian agencies gradually resumed partnering with FAO, which led to the opening of an FAO office in Brasília in 1980. The agency's renovated engagement in Brazil also coincided with a period of moderate political opening in the mid-1970s. It was in this context that FAO sent Wetterberg to Manaus in 1975 to help IBDF devise a plan for national parks in Brazilian Amazonia. In this plan, he would follow the approach first proposed by Kenton Miller, head of FAO's Latin American Program on

Wildland Management and a veteran of national park planning in other Latin American countries (Pizarro, 2012: 294–99). Both Wetterberg and Miller studied at the University of Washington under C. Frank Brockman, who had worked with the IUCN in the establishment of an international system of protected areas (Miller, n.d.; Brockman, 1962; Phillips, 2014). Their approach followed IUCN's late 1960s typology of protected areas, which emphasized national parks should be devoid of any human settlements and defined by habitat preservation (IUCN, 1970: 116–19, 148). Miller first developed his plan for a coherent national system of protected areas in Costa Rica, during his tenure at the Inter-American Institute for Cooperation on Agriculture (IICA), a research institute founded by the Organization of American States (OAS) in 1942 (Evans, 1999). Miller's specific innovation was a systematic approach to park management based on field surveys that informed the selection of areas to be preserved and standardized management plans to foster and regulate park development. That was a sharp contrast to the haphazard and idiosyncratic motivations behind national park creation in Latin American. Once at FAO, Miller used the organization's structure to spread his national park model throughout Latin America. By the mid-1970s, much had been done in the establishment of national parks throughout Latin America, but there was still an immense gap, Amazonia. With few charismatic species and little monumental landscape, the tropical rainforests of the Amazonian basin had been overlooked by international conservationists who, before the 1960s, concentrated their efforts elsewhere in South America. Miller viewed the absence of protected areas in Amazonia as an inexcusable omission and hired Wetterberg through FAO to help craft a coherent conservation plan for the region to be centered in Brazil, the country that harbored most of the biome (Foresta, 1991: 19–32; Miller, n.d.; 1976).

By 1975, Brazil had already created eighteen national parks in areas chosen for a variety of reasons—most commonly the preservation of monumental landscapes. The selection of areas for conservation did not follow field surveys and technical studies as proposed by Miller in other Latin American countries. Wetterberg's mission was to change that, proposing protected areas in Amazonia based on ecological studies. After moving to Manaus in 1975, he began gathering information to write a plan for a system of national parks in the region. His primary challenge was finding relevant technical criteria to determine which areas in Amazonia were of interest for conservation. He would later admit that, upon his arrival, the "Amazon rainforest . . . looked like a green blotch on a map, and I had no way to differentiate things that might exist in it" (Wetterberg, disc. with author, 2017). Wetterberg eventually happened upon a tentative classification system after meeting British botanist Ghillean Prance at the National Institute of Amazon Research (INPA) in Manaus. Prance's work on the geographic distribution of plant species revealed a forest that was far from homogenous. Amazonia had numerous pockets of endemic plant species, which Prance claimed had evolved in drier past periods when the forest retracted into different "islands" of dense vegetation separated by open fields (Prance, 1974; Wetterberg, Prance, and Lovejoy, 1981). Prance was not the first to notice this pattern, as zoologists had documented similar clusters of endemic animal populations (Haffer, 1969; Vanzolini and Williams, 1970). By the mid-1970s researchers started favoring a "Pleistocene refuge" theory that explained neotropical diversity as the result of refuge speciation. Wetterberg found in the Pleistoscene refuge theory an empirical geographic parameter that allowed portions of the Amazonian landscape to be ranked according to their biological relevance.

Between 1975 and 1977, Wetterberg produced a series of reports that laid out a coherent plan to protect fauna and flora refuges in Brazilian Amazonia. He based his ideas on studies of endemic populations of plants, lizards, birds, and butterflies, each producing maps with different clusters that he overlaid to create a joint base map of Amazonian Pleistocene refuges to be protected. The change from the PIN to POLAMAZONIA offered Wetterberg the possibility of using the spaces between development zones to propose national parks based on Pleistocene refuges. The goal, according to Wetterberg, was "to analyze potential compatibilities or incompatibilities between the POLAMAZONIA priority development poles and the preservation of key areas" (pp. 3–4). Wetterberg knew the Brazilian military regime prioritized development over conservation. However, by using a technical parameter that allowed for spatial flexibility, he intended to step ahead of development planning, eliminating the type of reliance on "the residuum of other land uses for parks and reserves" that had determined the creation of the Amazonia National Park in 1974. With his plan, "samples of biologically significant Amazon sites" could be selected for conservation before "the elimination of this option by other development" (p. 35). Wetterberg envisioned his Amazonian plan as the template for a Brazilian system of national parks that would take into account not only the biophysical elements of a country but also the "actual and potential soil use, transportation networks, national urban structure and development planning units." In this sense, he framed the creation of national parks as an essential part of frontier development, as it would "serve as a guide for subsequent action in this realm of land use planning" (p. 36).

Wetterberg's plan for national parks in Amazonia presented the region as a blank canvas, tacitly ignoring the possible existence of indigenous and peasant communities in the areas chosen for preservation. It aligned with the vision of park boosters in the IBDF, who historically believed humans should not dwell inside national parks (Freitas, 2017). It also reproduced the military understanding of the rainforest as a frontier—a space empty of people to be filled with rational uses of territory. In this sense, the parks proposed by the IBDF in the late 1970s should work as frontier development projects (Foresta, 1991: 79–80). Maria Tereza Jorge Pádua embraced Wetterberg's plan for protected areas in Amazonia, using it as the basis for an overhaul of the entire national park system (Pádua and Quintão, 1982; Urban, 1998: 145–51). A peer-reviewed version of the plan proposed a tiered system of conservation priorities, demanding the surveying of selected areas to identify optimal locations for the establishment of national parks (Wetterberg, 1978). Despite recent growth in staff and funding, the IBDF's protected areas department still lacked structural organization and personnel to survey areas of difficult access in Amazonia. Pádua solved this problem by contracting with the FBCN—which had a history of working with the agency as a nongovernmental organization—more than ninety external consultants to visit priority areas and suggest locations for future reserves (Wetterberg, disc. with author, 2017; Foresta, 1991: 67; Franco and Drummond, 2013; Urban, 1998: 145–51).

Once a consensus had been reached in the form of a definitive plan for national parks in Amazonia, Pádua started working hard to sell the proposal to military audiences, framing it as compatible with their vision of Amazonian development. Politically astute, Pádua began lecturing at the ESG, where Couto e Silva had worked a decade earlier, advocating the role of protected areas and conservation in the development of Amazonia. She cultivated a close relationship with generals in the Army, including João Figueiredo, who would eventually substitute Ernesto Geisel as president in 1979 (Wetterberg, disc. with author, 2017; FBCN, 1979: 145–47; Foresta, 1991: 92–94; Machado, COPLAN and IBDF, 1978). At the time he assumed office, Figueiredo was cognizant of the national park plan proposed by Pádua and Wetterberg, and in the months after assuming office he signed the decrees designating several new protected areas. They included four Amazonian national parks—Pico da Neblina, Pacaás Novos, Cabo Orange, and Jaú—along with three smaller biological reserves and several national forests.[18] Figueiredo also signed another decree implementing a national park system that introduced general technical procedures and norms for existing and future protected areas. The new national park system was inspired by Kenton Miller's systematic approach to national park development based on standardized management plans that emphasized the strict preservation of representative natural habitats.[19]

At first glance, the role of an intergovernmental agency such as FAO in providing the technical expertise for the creation of protected areas in Amazonia during a period of heightened nationalism might seem paradoxical. After all, one justification for promoting the development of Amazonia among the Brazilian military was the fear of territorial intervention by foreign actors. However, as the Brazilian geographer Otávio Velho noticed already in 1976, despite its nationalist discourse, the Brazilian government was willing to accept foreign capital and expertise in its plans to integrate Amazonia to Brazil's industrialized core (Velho, 2019: 201–2). Several Amazonian development projects in the 1970s—the Jari Project, the Manaus Free Trade Zone, or the Vale do Rio Cristalino Company, to name a few—were carried out with foreign capital and expertise. The production of geographic knowledge also, at times, relied on foreign actors, as in the case of the RADAM project, a remote sensing initiative supported by the U.S. Air Force and NASA. FAO, as pointed out by historian Oliver Dinius, not only was present in Brazil since the 1950s but had built an image of a "supra-national" nonideological agency that did not arouse suspicion of foreign meddling among Brazilian nationalists (Dinius, 2019: 140).

The technicians planning the new Amazonian parks believed themselves to be witnessing a change in how military planners viewed the role of the exploitation of natural resources in frontier development. POLAMAZONIA had increased, in the words of IBDF agents, the "sensibility and political will" toward nature protection, which materialized in the form of greater funding for the implementation of forestry and conservation projects in the second half of the decade and the creation of the new parks in 1979 and 1980 (IBDF, 1976; IICA and IBDF, 1976; Wetterberg et al., 1976: 1). Speaking years after the creation of the parks, Pádua argues that the military had agreed to embrace conservation because they respected the technical knowledge of a new field that appeared to be apolitical (Urban, 1998: 148). Wetterberg adds that the military in the 1970s recognized the international criticism triggered by frontier projects such as the Trans-Amazonian Highway and seized the opportunity to establish national parks in order to greenwash their development plans (Wetterberg, disc. with author, 2017). In fact, environmental protection never took precedence among officers, and it was only embraced when it did not interfere with other modernization priorities (Guimarães, 1991: 131). This was clear for the FAO and IBDF conservationists of the 1970s, who strategically avoided any overlap between new protected areas and existing development projects. POLAMAZONIA was, in fact, crucial in this as it treated Amazonia as a geographically diverse landscape, abandoning a blunt vision of a homogenous frontier in favor of a more sophisticated view of a patchwork of development possibilities (figure 3.1). Informed by surveys and technical reports, this new spatial imagining of Amazonia opened space for national parks and other types of protected areas. Framing the proposed parks in the language of spatial data and science allowed FAO and IBDF technicians to convince the military leaders of Brazil that conservation was, however small, a part of frontier development.

AMAZONIA AS A CONSERVATION FRONTIER

After the end of the military dictatorship in 1985, the new civilian administrations in Brazil continued to pursue Amazonian development, especially during the years that the Workers' Party was in power (2003–2016), when the federal government launched new development projects in the region as part of the Growth Acceleration Program (PAC). This program initiated the construction and renovation of thousands of miles of roads crisscrossing the forest, as well as dozens of energy-related projects, including the massive Belo Monte, Jirau, and Santo Antônio hydroelectric dams. These three projects alone generated considerable domestic and international backlash for their impact on the forest and local indigenous communities. In the post-dictatorship era, Brazil has also greatly expanded its number of protected areas, with the aggregate area of federal, state, and municipal reserves growing from about 45,000 square miles in 1980 to 558,000 square miles in 2017.[20] Eighty-two percent of all the protected areas in the country in 2017 were located within the confines of "legal Amazonia," a figure that had remained constant since 1980 (see figure 3.3). A crucial difference between the parks created by the military dictatorship and the ones established today is the role of the global paradigm of "sustainable development," as well as of the local concept of *socioambientalismo*, in justifying the creation of reserves that allowed the presence of dwellers and the development of some types of economic activities (Bratman, 2019; Campbell, 2015). Only about 6 percent of the 36,210 square miles of reserves created in Amazonia between 1974 and 1984 allowed for mixed-used activities such as selective logging. In contrast to this, 75 percent of the 24,962 square miles of reserves established in 2006 as part of the country's first "sustainable forest district" allowed for some economic activities and human inhabitation (Ministério do Meio Ambiente, 2006). This change in paradigm, from habitat protection to sustainable development and socio-environmental justice, also defined the geography of the new protected areas—instead of being plotted in the interstices of colonization projects, the new reserves overlapped areas receiving settlers since the 1970s such as the margins of the highway BR-163. Despite these differences, the establishment of parks and reserves in Brazilian Amazonia between 1974 and 1980 marked the inflection point when policy-makers and conservationists started to identify in Amazonia a frontier to be preserved from indiscriminate development through the establishment of protected areas. It did so by reimaging the rainforest as an archipelago of specific places worth of protection, places defined through a systematic approach based on field surveys and scientific literature.

What is the effect of these existing protected areas in preserving Amazonia? A spatial compromise was embedded in the creation of the Brazilian system of Amazonian parks in the late 1970s, one in which protected areas would

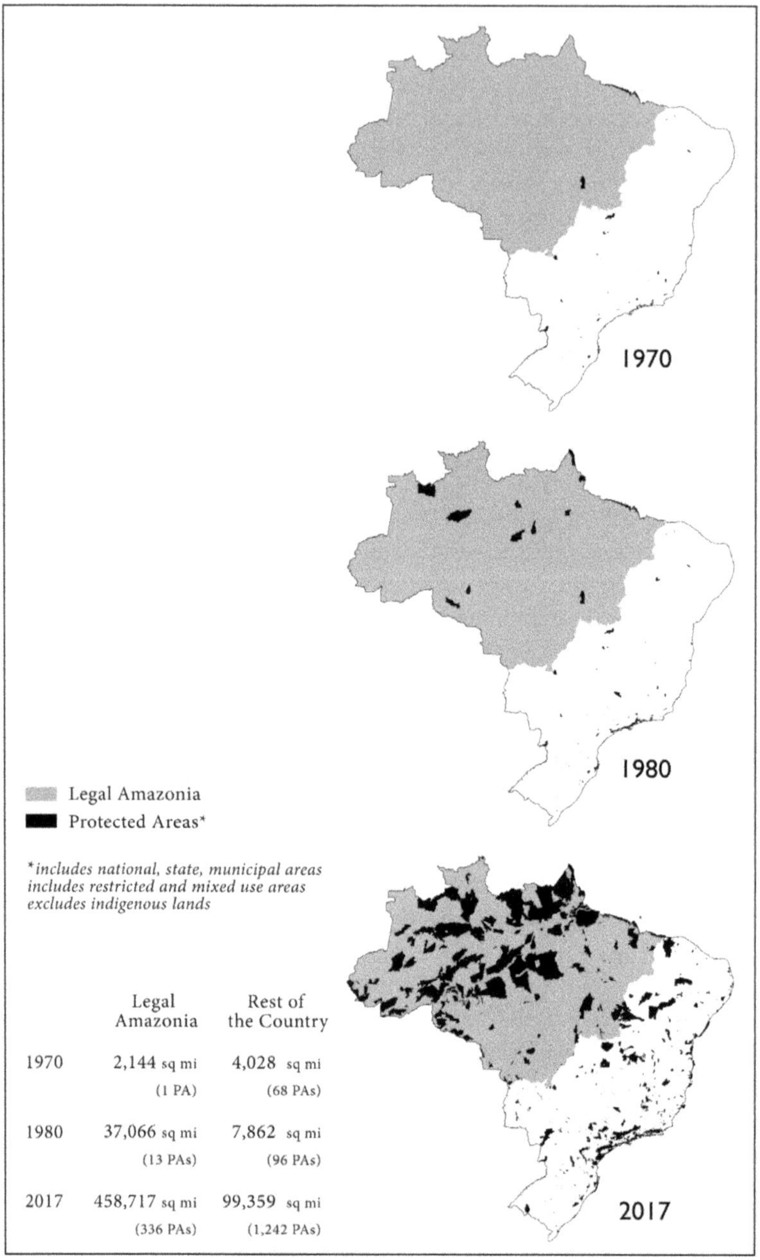

	Legal Amazonia	Rest of the Country
1970	2,144 sq mi (1 PA)	4,028 sq mi (68 PAs)
1980	37,066 sq mi (13 PAs)	7,862 sq mi (96 PAs)
2017	458,717 sq mi (336 PAs)	99,359 sq mi (1,242 PAs)

Figure 3.3 Evolution of Protected Areas in Brazil. *Source*: Ministério do Meio Ambiente, 2017.

not be established in sectors designated for development. As a result of this, many of the first-generation Amazonian parks and reserves were established far from the development frontier, in areas with lower demographic and economic pressure. There is a wide arc connecting the south and east of Brazilian Amazonia whose territory has never been put under environmental protection (figure 3.3), and it is this swath that has been most affected by land use change over the last fifty years.[21] On top of that, a good number of designated protected areas have yet to be fully implemented; in 2013, 96 percent of the reserves in Amazonia still presented low or medium implementation, and there are signs of a considerable increase in the deforestation rate inside reserves since 2017 (Araújo et al., 2017; Imazon, 2019). Overall, about 297,000 square miles in Brazilian Amazonia have been cleared since the launch of the PIN in 1970, an area twice the size of California, mostly to make way for cattle ranching and crop monoculture (Araújo et al., 2017; Buttler, 2017; INPE, 2016).

Still, four decades of establishing protected areas in Amazonia had a positive effect in reducing deforestation. This was especially notable during the Workers' Party's years in power, as figure 3.4 suggests. Between 2004 and 2013 the rate of deforestation in Amazonia dropped by 75 percent, and studies indicate that 84 percent of this reduction resulted from the existence of protected areas at all levels—municipal, state, and federal. As shown in figure 3.4, the deforestation rate in Amazonia stopped dropping in the 2010s, with 8,371 square miles of forests cut down between 2012 and 2015—12 percent of this number illegally, inside protected areas (Araújo et al., 2017; Barber et al., 2012; 2014; Buttler, 2017; INPE, 2016). The recent increase is still modest when compared to the years of high deforestation in the 1990s and early 2000s, but it reveals the dual nature of Amazonia today. On the one

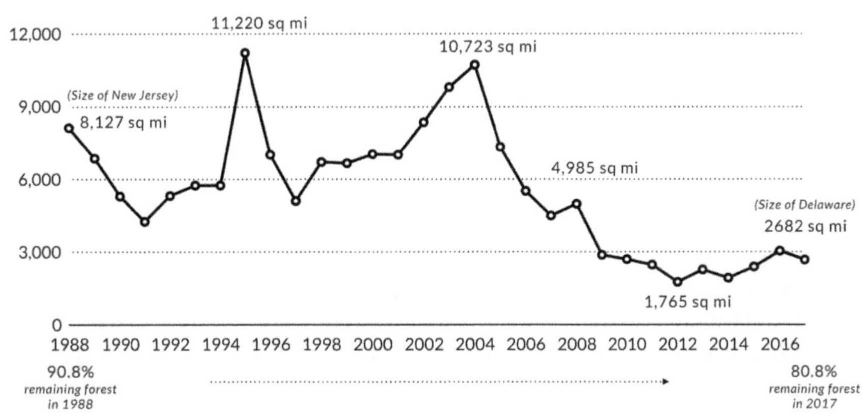

Figure 3.4 Annual Forest Loss (sq mi²) in the Brazilian Amazon. *Source*: INPE, 2018.

hand, the region went through a radical transformation in the late 1970s and 1980s, becoming a "conservation frontier" where the state acted to reform patterns of land use and prevent deforestation by changing the legal territorial regime covering extensive sections of the forest. The region's transformation into a conservation frontier followed the larger change, as appointed by Acker (2017), that reframed Amazonia as a world natural heritage area to be protected. On the other hand, Amazonia never stopped being a "development frontier"—deforestation was reduced but never halted or reversed and, more importantly, development projects continued, despite the accepted consensus around the need to protect the rainforest. Any initiative designed to address and understand the many challenges posed by Amazonia will have to acknowledge this perennial double nature of government policies in the region.

NOTES

1. These new Amazonian parks represented an increase of 450 percent in the total area of national parks in Brazil. In addition to national parks, the Brazilian government also established during this period in Amazonia 6,672 square miles of biological reserves, a category of protected area with similar characteristics, and 2,318 square miles of national forests, which allow extractive activities. In total, Brazil established 36,210 square miles of protected areas in Brazilian Amazonia between 1974 and 1984. Wetterberg, Prance, and Lovejoy (1981) put these numbers in perspective by including protected areas created by other Amazonian countries (i.e., Bolivia, Ecuador, and Venezuela) adding up to a total of 45,675 square miles of new protected areas in the Amazon basin by the early 1980s.

2. Two of the parks created between 1959 and 1961, the Araguaia and Xingu National Parks, were located in the transition zone between the savanna and the Amazonian rainforest. Xingu was a "national park" only in name. Initially planned as an "indigenous park," it was designated as a "national park" to give the federal government a mandate to expropriate the area owned by the state of Mato Grosso. Despite that, Xingu never fell under the umbrella of the Brazilian Forest Service or its successor environmental and forestry agencies. It was controlled by the Brazilian bureau for Indian affairs and the Air Force and eventually was reframed as an indigenous reservation (Brazil, Decree 47570, December 31, 1959; Brazil, Decree 50455, April 14, 1961; Brazil, Decree 51084, July 31, 1961; Brazil, Decree 82263, September 13, 1978; Barreto Filho, 1997: 4; Barreto Filho, 2004; Garfield, 2004; Maia, 2010; Menezes, 2000: 197–99, 304, 310–11, 317–20; Pádua, 2014b; Urban, 1998: 150; Wakild, 2016).

3. Following the usage of *conservação* [conservation] by government employees and nongovernmental organization in 1970s Brazil, I employ "conservation" here as a broad term encompassing both sides of the traditional U.S. dichotomy between conservation and preservation. For further discussion, see White (1985).

4. Brazil, Constitution of 1946, Article 1999, September 18, 1946; Brazil, Law 1806, January 6, 1953.

5. Brazil, Law 5173, October 27, 1966.

6. Brazil, Decree-Law 1110, July 9, 1970.

7. Brazil, Decree 1106, June 16, 1970; Brazil, Law 5727, November 4, 1971.

8. Brazil, Law 6151, December 4, 1974.

9. Military planners drew from "development pole theory," as developed by French economist François Perroux, to tackle the problem of the disparity between the industrialized southern Brazil and the underdeveloped north. On the different schools of development in Brazil, see Serra (2003), Love (1996), and Pereira and Van Dyke (1984).

10. Brazil, Decrees 51024 to 51030, and 51042 to 51043, July 25, 1961. In an interview to Foresta (1991: 33), Clara Pandolfo explained that "The idea of the reserves was a whim, as were many of Quadros's ideas, and the task of identifying suitable areas caught SPVEA unaware. Uncertain of the intended purpose of the reserves, the agency arbitrarily made river accessibility the key selection criterion."

11. Brazil, Law 4771, Forest Code, September 15, 1965; Brazil, Decree 62998, 1968; Brazil, Decree 68873, July 5, 1971; Brazil, Decree 71879, March 1, 1973; Brazil, Decree 82263, September 13, 1978. One of these 1961 forest reserves, Tumucumaque, was redesignated first as Indigenous National Park in 1968, then as an Indigenous Park in 1978 (Wetterberg and Pádua, 1978: 1; IBDF, 1975: 9).

12. The Forest Service, which had been reintroduced as an agency within the Ministry of Agriculture in 1938, lasted until 1962, when it was substituted by the *Departamento de Recursos Naturais Renováveis* (Department of Renewable Natural Resources, DRNR). The creation of the IBDF in 1967 combined in the same agency the DRNR and three other agencies created under Vargas: the *Instituto Nacional do Pinho* (National Pine Tree Institute, INP); the *Instituto Nacional do Mate* (National Yerba Mate Institute, INM); and the *Conselho Federal Florestal* (Federal Forest Council, CFF). Brazil, Delegated-Law 9, October 11, 1962; Brazil, Decree 52442, October 8, 1963; Brazil Decree-Law 289, February 28, 1967.

13. Over the years the FBCN, a civil society organization, developed close connections to the government, assisting government agencies with proposing conservation policy (Franco and Drummond, 2013).

14. Brazil, Decree-Law 1106, "Plano de Integração Nacional," June 16, 1970; Brazil, Decree 67557, November 12, 1970; Brazil, Decree 68443, March 29, 1971; Brazil, Decrees 73683 and 73684, February 19, 1974; Brazil, Decree 74607, September 25, 1974.

15. In 1972, UNESCO selected the impact of human activities on tropical forests as the subject of the first project in its Man and the Biosphere Program. A meeting of the tropical forest working group took place in Rio de Janeiro in 1974 (UNESCO, 1973, 1974).

16. As pointed out by von Hardenberg et al. (2017: 9) conservation is a tool for modernization whose stakes are low when compared to high-modernist attempts to reform society and nature.

17. Josué de Castro (1908–1973) was FAO's chairman between 1951 and 1955. He was also a member of the Brazilian lower house for two terms (1954–1962) and

later acted as Brazil's representative at FAO and UNESCO in Switzerland until 1964, when the new military regime stripped him of his political rights and forced him to live in exile.

18. Brazil, Decree 83550, June 5, 1979; Brazil, Decree 84019, September 21, 1979; Brazil, Decree 84913, July 15, 1980; Brazil, Decree 85200, September 24, 1980.

19. Brazil, Decree 84017, September 21, 1979.

20. The establishment of national parks in Amazonia at the end of the 1970s was also concomitant with the growth of protected areas in the Global South. Worldwide, between the 1950s and the 2000s, the number of national parks, to use a widely adopted category of protected area, grew at a steady pace of about 60 percent per decade (IUCN, 2017).

21. It is important to note that this area harbors few large indigenous territories, which historically have also presented a barrier for logging and large-scale deforestation (Nepstad et al., 2006).

BIBLIOGRAPHY

Acker, A. 2017. *Volkswagen in the Amazon: The Tragedy of Global Development in Modern Brazil.* Cambridge University Press: Cambridge.

Andreazza, M. 1972. *Programas Especiais na Área dos Ministério dos Transportes: Objetivos e Execução.* Ministério dos Transportes: Rio de Janeiro.

Araújo, E., Barreto, P., Baima, S., and Gomes, M. 2017. Most deforested Conservation Units in the Legal Amazon 2012–2015. Available at: https://imazon.org.br/en/publicacoes/most-deforested-conservation-units-in-the-legal-amazon-2012-2015.

Barber, C. P., Cochrane, M. A., Souza Jr., C. M., and Veríssimo, A. 2012. Dynamic Performance Assessment of Protected Areas. *Biological Conservation*, 149, 6–14.

Barber, C. P., Cochrane, M. A., Souza Jr., C. M., and Laurence, W. F. 2014. Roads, Deforestation, and the Mitigating Effect of Protected Areas in the Amazon. *Biological Conservation*, 177, 203–9.

Barreto Filho, H. T. 1997. Da Nação ao Planeta Através da Natureza: Uma Tentativa de Abordagem Antropológica das Unidades de Conservação na Amazônia. *Cadernos de Campo*, 222, 2–32.

Barreto Filho, H. T. 2004. Notas para uma História Social das Áreas de Proteção Integral no Brasil. In: *Terras Indígenas e Unidades de Conservação da Natureza: O Desafio das Dobreposições*, Ricardo, F. (ed.). Instituto Socioambiental: São Paulo, pp. 53–63.

Bratman, E. Z. 2019. *Governing the Rainforest: Sustainable Development Politics in the Brazilian Amazon.* Oxford University Press: New York.

Brazil. 1971. *I Plano Nacional de Desenvolvimento (PND) 1972/74.* Imprensa Nacional: Brasília.

Brazil. 1975. *II Plano Nacional de Desenvolvimento (1975–1979).* Serviço Gráfico do IBGE: Rio de Janeiro.

Brockman, C. F. 1964. Supplement to the Report to the Committee on Nomenclature. In: *World Conference on National Parks [1st], Proceedings of a Conference*

Organized by the International Union for Conservation of Nature and Natural Resources . . . Seattle, Washington, June 30–July 7, 1962. U.S. National Park Service, Department of the Interior: Washington, DC.

Buttler, R. 2017. Calculating Deforestation Figures for the Amazon. *Mongabay*, January 26, 2017. Available at: http://rainforests.mongabay.com/amazon/deforestation_calculations.html.

Campbell, J. M. 2015. *Conjuring Property: Speculation and Environment Futures in the Brazilian Amazon*. University of Washington Press: Seattle.

Castelo Branco, H. A. 1966. Na Instalação da Primeira Reunião de Incentivo ao Desenvolvimento da Amazônia. December 3, 1966. Biblioteca da Presidência da República. Available at: http://www.biblioteca.presidencia.gov.br/presidencia/ex-presidentes/castello-branco/discursos/1966/37.pdf/view.

Child, J. 1979. Geopolitical Thinking in Latin America. *Latin American Research Review*, 14(2), 89–111.

Correio da Manhã. 1964. Novos Horizontes para a Amazônia. January 12, 1964.

Correio da Manhã. 1966. Castelo Acha Melhor Situação do País. December 4, 1966.

Correio da Manhã. 1970. Médici: Amazônia Ainda não Achou sua Vocação. October 9, 1970.

Couto e Silva, G. 1957. *Aspectos Geopolíticos do Brasil*. Biblioteca do Exército Editora: Rio de Janeiro.

Couto e Silva, G. 1967. *Geopolítica do Brasil*. Coleção Documentos Brasileiros No. 126. José Olympio: Rio de Janeiro.

Diário de Notícias. 1964. Forum da Amazônia: Marco Inicial de Desenvolvimento. March 2, 1964.

Dinius, O. 2019. Transnational Development on the Frontier: The FAO's Fishery and Forestry Missions in the Brazilian Amazon. *The International History Review*, 41(2), 407–26.

Drummond, J. A. 2016. From Randomness to Planning: The 1979 Plan for Brazilian National Parks. In: *National Parks Beyond the Nation: Global Perspectives on America's Best Idea*, Howkins, A., Orsi, J. and Fiege, M. (eds.). University of Oklahoma Press: Norman, pp. 201–34.

Evans, S. 1999. *The Green Republic: A Conservation History of Costa Rica*. University of Texas Press: Austin.

FBCN. 1967. Revisão da Política Brasileira de Parques Nacionais. *Boletim Informativo* No. 2, 22–23.

FBCN. 1979. Criação de Novos Parques Nacionais. *Boletim Informativo* No. 14, 145–47.

Foresta, R. A. 1991. *Amazon Conservation in the Age of Development: The Limits of Providence*. University of Florida Press: Gainesville.

Foweraker, J. 1981. *The Struggle for Land: A Political Economy of the Pioneer Frontier in Brazil from 1930 to the Present Day*. Cambridge University Press: Cambridge.

Franco, J. L. A. and Drummond, J. A. 2013. Nature Protection: The FBCN and Conservation Initiatives in Brazil, 1958–1992. *Historia Ambiental Latinoamericana y Caribeña*, 2(2), 338–67.

Freitas, F. 2017. Ordering the Borderland: Settlement and Removal in the Iguaçu National Park, Brazil, 1940s–1970s. *The Nature State: Rethinking the History of Conservation*, von Hardenberg, W., Kelly, M., Leal, C., and Wakild, E. (eds.). Routledge: New York, pp. 158–75.

Garfield, S. 2004. A Nationalist Environment: Indians, Nature, and the Construction of the Xingu National Park in Brazil. *Luso-Brazilian Review*, 41(1), 139–67.

Garfield, S. 2013. *In Search of the Amazon: Brazil, the United States, and the Nature of a Region*. Duke University Press: Durham.

Glerum, B. B. and Smit, G. 1962. *Report to the Government of Brazil on a Combined Forestry-soil Survey along the Road BR-14 from São Miguel do Guamá to Imperatriz,* FAO Report No. 1483. UN-Food and Agriculture Organization: Rome.

Guimarães, R. P. 1991. *The Ecopolitics of Development in the Third World: Politics and Environment in Brazil*. Rienner Publishers: Boulder.

Haffer, J. 1969. Speciation in Amazonian Forest Birds. *Science*, 165(3889), 131–37.

Hardenberg, W. G., Kelly, M., Leal, C., and Wakild, E. 2017. Introduction. In: *The Nature State: Rethinking the History of Conservation*, von Hardenberg, W., Kelly, M., Leal, C., and Wakild, E. (eds.). Routledge: Oxford, pp. 1–15.

Hecht, S. and Cockburn, A. 1989. *The Fate of the Forest: Developers, Destroyers, and Defenders of the Amazon*. Verso: London and New York.

Heinsdijk, D. 1961. Forest Survey in the Amazon Valley. *Unasylva*, 15(4), 167–74.

IBDF. 1975. *Plano de Ação*. Ministério da Agricultura: Brasília.

IBDF. 1976. *Relatório Anual 1975*. Ministério da Agricultura: Brasília.

IBDF and IBRA. 1969. *Parques Nacionais e Reservas Equivalentes no Brasil: Relatório com Vistas a uma Revisão da Política Nacional nesse Campo*. Ministério da Agricultura: Brasília.

IICA and IBDF. *Reunião Internacional sobre Administração de Unidades de Conservação na Região Amazônica*. Informe de Conferências Curso y Reuniones No. 107. IBDF: Santarém.

IMAZON. 2019. *Our Heritage Endangered: Why Amazon's Conservation Units are at Risk*. Imazon: Belém. Available at: https://imazon.org.br/en/publicacoes/our-heritage-endangered-why-amazons-conservation-units-are-at-risk/.

INPE. 2016. Taxas Anuais do Desmatamento—1988 até 2016. Available at: http://www.obt.inpe.br/prodes/prodes_1988_2016n.htm.

INPE. 2018. Taxas de Desmatamento. Available at: http://www.obt.inpe.br/prodes/dashboard/prodes-rates.html.

IUCN. 1970. *Tenth General Assembly: Vigyan Bhavan, New Delhi, 24 November—1 December, 1969*. Vol. 2. IUCN Publications New Series 27. International Union for Conservation of Nature and Natural Resources/UNESCO: Morges.

IUCN. 1972. The Opening of Brazil. *IUCN Bulletin*, 3(5), 18.

IUCN. 2017. World Database on Protected Areas. Available at: http://www.wdpa.org.

Jornal do Brasil. 1979. Criação de Novos Parques Dobra Área Preservada. June 6, 1979.

Jornal do Comércio. 1966. Empresários Amazonenses Pediram a Castelo Isenção Total de Impostos para o Amazonas; Com os Instrumentos que o Governo Deu é Hora de Fazê-los Funcionar. December 4, 1966.

Keck, M. E. and Hochstetler, K. 2007. *Greening Brazil: Environmental Activism in State and Society*. Duke University Press: Durham.

Keck, M. E. and Sikkink, K. 1998. *Activists beyond Borders: Advocacy Networks in International Politics*. Cornell University Press: Ithaca.

Lourenço, J. S. 2009. Amazonia: Past Progress and Future Prospects. In: *Brazil: A Century of Change*, Sachs, I., Wilheim, J., Pinheiro, P.S., and Anderson, R.N. (eds.). University of North Carolina Press: Chapel Hill, pp. 253–70.

Love, J. 1996. *Crafting the Third World: Theorizing Underdevelopment in Rumania and Brazil*. Stanford University Press: Stanford.

Lovejoy, T. E. 1973. The Transamazonica: Highway to Extinction? *Frontiers*, 37(3), 18–23.

Macekura, S. J. 2015. *Of Limits and Growth: The Rise of Global Sustainable Development in the Twentieth Century*. Cambridge University Press: Cambridge.

Machado, E. F., COPLAN, and IBDF. 1978. *Diagnóstico do Subsistema de Conservação e Preservação de Recursos Naturais*. Ministério da Agricultura: Brasília.

Mahar, D. J. 1979. *Frontier Development Policy in Brazil: A Study of Amazonia*. Praeger: New York.

Maia, J. M. E. 2010. As Ideias que Fazem o Estado Andar: Imaginação Espacial, Pensamento Brasileiro e Território no Brasil Central. *Dados*, 53(3), 622–55.

McGrath, K., Gachot, R., and Gallant, N. M. 1953. The Amazon Valley. *Unasylva*, 7(3), 98–104.

Menezes, M. L. P. 2000. *Parque Indígena do Xingu: A Construção de um território Estatal*. Unicamp: Campinas.

Miller, K. 1976. *National Parks Planning: A Manual with Annotated Examples*. UN-Food and Agriculture Organization: Rome.

Miller, S. n.d. Kenton Miller: A Biography by Sue Miller. IUCN. Available at: http://cmsdata.iucn.org/downloads/kenton_miller_a_biography_by_sue_miller.pdf.

Ministério da Agricultura. 1972. IBDF na Amazônia, Anexo ao Relatório das Atividades do MA no PIN, 2º Trimestre de 1972. Ministério da Agricultura: Brasilia.

Ministério da Agricultura, IBDF, and POLAMAZONIA. 1978. *Plano de Manejo Parque Nacional da Amazônia (Tapajós)*. Ministerio de Agricultura: Brasília.

Ministério do Meio Ambiente. 2006. Amazônia ganha Unidades de Conservação e 1º Distrito Florestal Sustentável. Available at: https://www.mma.gov.br/informma/item/3046-amazonia-ganha-unidades-de-conservacao-e-1o-distrito-florestal-sustentavel.

Ministério do Meio Ambiente. 2017. Download de Dados Geográficos. Available at: mapas.mma.gov.br/i3geo/datadownload.html.

Moran, E. F. 1990. Private and Public Colonization Schemes in Amazonia. In: *The Future of Amazonia: Destruction or Sustainable Development?* Goodman, D., and Hall, A. (eds.). St. Martin's Press: New York, pp. 70–89.

Mougey, T. 2018. Tracing the Origins of Brazil's Great Acceleration: The SPVEA's Primeiro Plano Quinquenal and the Technoscientific Recovery of Amazonia, 1945–1959. *Varia Historia*, 34(65), 375–408.

Muller, A. C. 1998. *Unesco e o Parque Nacional do Iguaçu: Western and Southwestern Paraná*. Archive AIPOPEC.

Nepstad, D., Schwartzman, S., Bamberger, B., Santilli, M., Ray, D., Lefebvre, P., Alencar, A., Prinz, E., Fiske, G., and Rolla, A. 2006. Inhibition of Amazon

Deforestation and Fire by Parks and Indigenous Lands. *Conservation Biology*, 20(1), 65–73.
O Jornal do Rio de Janeiro. 1963. Lançamento de uma 'Operação Amazônia' para o Progresso Econômico da Região, December 22, 1963.
Pádua, M. T. J. 2014a. Bons Tempos e Maus Tempos. *O Eco* (blog), October 14, 2014.
Pádua, M. T. J. 2014b. Os Presidentes e os Parques Nacionais do Brasil. *O Eco* (blog), June 24, 2014.
Pádua, M. T. J. and Quintão, A. T. B. 1982. Parks and Biological Reserves in the Brazilian Amazon. *Ambio*, 11(5), 309–14.
Panagides, S. and Magalhães, V.L. 1974. Amazon Economic Policy and Prospects. In: *Man in the Amazon*, Wagley, C. (ed.). Gainesville, Florida: University of Florida Press: Gainesville, pp. 243–61.
Pereira, L. C. B. and Van Dyke, M. 1984. Six Interpretations of the Brazilian Social Formation. *Latin American Perspectives*, 11(1), 35–72.
Phillips, A. 2004. The History of the International System of Protected Area Management Categories. *Parks*, 14(3), 4–14.
Pizarro, R. 2012. The Global Diffusion of Conservation Policy: An Institutional Analysis. PhD Thesis. Stanford University: Stanford.
Prance, G. T. 1974. Phytogeographic Support for the Theory of Pleistocene Forest Refuges in the Amazon Basin, Based on Evidence from Distribution Patterns in Caryocaraceae, Chrysobalanaceae, Dichapetalaceae and Lecythidaceae. *Acta Amazonica*, 3(3), 5–28.
Ricardo, C. 1940. *Marcha para Oeste: A Influencia da "Bandeira" na Formação Social e Politica do Brasil.* José Olympio: Rio de Janeiro.
Sawyer, D. R. 1984. Frontier Expansion and Retraction in Brazil. In: *Frontier Expansion in Amazonia*, Schmink, M. and Wood, C.H. (eds.). University of Florida Press: Gainesville, pp. 180–203.
Serra, M. A. 2003. Development Pole Theory and the Brazilian Amazon. *Estudos Econômicos*, 33(1), 1–43.
Strang, H. E. and Veloso, H. P. 1970. Alguns Aspectos Fisionômicos da Vegetação No Brasil. *Memórias do Instituto Oswaldo Cruz*, 68(1), 9–76.
SUDAM. 1976. *II Plano Nacional de Desenvolvimento: programa de ação do governo para a Amazônia (1975–1979).* Ministério do Interior: Belém.
Tambs, L. A. 1974. Geopolitics of the Amazon. In: *Man in the Amazon*, Wagley, C. H. (ed.). University of Florida Press: Gainesville, pp. 45–87.
UNESCO. 1973. Report of the Expert Panel on MAB Project 1: Ecological Effects of Increasing Human Activities on Tropical and Subtropical Forest Ecosystems. UNESCO: Paris. Available at: http://unesdoc.unesco.org/images/0000/000028/002808eb.pdf.
UNESCO. 1974. (Report of the) Working Group on MAB Project 1: Ecological Effects of Human Activities on Tropical and Sub-tropical Forest Ecosystems. UNESCO: Washington/Rio de Janeiro/Paris. Available at: http://unesdoc.unesco.org/images/0000/000098/009865eb.pdf.
Urban, T. 1998. *Saudade do Matão: Relembrando a História da Conservação da Natureza no Brasil.* Editora da UFPR: Curitiba.

Vanzolini, P. E. and Williams, E. 1970. South American Anoles: The Geographic Differentiation and Evolution of the *Anolis Chrysolepis* Species Group (Sauria, Iguanidae). *Arquivos de Zoologia de São Paulo*, 19(1–2), 1–176.

Vargas, G. 1938. *A Nova Política do Brasil: O Estado Novo, 10 e novembro de 1937 a 25 de julho de 1938*. Vol. 5. José Olympio: Rio de Janeiro.

Vargas, G. 1941. *A Nova Política do Brasil: Ferro, Carvão e Petróleo, 7 de agosto de 1940 a 9 de julho de 1941*. Vol. 8. José Olympio: Rio de Janeiro.

Vargas, G. 2011 [1937]. *Marcha Para Oeste*—Rio de Janeiro, 31 de dezembro de 1937. Série Perfis Parlamentares No. 62. Câmara dos Deputados: Brasília, pp. 368–73.

Velho, O. G. 1972. *Frentes de Expansão e Estrutura Agrária: Estudo do Processo de Penetração numa Área da Transamazônica*. Zahar: Rio de Janeiro.

Velho, O. G. 2009. *Capitalismo Autoritário e Campesinato: Um Estudo Comparativo a Partir da Fronteira em Movimento*. Centro Edelstein de Pesquisas Sociais: Rio de Janeiro.

Velloso, J. P. R. 1974. O II PND e as Novas Oportunidades do Brasil; Nova Estratégia; Razões para não Perder o Sono. Imprensa Nacional: Brasília. Available at: http://www.biblioteca.presidencia.gov.br/publicacoes-oficiais/catalogo/geisel/ii-pnd_reis-velloso/view.

Vital, A. V. 2018. As "Florestas Sagradas" do Impasse: a Reserva Florestal do Território Federal do Acre (1911). *HALAC—Historia Ambiental, Latinoamericana y Caribeña*, 8(1), 42–66.

Wakild, E. 2016. Conservation on Tour: Comparing Nations, Scientists, and Parks in the Americas. In: *National Parks Beyond the Nation: Global Perspectives on America's Best Idea*, Howkins, A., Orsi, J., and Fiege, M. (eds.). University of Oklahoma Press: Norman, pp. 91–113.

Wetterberg, G. B. 1974. The History and Status of South American National Parks and an Evaluation of Selected Management Options. PhD thesis. University of Washington: Seattle.

Wetterberg, G. B. 1978. *Wildland Management and Planning in Brazil*. Technical Series No. 14. UNDP/FAO/IBDF: Brasília.

Wetterberg, G. B. and Pádua, M. T. J. 1978. *Preservação da Natureza na Amazônia Brasileira, Situação em 1978*. Technical Series No. 13. UNDP/FAO/IBDF: Brasília.

Wetterberg, G. B., Pádua, M. T. J., Castro, C. S., and Vasconcelos, J. M. C. 1976. *An Analysis of Nature Conservation Priorities in the Amazon*. Technical Series No. 8. UNDP/FAO/IBDF: Brasília.

Wetterberg, G. B., Prance, G. T., and Lovejoy, T. E. 1981. Conservation Progress in Amazonia: A Structural Review. *Parks*, 6(2), 5–12.

White, R. 1985. American Environmental History: The Development of a New Historical Field. *Pacific Historical Review*, 54(3), 297–335.

Wood, C. H. and Schmink, M. 1993. The Military and the Environment in the Brazilian Amazon. *Journal of Political and Military Sociology*, 21, 81–105.

Wozniewicz, V. D. 1974. The Amazonian Highway System. In: *Man in the Amazon*, Wagley, C.H. (ed.). University of Florida Press: Gainesville, pp. 291–314.

Chapter 4

Politics of the Floodplain Commons in the Amazon

Fábio de Castro

INTRODUCTION

Over the last decades, the Brazilian Amazon has undergone major socio-environmental transformations due to the introduction of new ideas, practices, and institutions. This process has reshaped economies, territories, and social interactions; however, with different trends over time and space. Fast forest cover change observed until mid-2000s, was followed by a drop in deforestation rate and boom of implementation of protected areas for a decade (Castro, 2012a) and, more recently, expansion of new agrarian frontier (Fearnside, 2015). The well-documented agrarian frontier expansion in the Amazonian upland contrasts with the limited analysis on the floodplain where the land-water nexus context is a key socio-environmental factor influencing the outcomes of the frontier expansion. Floodplains are dynamic ecosystems where a combination of farming, ranching, and fishing production systems are carried out by local residents, large cattle ranchers, and commercial fishers (Castro, 2009). Local patterns, however, are directly influenced by the expansion of land-based activities in the upland (McGrath et al., 1993). In this chapter, I provide an analysis of how socio-environmental transformation of a floodplain systems is shaped by a combination of local social processes and external economic and political drivers. My goal is to move beyond polarized theoretical divides of local/conservation/collaboration versus external/impacting/conflict by providing an integrative perspective to understand socio-environmental change in the region.

The Amazon literature can be roughly divided into two main theoretical perspectives. First, the agrarian frontier debate, grounded in political ecology frameworks, highlights land cover change and the struggles around land tenure, production systems, and global market mediated by power asymmetries and national and transnational actors (e.g., Bunker, 1985; Schmink and Wood, 1992; Ioris, 2017). Second, the commons debate, based on human ecology concepts, has emphasized forest conservation and the innovative capacity of local communities to adapt to their local ecological and social context through socially embedded collective institutional crafting (McGrath et al., 1993; Brondizio et al., 2009; Peruzzi et al., 2018). Both theoretical perspectives shed light on key aspects of socio-environmental changes—the former on power relations and conflictive interactions between peasantry and other actors and the latter on social capital and collaborative interactions among traditional communities. However, they both fall short in the articulation between collaborative and conflictive interactions (Castro, 2012b).

The Amazon floodplain occupies a particular relevance in both debates. On one side, the long history of human occupation and land use change (Harris, 2005, 2011) has allowed for the emergence of sharing experiences and sustainable local institutions (e.g., Castro, 2002). On the other side, fast changes in the upland have directly impacted riverine communities through new policies and demand of floodplain resources (McGrath et al., 1993). While both processes are usually addressed by different perspectives—sharing experience by the commons literature and socio-environmental impacts by the political ecology literature—in this chapter I offer an integrative analytical perspective, which combines collaborative and conflictive interactions at both local and regional scales. The analysis of the co-design and implementation of an innovative territorial model for the Lower Amazon floodplain communities—Agro-Extractive Project (AEP)—is based on three analytical pillars: a) the land-water nexus—the environmental foundation of a plural customary land tenure systems; b) historical legacies—the social foundation of local interactions and practices; and c) institutional ethos of the state—the cultural foundation of implementation of inclusive policies.

THE COMMONS AND THE LOWER AMAZON

In this section, I conceptualize community-based management system (CBMS) as a space of collaboration and tension, followed by the contextualization of the floodplain communities in the Lower Amazon, Brazil. The CBMSs were virtually invisible institutions in conservation research and practice until recently. Since the new millennium, however, the commons have become a key policy strategy to address the *tragedy of the commons* (McCay and Acheson, 1989; Ostrom, 1990). Grounded in models of collective action and sustainable production, numerous CBMS case studies worldwide have provided solid empirical evidence upon which the *theory of the commons* has been elaborated (Ostrom et al., 2002). The *commons* narrative resonated with advocacy and policy circles in the North and in emerging democracies in the South (Armitage et al., 2008; Pinkerton, 1989; Wilson et al., 2003). CBMSs have gradually been integrated into co-management systems in order to address power asymmetries between commoners and other stakeholders, such as private and state actors (Adger et al., 2005; Berkes and Pomeroy, 1997).

Researchers and practitioners have attempted to make CBMS more visible and to propose it as a suitable means of achieving justice and sustainability goals. However, in doing so, they have, often unintentionally, presented such customary institutions as the only local tenure system in place and as one that is characterized by symmetric relations. As a result, CBMS research and practice have been consolidated through an incomplete institutional, socio-ecological analysis. It is focused on commoners as the only local actor, common property as the only local tenure system, and the managed *commons* as the only contested matter driving collective action (e.g., Berkes and Folke, 2000; Ostrom, 1990). Despite major advances toward the recognition of local management systems in territorial and environmental governance, co-management systems have often yielded disappointing results. I argue that persistent inequalities and conflicting interests among local users are some of the issues that inhibit the expected devolution of power to local communities.

CBMSs are realities embedded in social-ecological contexts in which multiple interactions, ecosystems, and tenure systems are shaped and reordered over time and space according to local contexts (Adger et al., 2005). Local tenure, for instance, may encompass multiple institutional arrangements that are formed through contestation and negotiation over resource access and use among different local users (Peters, 1994). In this dynamic process, different local users exercise their agency by combining distinct norms, practices, and relationships according to their perceptions, social positions, and assets (Cleaver, 2002). Multiple-use common-pool resources (see Steins and Edwards, 1999) and co-management (see Armitage et al., 2008; Jentoft, 2005) perspectives are theoretical efforts to capture the multifaceted context in which CBMSs exist. However, while these perspectives often emphasize social relations (Berkes and Pomeroy, 1997) and asymmetries (Cash et al., 2006) between commoners and nonlocal actors (e.g., state and corporations), the everyday politics among different local users remains understudied.

In addition to the social embeddedness of CBMS, the ecological context plays major role in the design and implementation of local rules regulating resource use and monitoring (Bromquist et al., 1994). Although the *commons* are often classified according to a particular ecosystem/resource unit (e.g., forest, fishing, grasslands, water, game), in more heterogeneous and changing ecosystems the dynamics of landscape reconfiguration is particularly relevant to understand how CBMSs emerge and change. Wetlands are a case in point where annual river fluctuation creates permeable boundaries between water- and land-based resources (Junk, 1997). These dynamic ecosystems, characterized by multiple and changing landscape units, host a range of human activities throughout the year. The continuous environmental change set the stage for contestations of access to and control over multiple-use common pool resources among different local users.

In the Lower Amazon floodplain, two main local users—riparian residents and landholders—appropriate local resources. The riparian residents are an ethnically mixed peasants (*caboclos*) characterized by a combination of indigenous, European, and African sociocultural traits.[1] Their social position, distinct from the "authentic" indigenous Amazonian population and marginalized migrant settlers, has—until recently—made the *caboclos* an invisible population among researchers and policy-makers (Nugent, 1993). Their livelihood has been shaped by broader economic and political processes (e.g., economic cycles, political uprising and governmental development projects) as well as local social and ecological processes (e.g., environmental patterns and social interactions) (Adams et al., 2009). Residents' social life is particularly rooted in the mosaic and changing nature of the floodplains. The risky nature of the ecosystem creates some constraints but also fosters opportunities to build resiliency through diversity of production systems, institutions, and social relations (Castro, 2009; Harris, 2005; Lima, 2009). However, while much emphasis has been placed on social relations among riparian residents, little has been described about their social relations with the landholders with whom they have shared the floodplain for the last two centuries.

Landholders constitute a rural elite that, since the eighteenth century, has appropriated a large share of the floodplains to establish extensive production systems like cocoa and cattle ranching (Merry et al., 2004). Despite the asymmetric relations, both local users have kept their access to and use of the floodplain resources. This dual access is based on two layers of local property rights: the CBMS, emerged more recently, is controlled by the riparian residents while last-longing local social norms that are controlled by the landholders. In order to make the CBMS visible in the region, researchers and practitioners have overlooked the subtle social norms rooted in local asymmetric relationships. I contend that temporal, spatial, and social diversity has played a key role in the way this dual floodplain tenure arrangement has been crafted, exercised, and negotiated between the two parties. The assumption of CBMS as the single local tenure system has not only provided an incomplete picture of how access and control over local natural resources are negotiated among users, but also reinforced (invisible) power asymmetries between the two local actors.

In the following sections, I discuss how seasonality, human occupation history, and economic alternatives have shaped social relations between the two local actors and with the local tenure system. The analysis is based on a longitudinal research of over two decades, undertaken in three phases. Between 1990 and 1994, an ethnographic study was carried out in three communities with different levels of performance in their CBMS. Data sources included in-depth interviews with community residents, structured questionnaires to measure the production system in thirty randomly selected families, and participant observation. Between 1995 and 2003, irregular follow-up visits included interviews with NGO members, state agents, local leaders, and riparian residents engaged in discussions over the new territorial model for the floodplain. During 2008 and 2013, annual visits were carried out to assess the implementation of the new territorial model based on co-management principles. Data sources included interviews with key stakeholders, observations, and reports prepared by a local NGO that was directly involved in the implementation of the co-management system.

ENVIRONMENTAL, SOCIAL, AND HISTORICAL CONTEXT

The Lower Amazonian floodplain is a patchy and constantly changing landscape based on land-water connections. The annual river level oscillation—with approximately five meters' difference between the high- and low-water seasons—creates a rhythmic *pulsing system* that allows for the reloading of water nutrients and fertile soil sediments (Junk, 1997). During the peak of the high-water season in May, the whole landscape is considerably flooded. In the low-water season, a mosaic of four main sub-systems emerges (figure 4.1). *River streams* are channels that contour the floodplain islands. They are connected to *lake systems*, clusters of semi-open, interlinked water bodies located within the floodplain islands. *Lowlands* are areas of fertile bare soil formed by gradual sedimentation process, where natural grass grows every year. They may expand over time, giving rise to unclaimed land (known locally as *grown land*). *Natural levees* are higher terrains where residents usually build their houses. The higher the levee, the less exposed it is to the annual flood. However, levees may be unstable and prone to landslides (known locally as *fallen land*), a phenomenon that leads to the loss of earth. In extreme cases, fallen land may drive whole communities away.

The floodplain is therefore characterized by two temporal variations: an annual cyclical flooding season and a unidirectional, long-term process of "grown" and "fallen" land. In both cases there is some level of unpredictability and major risks. Although annual flooding is predictable, its length and extent vary year to year. For example, the highest and lowest water seasons in the last decades were registered in 2009 and 2010, respectively.

Figure 4.1 Ecological Zones of the Lower Amazonian Floodplain (Low-Water Season). *Source:* Image created by author.

Fertile soil, waterways, and diversity of natural resources have made the Amazonian floodplain an area long contested by different users. Earlier, densely populated settlements turned into virtually empty spaces after epidemic outbreaks, slavery persecution, and war during the early colonization stage (Roosevelt, 1989). Only in the eighteenth century was the human occupation resumed in the floodplain, under asymmetric agrarian structure defined by a few historical factors.

Floodplain farms were established with support from *sesmarias*, a policy that granted land titles to Portuguese immigrants during the Colonization period and was reinforced after the Independency by the Constitution of 1891 (Benatti et al., 2005). These landholders grew cocoa, which was suitable for the forest-shadowed levees and the floodplain's fertile soils, and brought cattle to graze on natural grasslands during the dry season (Gentil, 1988; WinkerPrins, 2006). Slavery abolishment coincided with drop in competitive prices of cocoa toward the end of the eighteenth century. As a result, landholders transitioned to cattle as their main economic activity. According to Funes (1995), cattle-ranching activity in the floodplain increased from 12 percent to 42 percent between the first and second half of nineteenth century. It was in this period that landless mixed peasants gradually moved to settle along the riverbanks. They developed a mixed subsistence economy based on annual crops, hunting, small-scale husbandry, fishing, wood collection for steam boats, and erratic work for landholders. However, the insurgent *Cabanagem* uprising that took place in the Amazon in the 1830s led to the massacre of a large number of the *caboclo* population including those living along the river (Harris, 2010). The private system of the floodplain combined with relatively low population density gave support to a land accumulation process (Harris, 2011) and the consolidation of the spatial configuration of floodplain farms surrounded by a few scattered riparian houses.

Throughout the nineteenth and twentieth centuries, several economic cycles in the Amazon—such as rubber (Weinstein, 1983), *jute* (Gentil, 1988), and gold mining (Cleary, 1990), along with governmental development policies (Bunker, 1985)—triggered major demographic, social, and territorial transformation. In particular, the debt peonage system emerged during the rubber boom created dependency and inequality between a rural elite and mixed peasants (Schmink and Wood, 1992). This system became pervasive in the region and was reproduced in different economic cycles. In the 1930s, for example, the jute boom brought landholders and floodplain residents to a similar economic dependency relation. Jute sharecropping and patronage systems consisted of land, input, and market access provided by the landholder and labor force supplied by the residents (Gentil, 1988). In 1934, the floodplain's private land regime was revoked by the national Constitution (Benatti et al., 2005). However, limited presence of the state combined with politically empowered landholders on one side, and poor local organization and a consolidated patronage system on the other, supported the continuity of private regime as a *de facto* land tenure system. Land transactions were maintained and even advertised in local newspapers (Castro, 1999). During this period, economic relations between residents and landholders kept conflicts at bay.

After the decline of jute production in the early 1970s (Winklerprins, 2006), landholders turned to cattle ranching while residents engaged in commercial fisheries. Fishing was a subsistence activity that gradually became more economically important due to the growing fish market fostered by urbanization in the region. At the same time, commercial fishermen from other regions entered the floodplain lakes with more efficient fishing technologies, leading to conflicts with residents (McGrath et al., 1993). These conflicts coincided with the consolidation of community-based settlements in the Amazon, orchestrated by an educational program of the Catholic Church (MEB) (Castro, 2009; Lima, 2009). This self-governing structure gave foundation to the notion of community-based property rights of floodplain lakes, known locally as *fishing accords* (Castro and McGrath, 2003).

Fishing accords, created in several communities during the 1980s and 1990s, are written documents in which rules regulating fishing activity and monitoring systems are drafted and brought to a vote in community meetings (Castro, 1999). In the 1990s, this CBMS attracted the attention of researchers and governmental and nongovernmental organizations as a genuine bottom-up solution for the commons' dilemma. Inspired by the *commons* research and activism, several initiatives were developed in the Amazon basin (Piñedo-Vasquez et al., 2011). In particular, local NGOs have initiated a gradual process to develop an integrated co-management of the floodplain system in Santarém in the 2000s. In 2005, as part of a national policy-oriented program for the Amazon governance (PPG-7), NGOs succeeded to push forward the implementation of a special territorial model designed for ethnic communities—the AEP (Castro, 2012a). The AEP grants residents exclusive rights to use their resources and to manage them jointly with the state. Although the AEP addressed compatibility with the CBMS—*fishing accords*—it overlooked another layer of the local tenure system, which defines resource allocation and negotiation among residents and landholders.

INSTITUTIONAL DIVERSITY AND LOCAL POLITICS

Until recently, floodplain communities in Brazil lacked legal rights to claim control over local resources. Nevertheless, community meetings were responsible for some decisions regarding access to and control over land and natural resources. At the same time, the powerful position of landholders guaranteed privileged access to some resources. Lakes and lowlands are of particular relevance in this analysis as they represent the two most economically important systems for local users. Residents use lakes for subsistence and commercial fishing, whereas lowlands are primarily used for cattle grazing by landholders and, more recently, by residents as well (McGrath et al., 2007). Lowlands and lakes physically overlap and their boundaries change according to the river level—one expands as the other retracts during the low- and high-water seasons. The continuous spatial reconfiguration of the two *commons* (lowlands and lakes), two contesting local actors (residents and landholders), and a range of economic activities are ingredients for an exceptionally complex tenure system for the floodplain.

According to formal legislation and policies, enacted by the Constitution of 1934, the national river floodplains are state property. Several laws and agencies are part of this legal framework. Land access is controlled by the State Heritage Office while the management and monitoring of floodplain resources is regulated by a set of national legislations (e.g., water, fishing, land, and forest) under the umbrella of environmental agencies. In particular, Water Legislation allows residents to use local resources as long as their practices do not threaten public interest (Vieira, 1992).

Due to the virtual absence of the state until recently, the private structure of the floodplain has remained unchanged. For instance, the clearing of floodplain forest and land transactions persisted and intensified during the jute boom despite the establishment of state property rights (Castro, 1999). State absence, on the other hand, opened space for local actors to shape their own ruling systems according to the socioenvironmental context. Only after national agencies increased their influence in environmental and territorial governance in Brazil during the last two decades, did the interplay between the local tenure systems and formal legislation collide. The CBMSs have been incorporated into co-management systems. However, another layer of the local ruling system that took shape during the reoccupation of the floodplain has been ignored.

Community-Based Management System

The CBMS makes clear distinctions across four floodplain sub-systems. *Streams,* particularly river channels where the excludability of transportation and fishing boats is troublesome, are treated as open access. *Natural levees*, where houses and crop fields are located, are privately owned and can be sold, rented, and inherited. Land transactions are locally formalized through hand-written documents in which the owner's name, property boundaries, and price are recorded. The façade limits are clearly defined along the riverbank; backside limits may become sources of conflict due to divergent interpretations. Land and material losses caused by natural events (e.g., flooded gardens, landslides) are treated as private misfortunes without rights to compensation.

Lake systems and lowlands are held collectively. The former is ruled by residents through the fishing accords (Castro and McGrath, 2003). Based on a concept of collective access and use rights, this CBMS emerged as a response to the perception of threat from intensified commercial lake fisheries (McGrath et al., 1993). Landholders sometimes supported fishing accords because they also benefited from the monitoring system that indirectly hindered cattle piracy in their land.[2]

Fishing accords have proliferated throughout the region. In some cases, they present well-elaborated rules of access, use, monitoring, and sanctioning (Castro, 1999). Until recently, however, they remained mostly fragmented, poorly formatted, often limited to single communities, and focused primarily on fishing resources (Castro and McGrath, 2003). Although lake systems are managed collectively by residents, they cover large areas that emerge as grassland during the low-water season and are therefore held privately by landholders. As a result, tension between collective and private access and use of these two overlapping systems leads to a subtle negotiation process between the two local users. This tension is reflected in a layer of social norms that mediate decisions regarding access.

Social Norms

The long-standing use of floodplain resources by residents and landholders forged cultural, social, and economic bonds between them. Under a virtually absent state, landholders took on the role of providing services and employment to residents in exchange for their control over the floodplain system. Oftentimes, landholders became godfathers of the residents' children and participated actively in community celebrations and meetings. Economic dependency established during the jute boom added to the social norms that mediated decisions about access to and use of the floodplain system for both local users. However, in contrast to the CBMS, which has been the key element in the residents' discourse, the underpinning social norms are hardly expressed and only surface when incidents over the interpretation of rights emerge.

In the past, local incidents were more frequent when social bonds between residents and landholders were relaxed. In the early 1900s, for instance, reports of local incidents over land tenure were commonplace in the local newspapers (Castro, 1999). Ironically, land contestations declined after landholders lost their legal land rights in 1934, when the patronage system supported private property regime to remain as the *de facto* tenure. Only after the 1970s, when community-based organization emerged, did the social relation between the two local users swing back to more conflictive interactions. Two factors contributed to this trend: 1) the increased political support of floodplain communities by religious groups, state and nonstate agencies; and 2) the increasing engagement of residents in cattle raising activities. Below I present a few local incidents observed in the 1990s, in order to illustrate how social norms (controlled by landholders) tried to defy the rules established by the CBMS.[3]

Incident 1: As described earlier, crop fields located on the *natural levees* are held privately under the CBMS. While damages caused by natural events are treated as individual misfortunes, material losses caused by third parties are expected to be compensated. Landholders, however, tend to refuse to compensate residents from recurrent crop damages caused by their cattle. Residents emphasize that cattle are not allowed in their private land, to which landholders retort that private gardens should be properly fenced. Therefore, these incidents tend to remain unsettled.

While unresolved incidents in the *natural levees* are more tolerated, cattle-related incidents on *lowlands* have led to more severe outcomes. Competition between landholders and residents over grasslands has recently increased. In particular, young residents have increasingly engaged in cattle partnership with outside ranchers. According to this informal arrangement, they host and take care of a number of cattle in the floodplain grasslands during the low-water season; in exchange, they receive half of the calves (Merry et al., 2004). Cattle partnership has made residents economically independent from landholders, and more prone to reclaim their collective rights to the lowlands. As a result, tension between private and collective claims to lowlands rose in the 1990s. One particular incident illustrates this issue.

Incident 2: Landholders proposed a community rule to regulate the grassland leasing business in Ilha de Sao Miguel. According to the rule, each resident could host a limited number of heads according to the size of the *lowlands* located behind his or her house. The approval of this proposal showed that the community residents acknowledge the private tenure of lowlands. In contrast, the collective decision to ban water buffalo in the floodplain—due to severe environmental impact—has been repeatedly ignored by landholders.[4]

Contestation over lowlands is particularly stringent in grown land where property rights are still unclear. Such lands usually emerge as new islands close to communities that are used by residents to farm and are eventually appropriated by landholders for cattle grazing. In three incidents observed in the 1990s, external support from a grassroots organization was instrumental in supporting residents' struggles to maintain their access to the island.

Incident 3: A landholder claimed his exclusive rights to an island formerly shared with fifteen residents. Most of the users left the area immediately in deference to the landholder's request. However, two years later—with support from the Rural Workers' Union—two residents won a court case and were granted right-of-use.

Incident 4: In 1959, a landholder was granted exclusive rights to an island of approximately two hectares, which he shared with several residents. Through the sedimentation process over the years, the island grew considerably to 250 hectares. In 1994, the landholder claimed exclusive rights-of-use to the whole island. Supported by the Fishermen's Union, the residents filed a suit against the landholder and won their rights-to-use to manage the island collectively while the landholder was only granted exclusive use rights to the original two hectares.

Incident 5: A landholder shared an island with a few residents for over two decades. After the death of the landholder, the property rights to the island became a source of conflict. The heir claimed exclusive access to the island and, similar to the cases described above, the residents—with support from the Fishermen's Union—were granted exclusive collective use rights.

The three cases above illustrate the continuous contestation over grown land. It also reveals that landholders strive to keep private control over lowlands while granting access to residents. Perception of private rights is sometimes shared by residents who comply with the landholders' decision. Their recent engagement in cattle ranching combined with external political support has empowered residents to reclaim their collective rights to lowlands.

In contrast, conflicts related to access to and control over *lake systems* are less severe since landholders have less economic interest in this sub-system. Nevertheless, although lake management is the flagship of CBMS in the region, claims by landholders for private ownership of lakes were observed on a few occasions, both in discourse and action.

Incident 6: A landholder allowed his guests to violate a community-based fishing rule. In his defense, he alleged that the lake in question was located within *his* property. In another case, a landholder claimed to have exclusive rights to make use of locally banned fishing gear in *his water*. In both cases, after several discussions the community finally accepted the landholders' claims, showing some level of common understanding of the private tenure of lakes that are enclosed in one's landholding during the low-water season.

However, less reconciliatory incidents emerge when landholders' claims have an impact on the economic interests of the residents, as illustrated by the incident described below.

Incident 7: One illustrative example is the request from a landholder for exclusive rights to commercially exploit the pirarucu (*Arapaima gigas*), a high-priced lake fish collectively managed in a lake located within *her* property. After a bitter negotiation process, the two parties agreed to keep the community rights to the managed species with the condition to sell the catch exclusively to the landholder in exchange for her support of the community-based management of *her* lake. Possessive pronouns (his, her) are frequent signs of boundaries for access, control, and management rights in the local discourse. Two incidents reveal how private ownership of aquatic system is also in residents' discourse.

Incident 8: In one fishing accord, it was decided that those who did not comply with the local rules would be allowed to fish only in *their* water (i.e., in the area adjacent to their landholding).

Incident 9: During a participatory mapping exercise, a focus group of residents drew landholding boundaries on a satellite image taken during the high-water season. Some of the boundary lines crossed over the middle of lakes. They explained that the lines represented the division of lowlands, which could be seen only during the low-water season. Therefore, although residents perceive the collective access to lakes for fishing purposes during the high-water season, they seem to be aware of private boundaries based on the landscape configuration during the low-water season.

The notion of private rights to aquatic system in the residents' discourse clearly illustrates that, while ecological factors influence the pattern of collective access and management rights to this sub-system, political factors define private control over decisions. This nuanced variation between the two local actors regarding the bundle of rights has major implications for the implementation of a new territorial model in the floodplain.

A NEW TERRITORIAL MODEL UNDER OLD SOCIAL RELATIONS

In 2006, a new territorial model was proposed for the Lower Amazon—the AEP[5]—in order to provide legal grounds for the CBMS. The implementation of the AEP was based on four main pillars: 1) social justice for local residents; 2) exclusive collective land use rights for local residents by means of a (renewable) concession contract of ten years; 3) a participatory collective Management Plan; and 4) the provision of small grants, credit lines, and technical assistance to establish basic infrastructure for sustainable production systems.

As an agrarian model, the AEP is defined by ecological, cultural, and socioeconomic criteria. In the Amazonian floodplain, the AEP boundaries are defined according to a lake system and its surrounding floodplain. All community residents, based on their ethnic background (*caboclo*), are entitled to live in an AEP and to all benefits therein.[6] Landholders, on the other hand, are disentitled from their (informal) landholdings. Their access to and use of the floodplain resources will depend on formal permission from the AEP Council, composed of a representative from each community. This way, the AEPs are expected to foster change in the local power structure that has taken shape over the last two centuries, and to strengthen the bottom-up CBMSs developed over the last few decades.

A Management Plan formulated in collaboration with local communities and state agencies is expected to regulate access to and use of local natural resources. The CBMS *fishing accords* are to be incorporated into the Management Plan as monitored by the AEP Council, a cross-community organization run by local representatives. As evoked by the settlement model, agro-extractive activities such as artisanal fishing, agroforestry, and small-scale farming systems are the core of the local socioeconomy foreseen for the residents. In contrast, cattle ranching is expected to be limited.

In summary, the management of the floodplain has evolved from a fragmented, informal ruling system with limited state support to a consolidated territorial governance that includes state and nonstate actors. In particular, the land tenure system of an AEP implies a major change in the political positions of the two local users. Under this new territorial model, the *de facto* private tenure system was to be replaced by a formal collective system that granted exclusive use concession to community residents. It was to be regulated by a participatory management plan, which would incorporate the fishing accords monitored and regulated by the AEP Council and the Management Plan. By revoking the *de facto* private land tenure in the floodplain, landholders' protagonist role in the decision-making process should change to one of authorized user. The formal recognition of residents to oversee the AEP was expected to give them the power to decide whether and how landholders could use the floodplain system. Therefore, the establishment of the AEP implied a major change in the political position of the two local actors. Residents became legally recognized as the territory overseers who decide how and by whom the natural resources should be accessed and used.

Needless to say, the implementation of the AEPs found major resistance among landholders. An informant from the Rural Producers Union at Santarém (SIRSAN), which represents cattle ranchers, held that the implementation of AEPs in the region characterized "a disguised authoritarian process to steal land from producers." In his opinion, landholders were mistakenly treated as land grabbers who had illegally appropriated the floodplain land. He argued that landholders had acquired their properties through the same *de facto* land market as residents. He also criticized the image of "large ranchers" that was used to refer to floodplain landholders. He contended that an internal report prepared by the state agricultural agency revealed an average herd size of eighty heads per landholding in the floodplain, and only 5 percent of landholdings with more than 300 heads. Despite their relatively small-scale ranching activity, landholders have been excluded from the new territorial model and disregarded as legitimate users.

After several failed attempts to reverse the process, landholders went quiet and withdrew from public meetings. Those involved in the implementation process interpreted landholders' behavior as a victory of the co-management system. Nevertheless, they admitted that landholders were not ready to hand over their power to residents, and that land incidents were expected to emerge at a later stage. Despite the landholders' dissatisfaction, the number of incidents registered between the two local users was surprisingly low in the first three years of the implementation process. Only a few official complaints were turned in to the government office. Although the undisturbed process might reflect agreements reached between residents and landholders—the outcome envisaged by the AEP—practitioners and state agents mentioned three other less desirable reasons driving this apparent smooth process. First, residents were busy dealing with their financial benefits (e.g., credits, grants, and infrastructure) and paid little attention to landholders' violation. Second, most of the AEP Councils were not organized enough to mobilize and confront violations from landholders. Third, some residents may still regard landholders as legitimate and powerful local users.

The last two reasons are particularly relevant to local social relations and the images of land tenure among local users, illustrated by one incident.

Incident 10: One of the most well-organized AEP Councils (Aritapera) tried to apply their new territorial rules to solve a last-longing land conflict between residents and landholders in one community. The state agent in charge of the AEP implementation explains with frustration the way the Council formulated the formal complaint:

> The most capable leader of the community sends to us [state agency] a formal proposal requesting that the *landowner* donate half of his land to the Council as a way to resolve the land conflict. The appropriate proposal should be framed the other way around, that the Council would grant the user (*not landowner!*) half of the land he had illegally appropriated. If the local residents' perception of property rights will not change, it will be difficult to see a real change on the ground. (my translation and my emphasis)

Despite the persistent image of "landowner," the outcome has shown some level of residents' empowerment. Initially, the landholder reacted to the residents' demands with threats and sought support from influential politicians. After failing to reach his goal, he turned to negotiation and agreed to allocate half of the area to several residents to farm. Therefore, in more organized communities, landholders had to change their formerly resistant position to a negotiating strategy in order to adapt to the new institutional arrangement. In less-organized communities, however, local politics remain a major barrier to change in land tenure.

The implementation of AEPs has also not been effective in completely stopping the *de facto* land transactions. Under the new land tenure arrangements, residents do not hold alienation rights to their territory. However, interviews with local leaders reveal that land trade persists as a local practice among both residents and landholders.

The practices observed in well-organized communities show that contradictions between the new formal territorial model and the old *de facto* tenure system do not only occur in communities that lack institutional capacity. Virtually all communities seem to face challenges in replacing the asymmetric *de facto* private tenure with the more just land tenure system installed by the AEP. In short, despite the promises of the new formal tenure system to empower residents, landholders rely on local perceptions to maintain access to their appropriated land under the new territorial model. In some cases, residents cannot break the social norms controlled by landholders due to limited organizational capacity. In other cases, residents and landholders still share the *de facto* tenure system grounded in asymmetric social relations.

MOVING BEYOND THE CBMS AND THE COMMONERS

The theoretical divide in the analysis of the Amazonian socio-environmental changes tends to mask more complex processes combining collaborative and conflictive interactions. The Lower Amazonian floodplain case reveals the limitation of both analytical perspectives to address the transformations observed in the region. In contrast to the focus on social impacts by the political ecology perspective, innovative territorial models such as the AEP can emerge in agrarian frontiers as part of a co-production process among local residents, state, and practitioners. At the same time, in contrast to the focus on collective action offered by the commons perspective, local power asymmetries can lead to unequal share of resources and limited local autonomy. A more integrative approach combining the adaptive process to local ecological and social context in CBMSs (e.g., McCay and Acheson, 1987; Netting, 1976; Ostrom, 1990) with historical legacies and broader s (e.g., Peters, 1994; Edwards and Steins, 1999; Dietz and Henry, 2008) is needed in order to reveal hidden factors driving the struggles over access and control of natural resource.

CBMSs are usually embedded at multiple socio-environmental levels, in which power relations play a major role (Ribot and Peluso, 2003). A close look at local social interactions with temporal and spatial depth is needed in order to better understand how asymmetric relations shape tenure arrangements among local users (Agrawal and Gibson, 1999). Political life in rural communities goes beyond formal and concerted collective efforts. It combines subtle everyday life interactions, ranging from support to resistance (Kerkvliet, 2009). The Amazonian floodplain case offers a unique opportunity to explore these conceptual shortcomings in the commons research. It adds temporal, spatial, and institutional diversity into the analysis of everyday life politics and the implications on the local tenure arrangement.

The analysis of the floodplain territorial governance in the Lower Amazon reveals different layers of land tenure. Under the official legislation, designed and regulated by national agencies that is influenced by large-scale political demands, two layers of tenure and management systems driven by historical, socioeconomic, and political factors overlaps and interact over space and time—the CBMS, focused on the local socio-environmental context, primarily related to fisheries and secondarily to other floodplain resources; and a more implicit ruling system mediated by economic and social relations between residents and landholders.

The two-layered local floodplain tenure system reveals patterns of resource appropriation that play out at different levels. The *fishing accord* is an explicit ruling system controlled by residents to claim collective rights to lakes and, more recently, to the lowlands. Grounded in the social capital built over time, this CBMS emerged as a local solution to external pressure from fishing encroachment. Supported by the political structure developed by the Catholic Church and later by NGOs, the CBMS was relatively successful in limiting resource access and use from outsiders. Underneath the CBMS lies an implicit ruling system controlled by landholders, which is triggered whenever their privileged access is under threat. Supported by asymmetric relations emerged from land tenure privileges in the eighteenth century and later deepened by economic dependency relations, these norms have been successful in regulating resource access among local users.

While conflicts with outside fishermen take place in the form of verbal and physical confrontations, contestations over control of and access to natural resources involving residents and landholders often take place in the form of subtle negotiations. Social bonds between the two local users make open conflicts costly for residents (Castro, 2002). Therefore, the *fishing accords* have been more successful when the rules do not jeopardize landholders' benefits; otherwise, they are overruled by social norms. In other words, although *access* to "collective" resources—for example, lakes and grasslands—may be maintained among both local actors, landholders strive to maintain *control* over use of these resources. Only when residents are empowered to confront and even to overcome local power asymmetries do the tensions between CBMS and the social norms become more evident and yields more frequent incidents.

This dual floodplain tenure system observed in the Lower Amazon challenges unidimensional perspectives of the *commons*. If treated as the only local management system in place, the CBMS conforms to the *cultural ecology* model, which addresses collective action as an adaptive process driven by the ecological and economic attributes of the ecosystem and resource units (Netting, 1976). The *fishing accord* is influenced by the ecological patchiness generated from the annual water fluctuation, leading to a spatially and temporally variable property rights system. In particular, the permeability of boundaries and highly dispersed (grass) or fugitive (fish) resources make individual rights of lake and lowlands troublesome (McGrath et al., 2008). Similar explanations have been proposed by other authors. Thomas (1996) describes how the floodplains tenure system in Nigeria spatially varies according to level of concentration, predictability, and mobility of resources. Vondal (1987) explains how seasonality influences changes in appropriation regime in the same physical area in the swampland in Borneo, where lakes that were collectively shared during the high-water season turned into privately owned cropland during the low-water season.

Although the CBMS fits well within this adaptive model, social norms mediating the reinterpretation of property rights at the local level can only be explained by local power asymmetry and shared local perceptions of floodplain tenure among different local users. The history of the floodplain occupation shows that landholders have succeeded in maintaining control over floodplain resources even after the revocation of private regime by accommodating a *de facto* ruling system. The implementation of AEP signals a new round in this process. This new territorial model represents a unique opportunity to change the local power structure through a co-management system to reconcile the national legal framework with the local ruling system. Although this strategy is in line with the theory of the commons (Berkes and Folke, 2000; Jentoft et al., 1998; Ostrom, 1990), it overlooks veiled social norms that have outlived several institutional innovations and helped to maintain landholders' privileges.

Co-management initiatives emphasize the local ruling systems that are expressed through collective action. On the other hand, little attention is devoted to more subtle social norms embedded in local social relations. The invisibility of such norms among policy-makers, practitioners, and researchers reflects an incomplete contextualization of the local tenure system. Residents have exerted agency through community organization, conservationist discourse, and attracting external support in order to exclude outsiders (Castro, 2012b). However, they have not been able to overcome local structural power relations that supports the local governance of floodplain resources among residents and landholders.

CBMS research and practice must go beyond the analysis of how commoners work together, and how collective institutional arrangements are crafted. CBMS is ingrained in multiple ecological, economic, and institutional repertoires (Cleaver, 2002). Therefore, in order to better understand the potential and limitations of such local institutions, a temporal, spatial, and social depth is needed. More importantly, it is fundamental to account for everyday life practices to reveal how local rules are continuously contested and negotiated between different actors under asymmetric relations in order to adapt to environmental, market, and institutional changes.

CONCLUSIONS: POLITICIZING THE COMMONS

CBMS research usually focuses on features of the commoners, attributes of single managed ecosystems and resource units, and single local tenure systems. However, CBMS represents only part of a more complex system shaped by the ecological and social contexts faced by different local actors. The limited focus on CBMS and on the commoners may hide the political process that shapes the appropriation of the *commons* among local users. These actors negotiate access to and control over natural resources under socio-environmental, socioeconomic, and institutional diversity. Revealing diversity of ruling systems and everyday life practices that mediate these interactions is fundamental for understanding how the *commons* are used and managed at the local level. A micro-political contextualization of the commons is needed not only for theoretical reasons but also to better unveil potentialities and shortcomings for the legal recognition of sustainable and robust CBMSs. The Lower Amazonian floodplain case reveals how the land tenure arrangement has been influenced by a range of external and local factors.

While external economic, demographic, political factors have strongly created new pressures on the floodplain resources, local environmental and social diversity and change have been particularly relevant in shaping interaction between two local actors and their agency. To assume that the integration of "visible" local management systems into a formal legal framework is sufficient to achieve an efficient co-management system is rather simplistic. Despite major structural changes in the formal tenure framework, power relations between different local users may remain unchanged if local perceptions and everyday life practices of power relations are maintained. The success of a co-management system depends on the degree of compatibility with other layers of property rights and on whether or not it addresses local power asymmetries. Therefore, a diachronic analysis of local social interactions beyond commoners and the CBMS is crucial to reveal how socio-environmental and institutional diversity may shape constraints and opportunities for a truly participatory and democratic local governance of natural resources.

NOTES

1. The term *caboclo* has been contested by some scholars as it often implies derogatory meanings of laziness, ignorance, and underdevelopment. In this article, I do not engage in this debate and use *caboclo* as a social category with particular sociopolitical and historical features (Adams et al., 2009).
2. On the other hand, increased ranching activity among landholders fueled local conflicts with floodplain peasants who suffered from invasions of cattle into their gardens and fishing grounds.
3. The analysis of incidents highlights local perception regarding access to and control of each sub-system, and the political position of each local actor. For a more detailed description of the incidents analyzed in this section, see Castro (1999).
4. Community residents argue that water buffaloes lead to soil compaction, crop damage, and the destruction of fishing grounds.
5. AEP from the Portuguese PAE—"Projecto Agroextrativista."
6. Community residents employed by the state (e.g., teachers, health agents, environmental agents) are not eligible for financial benefits such as small grants, credit lines, and technical assistance.

BIBLIOGRAPHY

Adams, C., Murrieta, R., Neves, W., and Harris, M. (eds.). 2009. *Amazon Peasant Societies in a Changing Environment: Political Ecology, Invisibility and Modernity in the Rainforest.* Springer: Dordrecht.

Adger, W. N., Brown, K., and Tompkins, E. L. 2005. The Political Economy of Cross-scale Networks in Resource Co-management. *Ecology and Society*, 10(2), 9.

Agrawal, A. and Gibson, C. 1999. Enchantment and Disenchantment: The Role of Community in Natural Resource Management. *World Development*, 27(4), 629–49.

Armitage, D., Berkes, F., and Doubleday, N. (eds.). 2008. *Adaptive Co-management: Collaboration, Learning, and Multi-level Governance* (Sustainability and the Environment). UBC Press: Vancouver.

Benatti, J. H., Surgik, A. C. S., Treccani, G. D., McGrath, D. G., and Gama, A. S. P. 2005. *A Questão Fundiária e o Manejo dos Recursos Naturais da Várzea: Análise para a Elaboração de Novos Modelos Jurídicos.* Provárzea: Manaus.

Berkes, F. and Folke, C. (eds.). 2000. *Linking Social and Ecological Systems: Management Practices and Social Mechanisms for Building Resilience.* Cambridge University Press: Cambridge.

Berkes, F. and Pomeroy, R. S. 1997. Two to Tango: The Role of Government in Fisheries Co-management. *Marine Policy*, 21(5), 465–80.

Bromquist, W., Schlager, E., Tang, S. Y., and Ostrom, E. 1994. Regularities from the Field and Possible Explanations. In: *Rules, Games, and Common-pool Resources*, Ostrom, E., Gardner, R., and Walker, J. (eds.). University of Michigan Press: Ann Arbor, pp. 301–18.

Brondizio, E. S., Ostrom, E., and Young, O. R. 2009. Connectivity and the Governance of Multilevel Ecological Systems: The Role of Social Capital. *Annual Review of Environmental and Resources*, 34, 253–78.

Bunker, S. 1985. *Underdeveloping the Amazon.* University of Illinois Press: Urbana.

Cash, D. W., Adger, W. N., Berkes, F., Garden, P., Label, L., Olsson, P., Pritchard, L., and Young, O. 2006. Scale and Cross-scale Dynamics: Governance and Information in a Multilevel World. *Ecology and Society*, 11(2), 8.

Castro, F. 1999. *Fishing Accords: The Political Ecology of Fishing Intensification in the Amazon.* CIPEC Dissertation Series, No. 4. Indiana University: Bloomington.

Castro, F. 2002. From Myths to Rules: The Evolution of the Local Management in the Lower Amazonian Floodplain. *Environment and History*, 8(2), 197–216.

Castro, F. 2009. Patterns of Resource Use by *Caboclo* Communities in the Middle-lower Amazon. In: *Amazonian Peasant Societies: Modernity and Invisibility*, Adams, C., Neves, W., Murrieta, R., and Harris, M. (eds.). Springer: Dordrecht, pp. 157–77.

Castro, F. 2012a. Between Cooperation and Conflict: The Implementation of Agro-extractive Settlements in the Lower Amazon Floodplain. In: *Human–Environment Interactions: Current and Future Directions*, Brondizio, E. and Moran, E. (eds.). Springer: Dordrecht, pp. 213–34.

Castro, F. 2012b. Multi-scale Environmental Citizenship: Traditional Populations and Protected Areas in Brazil. In: *Environmental and Citizenship in Latin America:*

Natures, Subjects, and Struggles, Latta, A. and Wittman, H. (eds.). CEDLA Latin America Studies 101. Berghahn Books: New York, pp. 39–58.
Castro, F. and McGrath, D. 2003. Moving towards Sustainability in the Local Management of Floodplain Lake Fisheries in the Brazilian Amazon. *Human Organization*, 62(2), 123–33.
Cleary, D. 1990. *Anatomy of the Amazon Gold Rush*. The Macmillan Press: London.
Cleaver, F. 2002. Reinventing Institutions: Bricolage and the Social Embeddedness of Natural Resource Management. *The European Journal of Development Research*, 14(2), 11–30.
Dietz, T. and Henry, A. D. 2008. Context and the Commons. *PNAS*, 105(36), 13189–90.
Edwards, V. and Steins, N. 1999. The Importance of Context in Common Pool Resources. Special Issue: *Journal of Environmental Policy and Planning*, 1(3), 195–294.
Fearnside, P. M. 2015. Highway Construction as a Force in the Destruction of the Amazon Forest. In: *Handbook of Road Ecology*, Ree, R., Smith, D. J., and Grilo, C. (eds.). Wiley Blackwell, pp. 414–24.
Funes, E. A. 1995. Nasci nas Matas Nunca tive Senhor: História e Memória dos Mocambos no Baixo Amazonas. PhD thesis. University of São Paulo: São Paulo.
Gentil, J. M. L. 1988. A Juta na Agricultural de Várzea na Área de Santarém—Médio Amazonas. *Boletim do Museu Paraense Emilio Goeldi, Série Antropologia*, 4, 118–99.
Harris, M. 2005. Riding a Wave: Embodied Skills and Colonial History on the Amazon Floodplain. *Ethnos: Journal of Anthropology*, 70(2), 197–219.
Harris, M. 2010. *Rebellion on the Amazon: The Cabanagem, Race, and Popular Culture in the North of Brazil, 1798–1840*. Cambridge University Press: New York.
Harris, M. 2011. The Floodplain of the Lower Amazon as a Historical Place. In: *The Amazon Várzea: The Decade Past and the Decade Ahead*, Pinedo-Vasquez, M., Ruffino, M., Padoch, C., and Brondizio, E. S. (eds.). Springer: Dordrecht, pp. 37–51.
Ioris, A. A. R. 2017. *Agribusiness and the Neoliberal Food System in Brazil: Frontiers and Fissures of Agro-neoliberalism*. Routledge: London.
Jentoft, S. 2005. Fisheries Co-management as Empowerment. *Marine Policy*, 29(1), 1–7.
Jentoft, S., McCay, B. J., and Wilson, D. C. 1998. Social Theory and Fisheries Co-management. *Marine Policy*, 22(4–5), 425–36.
Junk, W. J. 1997. *The Central Amazon Floodplain: Ecology of a Pulsing System*. Ecological Studies No. 126. Springer: Berlin.
Kerkvliet, B. J. T. 2009. Everyday Politics in Peasant Societies (and Ours). *The Journal of Peasant Studies*, 36(1), 227–43.
Lima, D. M. 2009. The Domestic Economy in Mamirauá, Tefé, Amazonas State. In: *Amazonian Peasant Societies: Modernity and Invisibility*, Adams, C., Neves, W., Murrieta, R., and Harris, M. (eds.). Springer: Dordrecht, pp. 131–56.
McCay, B. J. and Acheson, J. M. (eds.). 1987. *The Question of the Commons: The Culture and Ecology of Communal Resources*. University of Arizona Press: Tucson.

McGrath, D., Almeida, O. T., and Merry, F. D. 2007. The Influence of Community Management Agreements on Household Economic Strategies: Cattle Grazing and Fishing Agreements on the Lower Amazon Floodplain. *International Journal of the Commons*, 1(1), 67–87.

McGrath, D., Cardoso, A., Almeida, O. T., and Pezzuti, J. 2008. Constructing a Policy and Institutional Framework for an Ecosystem-based Approach to Managing the Lower Amazon Floodplain. *Environment Development and Sustainability*, 10, 677–95.

McGrath, D., Castro, F., Futemma, C., Amaral, B. D., and Calabria, J. 1993. Fisheries and the Evolution of Resource Management on the Lower Amazon Basin. *Human Ecology*, 21(2), 167–95.

Merry, F. D., Sheikh, P. A., and McGrath, D. 2004. The Role of Informal Contracts in the Growth of Small Cattle Herds on the Floodplain of the Lower Amazon. *Agriculture and Human Values*, 21, 377–86.

Netting, R. M. 1976. What Alpine Peasants Have in Common: Observations on Communal Tenure in a Swiss Village. *Human Ecology*, 4(2), 135–46.

Nugent, S. 1993. *Amazonian Caboclo Society: An Essay on Invisibility and Peasant Economy*. Berg: Oxford.

Ostrom, E. 1990. *Governing the Commons: The Evolution of Institutions for Collective Action*. Cambridge University Press: Cambridge.

Ostrom, E., Dietz, T., Dolsak, N., Stern, P. C., Stonich, S., and Webers, E. U. (eds.). 2002. *The Drama of the Commons National Research Council*. National Academic Press: Washington, D.C.

Peter, P. E. 1994. *Dividing the Commons: Politics, Policy, and Culture in Botswana*. University of Virginia Press: Charlottesville.

Pezzuti, J., Castro, F., McGrath, D. G., Saikoski Miorando, P., Sá Leitão Barboza, R., and Carneiro Romagnoli, F. 2018. Commoning in Dynamic Environments: Community-based Management of Turtle Nesting Sites on the Lower Amazon Floodplain. *Ecology and Society*, 23(3), 36.

Piñedo-Vasquez, M., Ruffino, M. L., Padoch, C., and Brondízio, E. S. (eds.). 2011. *The Amazon Várzea: The Decade Past and the Decade Ahead*. Springer: Dordrecht.

Pinkerton, E. (ed.). 1989. *Co-operative Management of Local Fisheries: New Directions for Improved Management and Community Development*. UBC Press: Vancouver.

Ribot, J. C. and Peluso, N. L. 2003. A Theory of Access. *Rural Sociology*, 68(2), 153–81.

Roosevelt, A. C. 1989. Natural Resources Management in Amazonia before the Conquest: Beyond Ethnographic Projections. *Advances in Economic Botany*, 7, 30–62.

Schmink, M. and Wood, C. 1992. *Contested Frontiers in the Amazon*. Columbia University Press: Oxford.

Steins, N. A. and Edwards, V. M. 1999. Platforms for Collective Action in Multiple-use Common-pool Resources. *Agriculture and Human Value*, 16, 214–55.

Thomas, D. H. L. 1996. Fisheries Tenure in an African Floodplain Village and the Implications for Management. *Human Ecology*, 24(3), 287–313.

Vondal, P. J. 1987. The Common Swamplands of Southeastern Borneo: Multiple Use, Management, and Conflicts. In: *The Question of the Commons: The Culture and Ecology of Communal Resources*, McCay, B. J. and Acheson, J. M. (eds.). University of Arizona Press: Tucson, pp. 231–49.

Weinstein, B. 1983. *The Amazon Rubber Boom 1850–1920*. Stanford University Press: Stanford.

Wilson, D. C., Nielsen, J. R., and Degnbol, P. (eds.). 2003. *The Fisheries Co-management Experience: Accomplishments, Challenges and Prospects*. Kluwer Academic Publishers: Dordrecht.

WinklerPrins, A. M. G. A. 2006. Jute Cultivation in the Lower Amazon, 1940–1990: An Ethnographic Account from Santarem, Para, Brazil. *Journal of Historical Geography*, 32, 818–38.

Part II

QUESTIONING THE BASIS OF PROGRESS, POWER, AND POVERTY

Chapter 5

What Is New in Agribusiness in Brazil?

The Long Path of Conservative Modernization in the Perpetual Country-of-the-Future

Rafael R. Ioris and Aaron Schneider

INTRODUCTION

The increased involvement of Brazil in the global economy in the last twenty years has been closely tied to the growing importance of agribusiness and associated activities in the country. The largest economy in Latin America has indeed become a global agriculture powerhouse and much of the public debate about this achievement has thus far been largely shaped by a laudatory narrative that inevitably obscures inherent problems involved in this transformative historical dynamic. In fact, while the expansion of agribusiness has deepened capitalist relations by means of the expansion of credit, consolidation of large-scale, high-tech, monoculture farming, and waged labor relations, it has also fostered a new type of legitimacy for oligarchic rule, commodity export dependence, state-dependent private enterprise, social exclusion, and environmental destruction.

The rise in political importance of the subnational political economy of agribusiness in Brazil, particularly by means of expanding the economic frontiers of the Amazon region, is the focus of this chapter. Most importantly, we scrutinize how the following three elements are closely related and mutually reinforcing: (1) the ability of local and national economic and political oligarchies to control the very process of economic success by establishing a multitude of innovative political alliances; (2) the unequal distribution of gains derived from the growing involvement of the Amazon region in global commodity chains; and (3) despite its alleged modern economic efficiency, the primitive nature of accumulation in the sector.

Toward these goals, the present chapter examines the much-praised economic success of agribusiness in Brazil by looking at its most dynamic and internationalized sectors, beef and soybean production. In doing so, the piece indicates that though important innovations, mostly on the technical and macro-economic sides, were manifest in the process of internationalization of the Brazilian economy over the last twenty-five years; the country is still far from being able to display a socially just, economically competitive, and environmentally sound record of economic modernization. On the contrary, the logic demonstrated in the recent path of efficient insertion of Brazil in ever more connected commodity chains on a global scale only confirms that the Brazilian society is still trapped in a decades-old path of conservative modernization that has consistently confirmed the status of Brazil as rich but deeply unjust nation.

Agribusiness within the Brazilian Tradition of Passive Revolutions

The pattern of boom and bust visible in developing countries such as Brazil is closely associated with the perils of commodity-led development. The disadvantages of commodity exports arguably lie at the heart of the entire field of development economics, propelled by the Prebisch-Singer thesis that commodities face declining terms of trade relative to manufactured goods, producing both greater fluctuations and decreasing relative returns for developing countries over time (Singer and Sarkar, 1991). While the existence of occasional commodity price booms casts some doubt on the thesis, the inevitable turn to bust affirms that commodity-led development is particularly difficult. Even in the presence of extended and extreme swings in price lasting several decades, labeled commodity "super-cycles," declining terms of trade remain evident, though slightly more difficult to detect (Erten and Ocampo, 2013).

The political economy implications of commodity price super-cycles are multiple. One significant area of literature emphasizes the resource curse, in which the existence of relatively easy rents from natural resources severs the relationships between state and society. States can earn revenues without the bureaucratic and political development required for taxation, and powerful elites can earn profits with limited innovation or competition. In the absence of dense state-society linkages or pressure to innovate, a weak bourgeoisie fails to provide impetus to democracy, authoritarian states sustain themselves through a mix of repression and patronage, and corrupt dictatorship handicaps development.[1]

Even as commodity-driven growth undermines development and democracy, it is further associated with drastic swings in policy. At periods of high commodity price, revenues are abundant, and state largesse can be extended through patronage and social programs that ameliorate popular demands (Mazzuca, 2013). Yet, when prices inevitably crash, fiscal and macroeconomic constraints reappear, and commodity-dependent elites abandon broader social and political appeals by imposing austerity.[2]

These swings in policy are further correlated with volatile swings and instability in political regime. Democracy can function, albeit imperfectly, when commodity prices are high and states possess resources to distribute to multiple constituencies. When prices fall, austerity policies correspond to a return to authoritarianism, as democracies propped up by handouts have difficulty surviving once they have to curtail spending (O'Donnell, 1972).

Development itself may depend in part on overcoming the political economy of commodities and the way they handicap long-term growth. To emerge from low-income status, countries count on large influxes of capital into their commodities sector. Yet, moving away from the easy rents of exploiting land requires structural transformations such as expanding and improving education, increasing savings and investment, and movement toward the technology frontier (Agenor and Canuto, 2012; Aiyar et al., 2013; Lin and Treichel, 2012).[3]

These transformations are more than technical challenges; they are fundamentally political and in direct opposition to the interests of commodity-producing elites. Factions of capital interested in moving to production using more qualified labor and technology favor policies such as expanding domestic markets, funding research and development, and managing more closely foreign access to domestic markets (Ghosh, 2018; Vernengo and Caldentey, 2017).[4] Yet, commodity exporters oppose these policies, who need only a coercive state with capacity to insure cheap access to labor and land and occasionally the infrastructure to get products to market.[5]

Commodity-dependent elites also constrain development by exhibiting combinations of modern and oligarchic features. Even as they integrate globally and adopt sophisticated financial and technological tools, they remain dependent on the most primitive strategies of accumulation, namely hyper-exploitation of labor, dispossession of public and communal lands, environmentally destructive production processes, and government subsidy and tax incentives (Reuschemeyer et al., 1992). Their political conservatism emerges from their structural position, dependent as they are on violence to contain the demands of labor and peasants, corruption to limit the regulatory intrusions of the state, and continued subsidies in the form of public expenditure and low taxes.[6]

One explanation for this bizarre combination of modern and traditional can be found in the Gramscian concept of passive revolution. For Gramsci,

Northern European countries had undergone a thorough transition in which a hegemonic industrial bourgeoisie fully displaced oligarchic, feudal elites.[7] By contrast, in places where the rising bourgeoisie could not fully displace feudal elites and lacked hegemony within civil society, there was an incomplete transition. As observed by Cox, "These last were caught in a dialectic of revolution-restoration which tended to become blocked as neither the new forces nor the old could triumph" (Cox, 1983: 166). Commodity-dependent elites in developing countries are precisely that faction of the domestic bourgeoisie whose accumulation prevents the emergence of a hegemonic modernizing bourgeoisie.

Gramsci and others also observed that conservative modernization occurs at least in part, because of the ambiguous relationship between dependent elites such as commodity exporters and the international system. Dependence on international markets identifies them closely with capitalist factions in developed countries,[8] and this close relationship positions them as the domestic allies to international capital. They both distort national development and also accelerate it, as described by Leon Trotsky.[9] Trotsky notes on the one hand the "privilege" of late development, "The privilege of historic backwardness—and such a privilege exists—permits, or rather compels, the adoption of whatever is ready in advance of any specified date, skipping a whole series of intermediate stages" (1977: 26–27).[10] At the same time, the compressed nature of late development carries with it certain distortions, as elements of advanced capitalism coexist and interact with pre-capitalist formations. "The development of historically backward nations leads necessarily to a peculiar combination of different stages in the historic process. Their development as a whole acquires a complex, combined character" (Trotsky, 1977: 27).[11]

The handicap presented by the ongoing role of commodity exporters is further evident in countries that experienced late development but somehow did not have to deal with a retrograde oligarchic sector, namely East Asian developmental states. The newly industrialized East Asian countries of Taiwan, South Korea, Hong Kong, and Singapore enjoyed particular combinations of state institutional characteristics, relations between critical class forces, and an international context that allowed them to move from low-income into middle- and then high-income status. Among the specific policies they pursued were establishing new activities, accessing finance, and upgrading technology (Evans, 1995).

Of critical importance was also the absence of oligarchic elite power, either because this had already been eliminated or because it never existed. Korea redistributed land following the end of Japanese colonialism; Taiwan transformed rural elites into industrialists with the arrival of mainland military and political leaders after the Chinese revolution; and Hong Kong and Singapore were city-states without an important oligarchic rural sector (Haggard, 1990). Thus, without retrograde oligarchic interests to block the transition from middle-income to high income, these countries were able to overcome downturns and develop rapidly.[12]

A more specific description of semi-peripheral power dynamics that incorporates the combination of advanced and retrograde characteristics was observed by Ruy Marini in his category of "subimperialism."[13] Subimperial countries pursue hyper-exploitative production processes, creating highly unequal societies that limit domestic demand. As a result, local capital must extend its operations to weaker regional neighbors, thus their subimperial character. While they benefit from their extractive relationships with the rest of the developing world, hegemonic powers benefit from the stability provided by the stewardship of middle powers. For world systems theorists, this intermediate role has played a legitimating function, suggesting international mobility even as it affirmed global hierarchies (Chase-Dunn, 1997).

The nature of subimperialism has somewhat updated as a result of the globalization of the 1990s (Katz, 2015). Recent decades of global integration produced by finance capital and booming commodity prices led most of all to the intertwining of financial elites and commodity-dependent elites (Robinson, 2014). This combination only further affirmed the limits of conservative modernization led by commodity-exporters. In the 2000s, with booming prices, conservative elites once again became willing to tolerate Leftist attempts at developmentalism and social inclusion, but the price they required was to constrain Leftist efforts at developmentalism and social inclusion. Leftist governments accepted modernization led by large-scale agribusiness and counted on commodity export earnings to balance macro-accounts and avoid conflict with the priorities of finance in price stability and fiscal management (Roberts, 2014).[14] The fall in commodity prices increased the relative power of financial interests, making national capital once again subservient to domestic and international finance (Boito Jr., 2018). In addition, the coalition of now-dominant finance and commodity exporters was loath to continue supporting Leftist developmentalism and social inclusion, securing a turn toward Right-led governments willing to pursue austerity and repression.

Brazil has experienced repeated episodes of conservative modernization, passive revolution as described by Gramsci. The core of this conservatism is the nature of development led by commodity exporters, whose relation to other factions of domestic and international capital and popular sectors repeatedly places obstacles in the path of the structural changes required by modernization from low-income to middle-income and from middle-income to high-income status.

Brazil's Economic Emergence amid a Continued Course of Dependent Development

Over the last 500 years Brazil has been deeply embedded in global economic cycles due to its international significance as an area of commodity export production, and as such, its deep dependence on international economic pressures, demands, and constraints. From the time of the first Portuguese incursions into the New World, imperial directives covering topics as wide ranging as territorial occupation, demographic policies, and geopolitical and administrative structures were always closely tied to successive economic cycles of production of global export commodities.

From Brazilwood to sugar, gold, coffee, rubber, and more recently soybean and beef, the very modernization of the Brazilian economy, even in terms of its highly significant path to remarkable levels of industrialization, has been deeply associated with and largely dependent on the profitability of extractivist and agriculture-based activities, which have tended to prosper, at least for a time, in the abundant, diverse and rich ecosystems of the Brazilian territory. One of the latest iterations of this historic pattern is vividly displayed in the case of the purported economic success of agribusiness expansion in the Amazon region, particularly in Brazil's so-called final agriculture frontier in the Amazonian parts of the mid-western state of Mato Grosso (Barros, 2009).

In effect, what is on the surface considered to be one of the country's most successful experiences of both economic modernization and insertion into global trade routes, has to be seen as still marred by ongoing social, economic, and ecological injustices, as well as an international insertion into the global economy defined by an associated or dependent model of development. The continuation of this historical trajectory has recently assumed new formulations, which in turn require us to broaden our interpretive lenses in order to make sense of the unique association between neoliberalism and neo-developmentalism unfolding at the dawn of the twenty-first century in Brazil (Morais and Saad-Filho, 2012). Unsurprisingly, this complex articulation of different development models involved several internal contradictions, which were intensified in the frontier context of the Amazon. Combined, these complex historical dynamics have produced two main consequences.

First, agribusiness expansion in state of Mato Grosso *modernized* the region, in terms of its physical infrastructure and economic output, along clear conservative lines that concentrate wealth and dispossess peoples of their traditional lands. The economy of the state has grown significantly but this process has deepened its dependence on a very small number of activities, namely soybean, maize, cotton, sugar cane, and cattle farming. Similarly, while enriching a few, this process has significantly enhanced regional socioeconomic inequalities, and its so-called success has increased the fragility of the economy by deepening its ties to outside markets. What is more, in addition to its social and economic side effects, the very process has diminished the environmental diversity of the region (Ioris, 2017).

The second major outcome is Mato Grosso's increased participation in the national economy, as Brazil's main economic frontier for generating much-needed export revenues, and the associated enhanced political leverage of the state's oligarchies in the national political scene (as will be detailed later in this chapter). It should be noted that these events unfolded amid a praising narrative surrounding agribusiness in the whole country built around the notions of entrepreneurialism, technological advances, efficiency, and free-market rationality (Ioris, 2015).

But if even embedded in the symbolism of pioneering success, the recent history of agribusiness in the country has served, once again, to consolidate oligarchic interests and the political hegemony of traditionally powerful interests and social groups (Bruno, 2009). To be sure, in the recent dramatic expansion of agribusiness into the Amazon region, the Brazilian federal government has once again acted to support large and influential private economic interests—for example, by generous loans provided by key public financial agencies, Bank of Brazil, and the National Bank of Development—thus serving as subsidizer of first resort, regulator of market operations, and defender against countervailing political pressures affecting corporate interests.

It begs remembering that these historical trends followed the traditional pattern of conservative modernization occurring in late-coming countries, where the central government played the role of planner, subsidizer, and implementer of last resort of projects of modernization that nonetheless continued enmeshed in traditional socioeconomic, oligarchic arrangements (Gerschenkron, 1962; Hirschman, 1965).

In Brazil, this process unfolded for much of the mid-twentieth century through political reforms and economic activities, which though efficient in producing high rates of economic growth, maintained traditional levels of economic inequality and did not reframe the dependent insertion of the country within the global economy (Cardoso and Faletto, 1973; Fernandes, 1968, 1973, 1975; Ioris, 2014; Werneck Vianna, 1999). In the countryside especially, the so-called "archaic" forms of production and social relations proved to be not obstacles to economic modernization and social accumulation, but rather key instruments toward reaching these same outcomes (Ianni, 1979; Martins, 1975; Oliveira, 1972).

Along these historical lines, the expansion of the agricultural frontier in Brazil has always been tied to government support, which assumed many forms ranging from historical legal and illegal appropriation of lands by ruling elites, governmental provision of subsidized credit and other factors of production (e.g., fertilizers, research and development), to price supports, currency devaluations, and favorable tax incentives to big business groups.

Added to these forms of corporate abetting, and especially relevant to agribusiness interests, Brazilian labor laws were largely prevented from functioning in the countryside for much of the country's history, while its political and diplomatic elites consistently worked to promote sales of monoculture crops abroad and repeatedly put in place financial mechanisms that favored continued rents from ownership of unused land. Much in the same way, the repressive apparatus of the federal and state governments has repeatedly been used to reinforce the idea that rural properties tied to export activities are sacrosanct, as they have historically praised in ideological terms as being essential for the very survival of the nation and thus undeserving of any type of criticism.

Together, these multiple policies underpinning the renewed economic expansion on the basis of socially unjust and ecologically unbalanced agrarian-based conservative modernization have implicitly and explicitly included pushing aside, sometimes even exterminating, local indigenous and poor mestizo populations, thus further concentrating wealth in the hands of business groups, especially those based in Brazil's Southeast region.

In the case of the Amazon region, much of this expansion of the agriculture frontier took place in the states of Goiás, Rondônia, Pará, and especially Mato Grosso, with significant consequences for both the region and the country as it has inserted itself more assertively into the commodity linkages of today's globalized economy. In Mato Grosso, the expansion of modern agricultural activities with close ties to multinational agribusiness conglomerates (e.g., Cargill, Monsanto, John Deere, etc.) is a relatively new phenomenon, which involved with public agencies to make these interactions go as smoothly as possible (Wilkinson et al., 2012). This process gathered momentum in the 1990s, and had its apogee, interestingly, during the rise of Brazil as an agriculture powerhouse in the last fifteen years during the period of economic expansion and globalization of the economy overseen by former president Luis Inácio Lula da Silva.

The reluctance of most left-leaning governments to move beyond orthodox, market-oriented and commodity-export economic policies has been a recurrent feature in Latin America in the early twenty-first century. In fact, although some authors have described these countries under various types of leftist government in the early 2000s as post-neoliberal states, others have poignantly made the point that regardless of professed political positions, these nations have still been export-oriented economies, deeply dependent and focused on ever more integrated global markets. Along these idiosyncratic historical lines, the neoliberal restructuring of Latin America since the 1980s has consolidated a pattern of export specialization that recreated the international role of the region as provider of basic products (Katz, 2015).

The goal of maintaining an orthodox macroeconomic stability though commodity exports in times of favorable global prices guided the PT (Workers' Party) government in Brazil, which thus showed no appetite to reduce export-oriented agribusiness policies. In fact, PT administrations significantly expanded public credit to large-scale, export-led monoculture farming activities. In most places throughout the region, this has required deep transformations in the agriculture base, moving away from food independence, as a reality or at least as a goal, to export crop promotion (Machado, 2009). This process required greater involvement of global capitalist forces, traditional oligarchic groups found creative new ways to associate themselves with these foreign actors, while small- and even medium-sized producers were increasingly pushed out.

The Brazilian capitalist classes, as well as their counterparts in other parts of the region, should thus not be seen merely as satellites controlled from outside. In fact, large local business groups closely connected to international actors with good access to financial support have successfully adapted their more traditional activities to successfully compete internationally. A concomitant result has been increasing exclusion of peasants from rural activities and agriculture becoming more capital-intensive, labor-saving, and exploitative.

This reality was sharply experienced in the Amazon region, as well as in key neighboring areas such as the central savannas (*cerrados*) of Goiás and Southern Mato Grosso. This historical trend was largely put in place during the dictatorial regime of the 1960s, 1970s, and 1980s; but it has accelerated in the last three decades due to technological advances, particularly in soybean production in the poor soils of both ecosystems (traditionally, soy production for global markets has been based in fertile regions such as the U.S. mid-west, the Argentinian pampas, and southern Brazilian states), as well as targeted governmental policies aimed at expanding territorial occupation inwards and promoting monoculture agro-exports.

All in all, contemporary claims of technical sophistication notwithstanding, the process of expansion of the agricultural frontier into the Amazonian region in Brazil is a phenomenon with clear roots in the country's past, echoing similar historical developments around continued economic expansion cycles seen throughout Brazilian history. This pattern of territorial occupation and frontier expansion in the Amazon is briefly reviewed next.

The Amazon as the Last Frontier: The Long Trajectory of Sustained Social, Cultural, and Ecological Inequality

The historical occupation of the Amazon region took place in gradual stages usually situated within the occurrence of key geopolitical and/or economic cycles. And even though actual European occupation would need time to take hold in this borderland portion of the territory, the Amazon was always part of the narrative of the ever-expanding Portuguese colony in the New World. This presence was at times more symbolic (i.e., centered on potentiality rather than significant numbers of actually existing white settlements), but its relevance was nonetheless clear, especially when subsumed within the path of gradual occupation of the mostly unknown, abundant but also untamed and dangerously challenging interior.

Much in the same way, the associated internationalization of the Amazon has been a protracted process, dating back to the nineteenth century, when private companies from the U.S. advocated the opening of the region to international private actors, to the so-called economic miracle of the 1970s, when state-sponsored policies paved the way for major global corporations to implement economic projects, to today's alliances between domestic and international agribusiness operators that have inserted the region to an unprecedented level in global economic cycles both in its input (production-based on hybrid seeds and fertilizers produced by global agribusiness corporations) and output dimensions (soybean and beef exports to global markets).

Thus, though late in coming, the economic integration of the Amazon region was always shaped by similar dynamics to those defining the colonial and postcolonial dependent and exploitative path of development in the country as a whole. To be sure, multiple modalities of coerced labor co-existed (and still co-exists) in the always-expanding frontier portions of the Amazon, broadly defined (Monteiro, 1994). Similarly, in tandem with broader historical developments, though continuously portrayed as at the margin of civilization, extractivist economic activities in the region—including the hunt for new enslaved hands to work in the coastal plantations—were always closely integrated and vitally important to colonial and later capitalist business ventures elsewhere in the country.

Closely related, this historical logic systematically amalgamated numerous different indigenous groups under the homogenizing label of "savage" (*índio bravo*), a process which, at the same time, served to consolidate the image of the region as threatening—and hence in need of control, taming, occupation—in short, a civilizing mission—while at the same time, by emptying the land of its traditional inhabitants, an open, abundant, and virgin frontier to be penetrated by powerful male figures, as historical events echoed similar developments in the western expansion of the United States as well as the southern expansion of Argentina.

In specific, from the Marquis of Pombal's eighteenth century officially sponsored territorial and demographic occupation of the region, European encroachment was always represented as a cultural duty. The vastness of the

Amazonian landscape was creatively associated with, simultaneously, abundance and scarcity, pristine wilderness and population deserts, both in need of civilizing efforts. This logic became even more prevalent when the region became more integrated into global economic cycles at the turn of the nineteenth century, under the aegis of global laissez-faire capitalism, a process also imbued with a migratory element tied to the goal of demographically and culturally *whitening* the region.

Over time, the image of the native populations evolved chronologically and conceptually. Moving from peoples *frozen in time* and in need of rescuing by allegedly more advanced civilizations, to savages refusing to accept the *white man's* civilizational *burden*, indigenous communities have been not only the object of public policies but also historical actors able to exert agency (Roller, 2014). From the romanticized image of the *bon savage* of the mid-nineteenth century, to the militaristic discourse of protectors of the national frontiers under the leadership of Marshal Rondon, early in the twentieth century, and then again in the contradictory nationalist rhetoric of occupying the territory via corporate international business activities in the 1970s, to today's rhetoric of environmental protection masking actual unprecedented territorial expansion of agricultural frontiers, change and continuity have defined the history of capitalist occupation in the Amazon.

Amid all these many transformations, it is easy to identify the persistence of the association between native populations and the very notion of frontiers, be they geographical, social, cultural, economic, or any others; always in flux, with one pole always expanding. Rather than physical markers, frontiers are above all both a logic of operating within a territory and a heuristic instrument of analysis that articulates the societal limits of civilizational presence within the broader totality, geographic and otherwise, of the whole society. Much in the same way, frontier zones should also be seen as both a central economic component of an always expanding path of exploitative economic development, playing the essential role of pushing aside externalities ever present in the process, as well as the very space wherein the inherent contradictions of many developmental processes clash in more intense, clear, and often violent historical dynamics (Oliveira, 2016).

The destruction of traditional forms of occupation of territory, particularly more democratic and inclusive ones, such as the small holdings of mestizo populations and, particularly, communally held, environmentally sound indigenous settlements, have always defined the process of frontier expansion in Brazil, particularly in the Amazon. In addition to being a constitutive element of capitalist economic expansion in general, when abetted by governmental policies (e.g., legal support for land grabbing, favorable tax and fiscal policies, subsidized credit, etc.), this process has proved instrumental both in creating a landless labor supply regionally and in portraying the technology-dependent (hence, less adapted to the local ecological reality) agricultural activities as embodiments of progress and social redemption.

As previously mentioned, governmental policies have been central to private incursions in the Brazilian Amazon. In effect, from the *March to the West* in the early 1940s, to the creation of the Superintendence for the Valorization of the Amazon (SPVEA) in 1953, to Juscelino Kubitschek's expansive network of roads into the region, and finally the military regime's focus on national security concerns and associated top-down resettlement programs; all have demonstrated the federal government's growing presence in planning, organizing and, especially, financing the encroachment of large, national, and foreign private economic actors in the region (Acker, 2017).

In the last four decades, this path of expansion has implicitly and explicitly pushed aside traditional (mestizo and indigenous) populations on an even greater scale, ironically coinciding with a period of deepening of formal democratic institutions in Brazil. Similarly, as the Amazon region has become more connected to global economic commodity cycles, it has also become more ecologically unbalanced and socially unjust. Conversely, traditional local elites have found creative ways to modernize while sustaining long-established socially and environmentally exploitative economic activities and labor conditions.

Regional oligarchies have indeed become ever better articulated with domestic and international powerbrokers, largely by presenting themselves as promoters of technocratic, top-down modernity, while remaining nonpassive but still deeply dependent junior players in global economic transactions. In doing so, they managed to remain in close association with successive governments, despite their different party affiliations and ideological positions, thus retaining some levels of control of the processes of territorial, economic and sociocultural occupation of the region.

The path of intensified agricultural production for export in Mato Grosso had made the state more technologically advanced and connected to global economic actors, while at the same time, perpetuating highly antiquated forms of social exclusion, along class, gender, and racial lines, veiled under the banner of efficiency. It has also impoverished the ecological diversity of production while concentrating economic gains in very few hands. This process enhanced the power of regional land-owning elites in national politics, while the country, as a whole, deepened the path of deindustrialization of the last forty years, thus becoming more dependent on a small number of commodity exports. In fact, between 2004 and 2013, the percentage of manufactured goods in the export list of items from Brazil declined from 55 percent to 38 percent, while the role of commodities climbed from 29.5 percent to 46.7 percent (MDIC, 2013).

In addition to the rapid deindustrialization of the national economy, land and capital have become ever more concentrated. Around 10 percent of large property-owners control 85 percent of the total value of agricultural production, involving only around fifty economic conglomerates in the process (Katz, 2015). Thus, whereas between the 1950s and the 1970s Brazil's insertion into the global economy increasingly included exports of manufactured goods, mostly produced by multinational corporations that had established themselves in the country via attractive fiscal and monetary subsidies, in the last quarter century the country has witnessed rapid deindustrialization, whereby industrialized goods have been largely replaced by commodities in terms of export revenues (Evans, 1979; Mendes de Paula, 2017).

This process is never linear and involves both countermeasures on the pat local communities that appropriate encroaching developmental trends for their own interests by making new economic activities their own (Hoelle, 2015), or by finding ways to actively engage with new public policies such as environmental conservation (Bratman, 2019; Schmink et al., 2019). In any case, this process has represented both a more fragile (harking back to classic one-crop dependence) and substantially foreign-determined (in terms of global demand and prices) path of participation in global economic cycles, which has also significantly enhanced the domestic power and influence of agribusiness in the country, a trend that is further detailed below.

Before that can be done, in the next section, we review research on how middle-income economies have sought to reorient their historical patterns of insertion in global commodity chains as a way to assess the dependent path of globalization of the Brazilian economy in the last twenty-five years.

Purported Efficiency amid Continued Public Support: The Modernization of Amazonian Oligarchies in the Twenty-first Century

Presenting itself as one of the most effective cases of economic success in the process of globalization in Brazil—during the last few years, the agribusiness lobby has sponsored an aggressive media campaign under the slogans "the Brazil that does it right!" [*O Brazil que dá certo!*][15] and "Agro is pop, Agro is everything" [*Agro é pop, agro é tudo*][16]—the modernization and internationalization of Brazilian agro-economy should be viewed through the lens of a long-established tradition of conservative modernization, under which the economic base of the country is overhauled with the aim of meeting the new needs of international and domestic capitalist groups while wealth inequality, and in this case land concentration, significantly increases.

Along this path of contradictory transformations, it is again important to highlight how much the narrative of success of market forces, efficiency, technology, entrepreneurial spirit, etc. has been marred by the reality of continued and even expanded government support to large-scale, export-focused farming in the last fifteen years. While in 1990, federal subsidies to agribusiness amounted to R$ 15 billion, in 2014 they amounted to R$ 156 billion; interestingly a process occurring during a period of rising international commodity prices.[17] Similarly contradictory, this path of public support to large corporate business interests was deepened during the administration of Brazil's most important left-center party, PT, the most influential proponent of the agenda of land reform since the democratization of the mid-1980s. While total credit to farmers and cooperatives hovered between R$20bi and R$30bi during the 1990s, credit rose to almost R$80bi during the 2000s, under the PT (Santana and Nascimento, 2012).

To be sure, PT's reign in power was marked by a succession of prominent defenders of traditional agribusiness elites in the cabinet position of Minister of Agriculture, such as Roberto Rodrigues, Katia Abreu, and Blairo Maggi. The first, an academic, could perhaps be included in the somewhat ambiguous category of technocrats, thus diminishing Lula's association with landed oligarchs, but the latter two were themselves large land-owners and leaders of conservative agribusiness associations. Much in the same way, powerful lobbying groups such as CNA (the National Agriculture Confederation), as well as food-processing conglomerates such as Sadia and JBS, successfully made the case that their competitive advantage in terms of producing for global markets was vital to maintaining surpluses in the country's current accounts. Along these lines, agribusiness corporations managed to obtain extensive government subsidies and preferential treatment from the president himself, who quickly toned down the focus of land reform except on the margins of actual agriculture frontiers, such as in the Amazon.[18]

Similarly, the narrative of operating a technologically advanced and highly competitive agriculture, said to have been achieved on the merits of national talent and effort was repeatedly deployed by powerful business elites in order to obtain continued levels of support from the national government. These claims did find welcoming grounds in the national administrations of PT as they were well positioned to fit within Lula's (and later Dilma's) own conceptions of fast-paced economic development.

In fact, having both matured politically during the so-called Brazilian miracle, when the country was growing at remarkable rates, and later when a domestic sugarcane based energy program was being pursued by the federal government, and despite being political opponents of the dictatorial regime in power at the time, Lula and Dilma have been known for espousing the supposed successes of the period as something to be emulated, though this time in a democratic context. In effect, it did not take long for Lula himself to become one of the main supporters of Brazil's supposedly efficient agribusiness sector, and thus to consistently extol Brazil's agricultural potential around the world, while the same elements of the country's foreign policy were closely tied to opening global markets for agriculture exports.[19] Lula indeed boasted about the country's agribusiness achievements, including parading with Blairo Maggi, then the governor of Mato Grosso, in presidential events tied to agribusiness expansion.[20]

Known as the largest producer of soybean in the world, and with a family corporate worth of over US$ 1.5 billion, Maggi managed to reinvent himself in remarkable ways, though always sustaining a shrewd course of activism in favor of large, monoculture export-oriented agriculture.[21] In 2005, Maggi received Greenpeace's Golden Chainsaw Award for his support for deforestation in the Amazon, justified negative publicity that the agribusiness leader turned formal politician quickly brushed off through his growing proximity to Lula himself. In 2006, Maggi campaigned for Lula's reelection despite the official support of his party (PPS) for Lula's opponent, a move that involved the federal government writing off of longstanding debts held by large farmers in Mato Grosso to the tune of close to US$ 1 billion.[22]

This close association between the two political leaders continued even after the end of both of their mandates in the respective state and federal executives. In the summer of 2014, both visited a Cuba-based soy farm that had been set up in 2006 with support from Embrapa.[23] Similar synergy had also been witnessed in 2009, when they made a joint statement rejecting criticism from foreign environmental groups concerned about the path of deforestation in the region, stating "do not stick your noses in our business! The Amazon is ours."[24]

As a final sign of the malleability of agribusiness leaders in repositioning themselves, at least in terms of narrative and political alliances, for the sake of proximity to the realms of power of a governmental structure that has consistently provided generous support to large-scale export-oriented monoculture farming, Maggi served as Minister of Agriculture for two and a half years in the government of Michel Temer, an administration that otherwise sought to reverse many policies put in place during Lula's administration.

In this new ministerial role, the former governor not only sustained the course of federal disbursements to agribusiness throughout Brazil, especially in the new agriculture frontiers of the Amazon, but also worked assiduously to deepen the path of cooperation with global agribusiness corporations and foreign markets. More recently, Maggi has been involved in corruption scandals involving Brazil-based meat conglomerate JBS, though thus far he has managed to avoid any form of sentencing due to his legal privileges.[25]

JBS's role in the path of internationalizing Brazil's agribusiness along new but still dependent lines is examined next.

The Boom and Bust of Agribusiness within Brazil's Long Path of Curtailed Development

The Brazilian agribusiness conglomerate JBS began as a family-run butchery operation in the central-west Brazilian state of Goiás in 1953. In 2016, JBS earned the highest revenue of any single private Brazilian company at around R$ 116 billion (approximately US$ 37 billion at the end of 2018 exchange rate). At the start of 2018, the two principal heirs and executive officers to the company were in jail for corruption. By the end of the year, prospects again looked bright for JBS as the newly elected government named as Minister of Agriculture a politician who had both subsidized and profited from close association with the company in her home state of Southern Mato Grosso.[26]

The boom and bust (and perhaps boom again?) trajectory of JBS represents a microcosm of agribusiness in a developing country like Brazil. The JBS empire was built through combining the most advanced portions of the agribusiness sector with the most retrograde. Advanced elements include savvy financial dealing, technological upgrading, and strategic market positioning in an integrated global market for proteins. Retrograde elements include land-grabbing from the indigenous and ecological frontier, labor abuse, corruption, and dependence on state subsidy and largesse.

In addition to this combined advanced and retrograde structure, the political economy of the sector can be seen in the political alliance behavior of the company, joining and benefiting from a Left-led, inclusionary government during boom years but eagerly throwing its support to Right-led austerity governments in lean times. The explanation for this behavior in boom and bust periods as well as the combined advanced and retrograde strategy of accumulation can be found in the position of Brazilian agribusiness in the domestic class structure and the international system.

The conditions for agribusiness are propitious in Brazil. The country enjoys low production costs, especially low cost land and labor, and an ample supply of raw materials, namely feed. Costs of raising cattle in Brazil are 40 percent lower than in the United States. Brazil has the largest commercial cattle stock in the world, with 232 million head in 2018, and the particular conjuncture of the mid-2000s propelled Brazilian meat producers to the forefront of global competition. According to the European Livestock and Meat Trading Union report entitled "Brazilian Beef Imperialism," the country benefited from "increased productivity and lower costs, expanded marketing and advertising, wider export destinations, reductions in trade barriers, and a drought that hit Australian producers in the early 2000s, especially affecting competition in Asia" (European Livestock and Meat Trading Union, 2008: 5).

The beginnings of growth for JBS came in the late 1990s, as the company, then called Friboi, took advantage of the collapse of export markets under

Brazil's strong exchange rate regime to absorb failing exporters. In the early 2000s, JBS received funds from the public Banco de Brasil to purchase two of Brazil's largest exporters of frozen and processed meats, Anglo and Bordon, the latter linked to Bertin, a company JBS would also acquire in 2009.

In 2005, again with public support, JBS began its international expansion,[27] acquiring Swift Armour in Argentina for US$ 200 million, of which 40 percent was financed by the Brazilian National Development Bank (BNDES).[28] In 2007, JBS launched the then biggest IPO in the history of the São Paulo stock exchange, netting R$ 800 million (approximately US$ 450 million) by selling 20 percent of the company, of which 12.95 percent was purchased by the BNDES.[29]

The funds from the IPO went immediately toward the company's biggest purchase to date, SwiftFoods in the United States. Swift was facing a downturn on the eve of the U.S. financial crisis, and the purchase immediately launched JBS as the biggest beef producer in the United States.[30] In the same year, JBS purchased Inalca in Italy, SB Holdings in the United States, Rio Platense in Argentina, and JV Beef Jerky in the United States and Brazil. The year 2008 saw additional purchases, including Smithfield Beef, Five Rivers, and National Beef in the United States, as well as the Tasman Group in Italy and Gruppo Cremonini in Italy. These purchases were again supported publicly, with US$ 200 million from BNDES, as well as an undisclosed amount from a BNDES-related investment fund. Public support continued in 2010, with the BNDES purchase of US$ 1.5 billion of JBS debt, later to be turned into shares that increased BNDES ownership of JBS to 31 percent in 2011.

The JBS international purchasing spree continued, with acquisition of Pilgrim's Pride, the second largest poultry producer in the United States and the absorption of Bertin, their one-time partner company in Brazil and also a leading leather exporter, including to factories in China. Additional acquisitions included Independencia (2012—Brazil), Frangosul (2013—a French-owned company in Brazil), and Marfrig (2013—the second-largest meat company in Brazil with leather interests in South Africa, Uruguay, and Mexico). In 2015, in its most recent round of major acquisitions, the company purchased Tyson (Brazil and Mexico), Cargill Pork (United States), Moy Park (United Kingdom, Ireland, France, and Holland), and Primo (Australia).

The international insertion of JBS invites competing interpretations. JBS executives claimed their strategies were driven by a desire to upgrade in at least three ways. First, by absorbing technology and management expertise from international competitors, including research and development facilities; second, to gain insider knowledge of the tastes of consumers in different markets; and third, to evade phyto-sanitary and other trade barriers erected particularly in developed country markets (Interview, JBS U.S. headquarters, 2016). On the other hand, JBS acquisitions were hardly moving Brazilian producers to the front of the technology frontier. In the words of Wesley Batista, "Meat processing is the same all over the world. Killing cows is the same" (as quoted in Dieguez, 2015).

A more accurate understanding of the global strategy of JBS notes the reality of the international regime. WTO, TRIMS, and TRIPS rules favor international investors, and firms in developing countries can only take advantage of the regime if they too can turn themselves into international investors (Serra, 2017). Accomplishing this feat, however, requires public help, as capital in developing countries operates at a disadvantage compared to that in developed countries. While firms based in developed countries can access deep home-country capital markets and advanced technology and innovation systems, sectors in developing countries tend to be fragmented among a few oligopolistic and uncompetitive firms, described by some as "hierarchical capitalism" (Schneider, 2013).

For such firms to compete globally, public subsidy allows them to absorb domestic competitors and secure a dominant market position. Monopoly profits and further public subsidy (and for JBS the overvalued Real of the period) allowed the absorption of international competitors and increasing operations in scale. Similarly, international acquisitions provided the benefits named by JBS executives in terms of market access, knowledge, and innovation, affirmed as development bank strategy by the president of BNDES, Luciano Coutinho (2012–2016): "We cannot do without a set of national groups with global capacity. Without them, we will not develop endogenous cores of technological progress capable of affirming Brazilian brands, creating new markets, and generating high quality activities and jobs" (Coutinho, 2002: 205).

In addition to the benefits to the company bottom line, international insertion shifts the relative power of interests within firms like JBS. One indicator of whom JBS increasingly sought to satisfy can be found in the change of name from Friboi to JBS. According to Wesley Batista, "JBS is easier to assimilate in international markets, 'J' 'B' 'S'" (Dieguez, 2015). Critics noted the shift away from productive capital, and Mansueto Almeida, of the government Institute for Policy and Economic Analysis (IPEA), even noted that "selling cows will not impact the collective welfare or modernize national industry" (Dieguez, 2015).

In fact, in at least one of the major deals, financial interests would appear to have driven decision-making over and above productive interests. The purchase of Pilgrim's Pride in 2008, for US$ 2.8 billion, occurred only after the BNDES had purchased approximately R$ 3.5 billion of convertible debentures, and the condition for the infusion of cash appeared to have been JBS absorbing the failing Bertin meat processing company, along with the R$ 4 billion in debt that the company owed BNDES (Dieguez, 2015).

Even as JBS globalized with public assistance, gained access to advanced technologies, and increasingly organized its global strategy to benefit domestic and international financial interests, the meat-processing sector, and JBS in particular, retained some of its most unsavory elements. In fact, the cattle sector, including JBS and its network of suppliers in Brazil, were responsible for some of the worst excesses of land grabbing, environmental degradation, land concentration, labor abuse, and corruption.

Cattle producers consistently seek ever larger areas of land for pasture, particularly in regions considered undeveloped, and therefore available at lower prices. This logic has repeatedly been applied to environmental reserves and indigenous reservations, as well as areas like much of the Amazon, where there is limited government oversight and insecure titling. Once timber has been cleared, cattle producers move in, and the preparation of the land increases the market value of land in general, leading to bids from more intensive users, such as grain (especially soy) and eventually sugarcane producers, driving up prices further and forcing cattle producers deeper into areas defined as frontier, restarting the cycle anew.

The attractiveness of land as an investment created a globalized land market, with domestic and international actors speculating in land for a variety of financial and productive objectives (Wilkinson, Reydon, and DiSabbato, 2012).[31] The result of these processes has been an increase in land concentration. From 2003 until 2010, the percentage of large properties increased from 51.6 percent to 56.1 percent of the total. Small farms of less than fifty hectares accounted for 12.8 percent of the total productive land in 2017, down from 13.3 percent in 2006, and extremely small farms of less than ten hectares, which make up half of all farms in Brazil, accounted for only 2.2 percent of productive land. Large farms of over 10,000 hectares number only 0.04 percent of all establishments, but occupy 14.8 percent of all productive land (Davalos et al., 2014). The state with the largest farms was Mato Grosso, with 868 large estates, followed by Southern Mato Grosso do with 341, and another state including a significant amount of Amazon territory, Pará, with 188 large estates. The Gini index for land distribution in Brazil rose from 0.856 to 0.872 between 1996 and 2006.[32]

The profitability of agribusiness comes not only from low-cost land obtained through questionable processes and environmentally destructive production, but also from the hyperexploitation of labor. "Since 2003, over 21 thousand workers have been freed from slavery in the Brazilian Amazon; 70% of those cases were related to the cattle sector" (Greenpeace, 2015: 14). JBS was fined R$ 3 million for use of slaves in its facility in Rondônia and R$ 1 million in Mato Grosso. Furthermore, the company was found to have purchased cattle from multiple providers who had been punished for keeping workers in conditions of forced labor.[33]

It is in penetrating the Amazon that much of the environmental degradation associated with cattle farming occurs.[34] According to the 2009 Greenpeace study, "JBS' cattle supply chain includes significant supplies from within the Amazon rainforest. The company sources from suppliers whose ranches are grossly in breach of deforestation limits and who actively continue to clear remaining forest within ranch boundaries" (Greenpeace, 2009: 97). In fact, the cattle sector in the Brazilian Amazon is responsible for 14 percent of annual deforestation worldwide, and 62 percent of Brazilian greenhouse gas emissions. While JBS and other major slaughterhouses entered a voluntary "Cattle Agreement" committed to cleaning their supply chain of providers engaged in deforestation, JBS has been fined R$ 264 million for purchasing cattle from farms barred from selling for their deforestation practices, especially in Pará.[35] More than 750,000 square kilometers of Amazon rainforest have been destroyed since the 1980s, with about 60 percent replaced by cattle pasture, as cattle are the most extensive use of large areas of land (Greenpeace, 2015: 12).[36]

At least part of the profitability in agribusiness, including that of JBS, comes from corruption. Some corruption seeks to evade the few regulations that exist to control the sector. In Operation Weak Flesh [*Carne Fraca*], JBS and other major meat processors were exposed as having paid monthly bribes to officials to lower food safety standards.[37] Other forms of corruption seek to overturn or reverse regulation entirely, allowing agribusiness firms a free hand. Shortly after Operation Weak Flesh, one of the Batista brothers revealed that he had taped encounters with the sitting president Michel Temer requesting bribes.[38] Further, he revealed additional tapes with a leading Right-wing politician, Aécio Neves, in which he requested a R$ 2 million payment.[39]

To wield their influence through more quasi-legal mechanisms, the Rural Caucus of the Brazilian Congress [*Bancada Ruralista*] accounts for as many as 261 deputies out of a total of 513. They operate in almost every political party except those on the Left, and they mobilize their members specifically to control legislation of interest to agribusiness. In 2011, for example, they secured a long-debated new forestry law, decentralizing control over forest management and reducing federal protections for indigenous and environmentally sensitive lands, playing into the hands of agribusiness elites and their political allies (Ferraz da Fonseca and Moreira da Silva, 2011).

In sum, JBS displays the most modern and the most retrograde elements of Brazilian agribusiness. The company positioned itself strategically to capitalize on public subsidy, beating out competitors in the domestic market and purchasing rivals internationally. With a global presence, the company increasingly operated according to the financial interests of shareholders, both domestic and international, upgrading its production using the technology, know-how, innovation systems and market access it could obtain by operating in multiple markets.

Yet, even as the company became dominant in global protein markets, it continued to enhance its profits using land grabbing, slave labor, environmental degradation, and corruption. The apparent fall of the firm's owners, the Batista brothers, who served time in jail before being released, has barely damaged the company, which remains one of Brazil's largest and is still dominant internationally.

Bringing it All Together: Winners, Losers, and Possible Alternatives

The reflection presented here sought to examine through critical and interdisciplinary lenses the economic transformations taking place in the agribusiness expansion in the Amazon frontier in Brazil, and the associated political dimensions of the recent agriculture-led economic expansion of the Brazilian economy. We have reviewed how, over the last quarter century, Brazil has inserted itself more deeply in the global commodity chains by means of its booming agribusiness sector, and how subsequent busts have revealed the distortions and vulnerabilities of combining technologically advanced agriculture with oligarchic-style political economy.

In effect, as the country ceded its hard-won status as a manufactured-goods exporter and stepped back into its traditional role as a major commodity exporter, in the last twenty-five years, Brazilian agricultural elites, both, modernized their activities by means of technological innovations and more complex global trade and international agribusiness alliances, while also reasserting their prominence as central economic and political players. This process included a huge expansion into new agricultural frontiers, this time more than ever into the Amazon region. These developments present several tragic ironies, as suggested here.

To be sure, oligarchic emergence occurred amid the strengthening of the country's formal democratic institutions and during the rule of PT center-left political coalitions. In this sense, rather than being forced to accept a more sustainable path of territorial occupation and a more labor-inclusive mode of production, latifundia and traditional land-based oligarchic elites used their importance as generators of export revenues and their political influence to position themselves as essential coalition partners for the PT.

Much in the same way, when economic fortunes and PT electoral dominance declined, starting in the 2014, these oligarchic elites were the first to abandon the government, gradually pursuing undemocratic forms of controlling power while also seeking to reverse even the few social advances they had conceded during the boom years. In effect, with the political reversal unfolding in Brazil in the last five years, the developmental path of conservative modernization has intensified the already socially, politically, and environmentally unsustainable reality of the country's most important economic base.

Thus, though purportedly successful, the recent path of global insertion of the Brazilian economy by means of commodity exports is not only unsustainable in the long-term given its environmental costs and close dependence on uncontrollable international variables such as global demand and prices. It is indeed a process deeply ridden with deep social, political, and cultural fractures, which are very likely to grow in the short, medium and long term. What is more, the recent Brazilian experiences of predatory commodity export economic boom and global economic insertion have the potential to illuminate analysis of related development taking place beyond the Latin American context.

In short, as occurring in similar dynamics currently unfolding elsewhere in the world, it is ever more noticeable that socioeconomic, political, and cultural conflictive dynamics are increasingly reshaping domestic and global developmental trends (Burkhardt and Dietz, 2014). In all of these, nature itself is becoming, on a much deeper level and accelerated fashion, a realm for the implementation of the centuries-old conflictive developmental patterns, such as those examined here in the case of the always socially exclusionary and environmentally predatory experiences inherent to the logic of perennial frontier-making in the vast Amazonian tracts of Brazil.

NOTES

1. For a review of resource curse arguments, see Ross (2015).
2. Alternative economistic approaches have recently focused on the middle-income trap. Taken literally, such approaches link explicitly to GDP per capita, creating a causal claim out of the way average income raises wages without necessarily increasing productivity, characterizing middle-income states as trapped, "unable to compete with low-income (LI), low-wage economies in manufactured exports and unable to compete with advanced economies in high-skill innovations" (Kharas and Kohli, 2011: 282).
3. In addition to these more complicated economic policy requirements, observers point to institutional upgrading with respect to things like property rights, labor flexibility, anti-corruption, and deepening financial systems (Gill and Kharas, 2007).
4. "New developmentalist" variants advocate correcting overvalued exchange rates and emphasize industrial policy to upgrade capacity for export competitiveness (Paus, 2014; Ohno, 2009).
5. Popular sectors also remain divided, as labor markets in middle-income countries remain fragmented between the formal sector and a vast and growing informal sector. As a result, "growth trajectories on the way to middle-income status fractured the groups, especially business and labor, which are the core potential constituencies for a coalition that could take the big leap" (Schneider and Doner, 2016: 611).

6. Marx identified the combination of extra-economic coercion in the violently disruptive nature of "primitive accumulation," necessary in early stages of capitalism to concentrate wealth while displacing peasants and transforming them into a mass proletariat with only their labor power to sell. "The parliamentary form of the robbery is that of Acts for enclosures of Commons, in other words, decrees by which the landlords grant themselves the people's land as private property, decrees of expropriation of the people" (Marx, 1992: chapter 27). Luxembourg extended this combination to all stages of capitalism, "The relations between capitalism and the noncapitalist modes of production start making their appearance on the international stage. Its predominant methods are colonial policy, an international loan system—a policy of spheres of interest—and war. Force, fraud, oppression, looting are openly displayed without any attempt at concealment, and it requires an effort to discover within this tangle of political violence and contests of power the stern laws of the economic process" (Luxembourg, 1968: 396).

7. Even where they allowed such feudal elites to retain some role in government, as in the landed gentry of England or the Junkers in Germany, a hegemonic bourgeoisie could be confident that the policies of the state would obey their requirements for industrial development (Gramsci, 1971: 366).

8. Patnaik argues that the international system requires intervention in developing countries, labeled "imperialism" in its most honest descriptions (Patnaik and Patnaik, 2018). Capital requires the raw materials and labor power of developing countries precisely at moments of decline in core countries, who can renew their growth by displacing their crises to developing countries through empire.

9. Justin Rosenberg and colleagues at the University of Sussex have been particularly central to recuperating Trotsky's theory of uneven and combined development (Rosenberg, 2006).

10. The full quotation is, "Although compelled to follow after the advanced countries, a backward country does not take things in the same order. The privilege of historic backwardness—and such a privilege exists—permits, or rather compels, the adoption of whatever is ready in advance of any specified date, skipping a whole series of intermediate stages. Savages throw away their bows and arrows for rifles all at once, without traveling the road that lay between those two weapons in the past. . . . The fact that Germany and the United States have now economically outstripped England was made possible by the very backwardness of their capitalist development." Writing around the same time, Hilferding notes, "capitalist development did not take place independently in each individual country, but instead capitalist relations of production and exploitation were imported along with capital from abroad, and indeed imported at the level already attained in the most advanced country. . . so capitalism is now imported into a new country in its most advanced form and exerts its revolutionary effects far more strongly and in a much shorter time than was the case, for instance, in the capitalist development of Holland and England" (Hilferding, 1981: 322–23).

11. "The law of combined development—by which I mean a drawing together of the different stages of the journey, a combining of separate steps, an amalgam of archaic with more contemporary forms" (Trotsky, 1967: 23).

12. Other elements of the East Asian developmental states included the capacity to "govern markets," whereby strong executives preserved autonomy from powerful

interests, delegated authority to meritocratic Weberian bureaucracies, enjoyed close relationships with the private sector, and subordinated labor (Wade, 1990; Johnson, 1982). At least in part, successful movement into high-income states was possible also because the international context provided both a market for surplus output and sufficient threat to discipline potentially predatory state actors (Amsden, 1989).

13. He applied the concept to Brazil, especially in its rapid industrialization during the 1970s, outlining economic and geopolitical dimensions (Marini, 1973: 99–100).

14. The limits on developmentalist policies led some to characterize the period in Brazil as "neoliberal developmentalism" (Ban, 2013).

15. https://fotografia.folha.uol.com.br/galerias/37065-o-brasil-que-da-certo-agronegocio.

16. https://noticias.r7.com/jornal-opcao/o-agro-e-tech-o-agro-e-pop-o-agro-e-tudo-08022019.

17. "Credito ao Setor Rural." *O Estado de São Paulo*, May 20, 2014.

18. https://www.cartacapital.com.br/politica/politica-agraria-do-governo-lula-valorizou-o-agronegocio.

19. https://www.nytimes.com/2007/09/23/world/americas/23iht-23lula.7609773.html.

20. http://www.brasildamudanca.com.br/agronegocio/agronegocio.

21. R. Geromel, Meet Blairo Maggi: From Soybean King to Politician to One of the World's Most Powerful People to Tree Hugger. *Forbes,* November 8, 2013.

22. G. Guerreiro, Lula Libera Recursos para MT e Blairo Maggi Formaliza Apoio a Petista. *Folha de São Paulo,* October 11, 2016.

23. http://www.institutolula.org/blairo-maggi-e-lula-em-cuba-para-ampliar-producao-de-alimentos/.

24. http://www.institutolula.org/blairo-maggi-e-lula-em-cuba-para-ampliar-producao-de-alimentos/.

25. https://news.mongabay.com/2017/07/soy-king-blairo-maggi-wields-power-over-amazons-fate-say-critics/.

26. https://www1.folha.uol.com.br/mercado/2018/11/parceira-da-jbs-ministra-de-bolsonaro-deu-incentivos-fiscais-a-empresa-em-ms.shtml.

27. Many of these dates and figures are from Serra (2016).

28. They would go on to buy the next-biggest Argentine producer, CEPA, in 2006. https://www.lanacion.com.ar/863117-otro-frigorifico-paso-a-manos-del-grupo-brasileno-jbs.

29. BNDES would eventually hold as much as a 31 percent stake in the company.

30. https://www.marketwatch.com/story/brazils-jbs-buys-swift-foods-for-14-bln. One US executive called it "a case of the guppy swallowing the whale" (Interview, US executive of JBS, 2016).

31. Wilkinson, Reydon and DiSabbato typologize the interests of agribusiness, synergistic and convergent sectors, newly convergent sectors, real estate companies, food security, investment funds, environmental services, mining, and oil (2012: 427–28). They find that "The limitations currently imposed upon access to land by foreigners in Brazil are, in many cases, innocuous due to the use of Brazilian figureheads. On the other hand, they stimulate strategies geared to forms of indirect control over land" (436).

32. https://www.brasildefato.com.br/2018/07/26/no-brasil-2-mil-latifundios-ocupam-area-maior-que-4-milhoes-de-propriedades-rurais/.
33. https://reporterbrasil.org.br/2017/06/jbs-comprou-de-fazendas-flagradas-com-trabalho-escravo-e-desmatamento-ilegal/.
34. There is an association between land grabbing, slavery, and environmental destruction, "Rampant slave labor and land-grabbing (such as the theft of indigenous lands) often goes hand-in-hand with environmental destruction" (Greenpeace, 2015: 12).
35. https://reporterbrasil.org.br/2017/03/jbs-compra-gado-de-areas-desmatadas-ilegalmente-e-leva-multa-de-r24-milhoes/.
36. As land values increased, more intensive uses of land took over, primarily soy in places like Mato Grosso, which increased production of soy from 3 million tons to 26 million tons from 1990 to 2014 (Flexor and Leite, 2017: 395).
37. To provide the specific, and disgusting, details, JBS and another processor, BRF, were found to have mixed rotten meat, meat doctored with carcinogenic chemical substances, and low-quality meat into meat for export. https://www.bbc.com/news/world-latin-america-40377203.
38. https://oglobo.globo.com/brasil/dono-da-jbs-grava-temer-dando-aval-para-compra-de-silencio-de-cunha-21353935.
39. https://www1.folha.uol.com.br/poder/2017/05/1884947-senador-aecio-neves-psdb-mg-pediu-r-2-milhoes-a-joesley-diz-jbs.shtml.

BIBLIOGRAPHY

Acker, A. 2017. *Volkswagen in the Amazon: The Tragedy of Global Development in Modern Brazil*. Cambridge University Press: Cambridge.

Agénor, P. -R. and Canuto, O. 2012. *Middle-income Growth Traps*. Policy Research Working Paper No. 6210. World Bank: Washington, DC.

Aiyar, S., Duval, R., Puy, D., Wu, Y., and Zhang, L. 2013. *Growth Slowdowns and the Middle-Income Trap*. IMF Working Paper 13–71. International Monetary Fund: Washington, D.C.

Amsden, A. 1989. *Asia's Next Giant: South Korea and Late Industrialization*. Oxford University Press: New York.

Ban, C. 2013. Brazil's Liberal Neo-developmentalism: New Paradigm or Edited Orthodoxy? *Review of International Political Economy*, 20(2), 298–331.

Barros, G. 2009. *Brazil as an Economic Superpower? Understanding Brazil's Changing Role in the Global Economy*. Brookings Institution Press: Washington, D.C.

Boito Jr., A. 2018. *Reforma e Crise Política no Brasil*. Editora UNESP: São Paulo.

Bratman, E. 2019. *Governing the Rainforest: Sustainable Development Politics in the Brazilian Amazon*. Oxford University Press: Oxford.

Bruno, R. 2009. *Um Brasil Ambivalente: Agronegócio, Ruralismo e Relações de Poder*. Mauad: Rio de Janeiro.

Buckhardt, H. -J. and Dietz, K. 2014. (Neo)extractivism: A New Challenge for Development Theory from Latin America. *Third World Quarterly*, 35(3), 468–86.

Cardoso, F. H. and Faletto, E. 1973. *Dependência e Desenvolvimento na América Latina. Ensaio de Interpretação Sociológica.* Zahar: Rio de Janeiro.
Chase-Dunn, C. 1997. *Rise and Demise: Comparing World Systems.* Westview Press: Boulder.
Coutinho, L. 2002. Marcos e Desafios de uma Política Industrial Contemporânea. In: *Desenvolvimento em Debate*, Castro, A.C. (ed.). Editora Mauad: Rio de Janeiro, pp. 191–209.
Cox, R. W. 1983. Gramsci, Hegemony, and International Relations: An Essay in Method. *Millennium: Journal of International Studies*, 12(2), 162–75.
Cox, R. W. 1987. *Production, Power, and World Order: Social Forces in the Making of History.* Columbia University Press: New York.
Davalos, L. M., Holmes, J., Rodriguez, N., and Armenteras, D. 2014. Demand for Beef is Unrelated to Pasture Expansion in Northwestern Amazonia. *Biological Conservation*, 170, 64–73.
Dieguez, C. 2015. O Estouro da Boiada: Como o BNDES Ajudou a Transformar a Friboi na Maior Empresa de Carnes do Mundo. *Revista Piauí*, 101, 16–24.
Erten, B. and Ocampo, J. A. 2013. Super Cycles of Commodity Prices Since the Mid-Nineteenth Century. *World Development*, 44(C), 14–30.
European Livestock and Meat Trading Union. 2008. *Globalization of the Meat Industry = Brazilian Meat Imperialism?* EMTU: Brussels.
Evans, P. 1979. *Dependent Development: The Alliance of Multinational, State and Local Capital in Brazil.* Princeton University Press: Princeton, NJ.
Evans, P. 1995. *Embedded Autonomy: States and Industrial Transformation.* Princeton University Press: Princeton.
Fernandes, F. 1968. *Sociedade de Classes e Subdesenvolvimento.* Zahar: Rio de Janeiro.
Fernandes, F. 1972. *Capitalismo Dependente e Classes Sociais na America Latina.* Zahar: Rio de Janeiro.
Fernandes, F. 1975. *A Revolução Burguesa no Brasil.* Zahar: Rio de Janeiro.
Ferraz da Fonseca, I. and Moreira da Silva, A. P. 2011. Mudança Institucional: O Novo Código Florestal e a Descentralização da Política Ambiental. *Boletim de Análise Político-Institucional*, 1, 43–53.
Flexor, G. and Leite, S. P. 2017. Land Market and Land Grabbing in Brazil during the Commodity Boom of the 2000s. *Contexto Internacional*, 39(2), 393–420.
Gerschenkron, A. 1962. *Economic Backwardness in Historical Perspective.* Belknap: Cambridge, MA.
Ghosh, J. 2018. Global Instability and the Development Project: Is the Twenty-First Century Different? *European Journal of Economics and Economic Policies*, 15(2), 193–207.
Gill, I. and Kharas, H. 2007. *An East Asian Renaissance: Ideas for Economic Growth.* World Bank: Washington, D.C.
Gramsci, A. 1971. *Selections from the Prison Notebooks.* International Publishers: New York.
Greenpeace. 2009. *Slaughtering the Amazon.* Greenpeace: São Paulo.
Greenpeace. 2015. *Grilling Away the Amazon.* Greenpeace: São Paulo.

Haggard, S. 1990. *Pathways from the Periphery*. Cornell University Press: Ithaca.
Harvey, D. 2005. *The New Imperialism*. Oxford University Press: New York.
Hilferding, R. 1981 [1910]. *Finance Capital: A Study of the Latest Phase of Capitalist Development*. Trans. M. Watnick and S. Gordon. Routledge: London.
Hirschman, A. O. 1965. *The Strategy of Economic Development*. Yale University Press: New Haven.
Hoelle, J. 2015. *Rainforest Cowboys: The Rise of Ranching and Cattle Culture in Western Amazonia*. University of Texas Press: Austin.
Ianni, O. 1979. *Colonização e Contrareforma Agrária na Amazônia*. Vozes: Petropolis.
Ioris, A. A. R. 2015. Cracking the Nut of Agribusiness and Global Food Insecurity: In Search of a Critical Agenda of Research. *Geoforum*, 63, 1–4.
Ioris, A. A. R. 2017. Encroachment and Entrenchment of Agro-Neoliberalism in the Centre- West of Brazil. *Journal of Rural Studies*, 51, 15–27.
Ioris, R. 2014. *Transforming Brazil: A History of National Development in the Postwar Era*. Routledge: London and New York.
Johnson, C. 1982. *MITI and the Japanese Miracle: The Growth of Industrial Policy*. Stanford University Press: Stanford.
Katz, C. 2015. Dualities of Latin America. *Latin American Perspectives*, 42(4), 10–42.
Kharas, H. and Kohli, H. 2011. What Is the Middle Income Trap, Why Do Countries Fall into It, and How Can It Be Avoided? *Global Journal of Emerging Market Economies*, 3(3), 281–89.
Lin, J. and Treichel, V. 2012. *Learning from China's Rise to Escape the Middle-Income Trap: A New Structuralist Economics Approach to Latin America*. Policy Research Working Paper No. 6165. World Bank: Washington, D.C.
Luxembourg, R. 2003. *The Accumulation of Capital*. Routledge: New York.
Machado, E. 2009. Governo Lula, Neoliberalismo e Lutas Sociais. *Lutas Sociais*, 21/22, 23–34.
Martins, J. S. 1975. *Capitalismo e Tradicionalismo*. Pioneira: São Paulo.
Marx, K. 1992. *Capital*. Penguin Classics: New York.
Mazzuca, S. 2013. The Rise of Rentier Populism. *Journal of Democracy*, 24(2), 108–22.
Mendes de Paula, G. 2017. Deindustrialization in Brazil? In: *The New Brazilian Economy: Dynamic Transitions into the Future*, Grivoyannis, E. (ed.). Palgrave Macmillan: New York, pp. 63–85.
Montero, J. 1994. *Negros da Terra*. Companhia das Letras: São Paulo.
Morais, L. and Saad-Filho, A. Neo-developmentalism and the Challenges of Economic Policy- Making under Dilma Rousseff. *Critical Sociology*, 38(6), 789–98.
Ohno, K. 2009. Avoiding the Middle Income Trap: Renovating Industrial Policy Formation in Vietnam. *ASEAN Economic Bulletin*, 26(1), 25–43.
Oliveira, F. 1972. *A Economia Brasileira: Critica `a Razão Dualista*. Vozes: Petropolis.
Oliveira, J. P. 20016. *O Nascimento do Brasil e Outros Ensaios: 'Pacificação', Regime Tutelar e Formalção de Alteridades*. Contracapa: Rio de Janeiro.
Patnaik, P. and Patnaik, U. 2018. *A Theory of Imperialism*. Columbia University Press: New York.

Paus, E. 2014. *Latin America and the Middle Income Trap*. CEPAL: Santiago.
Reuschemeyer, D., Stephens, J. D. and Stephens, E. H. 1992. *Capitalist Development and Democracy*. Cambridge University Press: New York.
Roberts, K. 2014. *Changing Course in Latin America: Party Systems in the Neoliberal Era*. Cambridge University Press: New York.
Robinson, W. I. 2014. *Global Capitalism and the Crisis of Humanity*. Cambridge University Press: New York.
Roller, H. 2014. *Amazonian Routers: Indigenous Mobility and Colonial Communities in Northern Brazil*. Stanford University Press: Stanford.
Rosenberg, J. 2006. Why is There no International Historical Sociology? *European Journal of International Relations*, 12(3), 307–40.
Ross, M. 2015. What have We Learned about the Resource Curse? *Annual Review of Political Science*, 18, 239–59.
Santana, Carlos Augusto M. and Rente Nascimento, José. 2012. *Public Policies and Agricultural Investment in Brazil*. Food and Agricultural Organization (FAO): Brasilia.
Schmink, M., et al. 2019. From Contested to "Green" Frontiers in the Amazon? A Long-Term Analysis of São Felix do Xingu, Brazil. *The Journal of Peasant Studies*, 46(2), 377–99.
Schneider, B. R. 2013. *Hierarchical Capitalism: Business, Labor and the Challenges of Equitable Development*. Cambridge University Press: New York.
Schneider, B. R. and Doner, R. 2016. The Middle Income Trap: More Politics than Economics. *World Politics*, 68(4), 608–44.
Serra, J. 2017. Global Champions Are Made at Home: The Brazilian Development Bank and Domestic MNCs. Paper Presented at American Political Science Association Annual Meeting, San Francisco.
Singer, H. and Sarkar, P. 1991. Manufactured Exports of Developing Countries and Their Terms of Trade since 1965. *World Development*, 19(4), 333–40.
Trotsky, L. 1928. *Third International After Lenin*. The Militant: New York. Available at: https://www.marxists.org/archive/trotsky/1928/3rd/index.htm.
Trotsky, L. 1969. *The Permanent Revolution and Results and Prospects*. Red Letter Press: New York.
Trotsky, L. 1977. *History of the Russian Revolution*. Haymarket Books: Chicago.
Vernengo, M. and Caldentey, E. P. (eds.). 2017. *Why Latin American Nations Fail*. UC Press: Berkley.
Wade, R. 1990. *Governing the Market: Economic Theory and the Role of Government in East Asian Industrialization*. Princeton University Press: Princeton, NJ.
Werneck Vianna, L. 1999. *Liberalismo e Sindicato no Brasil*. UFMG: Belo Horizonte.
Wilkinson, J., Reydon, B., and Di Sabbato, A. 2012. Concentration and Foreign Ownership of Land in Brazil in the Context of Global Land Grabbing. *Canadian Journal of Development Studies*, 33(4), 417–38.

Chapter 6

The Socioenvironmental Contradictions of Contemporary Developmentalism

Urbanization, Food Sovereignty, and Sustainable Development in the Amazon

Tatiana Schor and Gustavo S. Azenha

INTRODUCTION

Since the 1980s, with Brazil's return to democracy, there has been numerous advancements in socio-environmental policies in the country, including the strengthening of federal, state, and municipal environmental governance institutions, the creation and expansion of formalized channels for civil society participation in decision-making, and more rigorous environmental licensing procedures. Brazil has also developed more robust environmental monitoring (e.g., satellite and remote sensing) and data collection and increased the technical capacities of public servants to support evidence-based planning and the implementation of environmental policies. These changes have contributed to various gains for environmental conservation, resource management, and rural livelihoods and resource rights. Among these gains have been decreases in deforestation rates during the mid-2000s (reversing previous, decades-long trends); expansion of conservation and sustainable use areas; new programs for traditional resource rights and management; and increased sustainability of some agricultural systems. Although not expressly connected with environmental matters, social programs and economic policies in the 1990s and early 2000s targeting the poor and promoting, inclusion diminished poverty and decreased inequalities, contributing in some respects to sustainable development in a broad sense.

Despite these apparently positive trends in policies, institutional innovations, and socioenvironmental impacts, advances have often been inconsistent and uneven. At the same time, Brazil has seen increasing demands for territorialized resources due to a combination of population growth, and, especially, changing patterns of consumption including among those in the poorest strata and the expanding middle class. Increased domestic demand for energy, water, food crops, animal protein, cars, construction materials, and consumer goods have placed new demands on Brazil's ecosystems and their resources. Increased monetization of social relations in rural communities through the expansion of markets in goods and services, increased access to public services (e.g., electricity), and the influx of money through government social services (e.g., rural pensions and conditional cash transfer programs) and novel forms of development (e.g., government- and NGO-sponsored sustainable development and ecosystem service projects), especially in socioeconomically and politically peripheral areas such as the Amazon, has transformed consumption patterns, resource use, and the social fabric of rural communities.

Beside changes in internal demand and socioeconomic conditions in rural communities from public and nonprofit interventions, evolving engagements with external markets also contributed to shifting socio-environmental dynamics. Brazil—like many other Latin American countries—made export-oriented commodity production central to its economic growth, which has often been accompanied by problematic socio-environmental consequences. Agricultural and extractive enterprises that have been an important priority in Brazil—such as soybean production, sugarcane cultivation, iron ore extraction, petroleum extraction, beef production, and eucalyptus cultivation—have enabled and been accompanied by a variety of socioenvironmental changes with potential and actual detriments to the environment and its sustainable use, as well as to the livelihoods of traditional peoples and the rural poor (e.g., pollution, dispossession, and workplace hazards).

The country has increasingly pursued what some refer to as a "neodevelopmentalist" approach, which draws from older forms of state-driven development from the mid to late 1900s, reimagining them in the neoliberal era to yield a novel hybrid form of rural development and natural resource exploitation. This paradigm was exemplified by the Growth Acceleration Program [*Plano de Aceleração de Crescimento*, or PAC], which emphasized state-subsidized or financed large-scale infrastructural projects including roads and hydroelectric plants. These development initiatives have been criticized for bypassing environmental licensing processes and environmental impact assessment requirements, lacking adequate public consultation and civil society participation, and violating the human, land, and resource rights of indigenous peoples, traditional peoples, and the rural poor.

Thus, despite advances, as a combined result of economic growth trajectories and the fragilities and limitations of governance structures and public policy processes, socio-environmental challenges have persisted as key barriers to more sustainable and equitable socioeconomic growth. Brazil continues to grapple with a variety of critical social and environmental issues that inhibit sustainable development including droughts, water rationing, water pollution, air pollution, displacement of people for large infrastructure projects, violent conflict in rural areas, the persistence and exacerbation of varied environmental health issues, the recent reversal of temporarily favorable trends in deforestation rates, and the persistence of social and economic inequalities with large portions of the population living in conditions of poverty.

The contradictions and limitations described above existed throughout the 1990s and early 2000s, but have become increasingly apparent and exacerbated in the last five to ten years. In recent years, there has been an erosion of traditional peoples' influence in socio-environmental governance and an undermining of traditional land and resource rights. Delineating the specific timeline for this backsliding of the socio-environmental citizenship of traditional peoples (i.e., their political agency to exert influence on the design and implementation of land and resource management policies) is a tricky and politicized affair, considering that the political agency of traditional peoples remained limited, circumscribed, and spatially heterogeneous even during the periods of greatest advancements in rights and participatory frameworks (roughly the 1990s and the first decade of the twenty-first century), but a shift happened around 2010, with a further accentuation of backsliding following the impeachment of President Dilma Rousseff in 2016. This shift is linked to (and contributes to) the recent and ongoing economic and political crisis. With this crisis, more conservative political forces have come to power that have increasingly privileged agribusiness, energy, and mining interests, while being hostile and committed to scaling back, social programs, conservation measures, land and resource rights of the rural poor and traditional peoples.

More concretely, through ministerial appointment choices, budget cuts, and attempts at institutional reconfigurations, recent administrations have weakened and subordinated the government agencies that oversee environmental protection, indigenous rights, and land reform to competing interests. Some of these attempts have been thwarted by social movements, public pressure, international pressure, and some private sector actors, but important changes have gone through, with these agencies being undermined in important ways since 2016. A notable example of a change to socio-environmental governance that has undercut environmental interests and the influence of indigenous and rural peoples is the Bolsonaro administration´s decision to dissolve or diminish the influence of government councils for civil society participation. Changes to the *Conselho Nacional do Meio Ambiente* (Conama) have eliminated a vital channel for environmentalists, scientists, and traditional peoples to contribute to policy-making and monitor government actions, such as environmental licensing procedures.

These trends toward increased limitations in the rights of traditional peoples and the effective influence of their political agency occur alongside seemingly contradictory efforts at promoting poverty alleviation and socioeconomic inclusion under the guise of the rhetoric of sustainable development. These efforts are based on narrow conceptions of inclusion and citizenship that are modest in scope and focused on the short term. These have insufficiently addressed participation in policy-making and the safeguarding of established rights, yielding the so-called "sustainable development" practices in which environmental and social concerns are superficial, and that echo historical exclusionary, assimilationist, and developmentalist efforts to promote "progress" and integration of rural peripheries into the national economy. Even considering the limits of social participation in policy-making processes[1] the lack of institutional frameworks for participation, where civil society groups and local populations can and many times have contributed to the debate, has weakened the democratic impetus of policy-making since the 1988 National Constitution.

Considering this broad diagnosis of the ambivalent and worrisome state of Brazil´s socio-environmental politics and the government´s weak commitment to environmental justice and poverty alleviation, in this chapter we highlight and critically examine the issue of food sovereignty and sustainable development in the Amazon. Exploring the links between trends in urban-rural dynamics, food sovereignty, and livelihoods in the region, and the broader scenario of crisis within Brazil´s socio-environmental politics. More specifically, we examine recent patterns in urbanization and food consumption in the Western Amazon. Employing the transforming issue of food security as a lens to: 1) understand some potentially problematic implications of economic inclusion policies; 2) critique the sustainability of agribusiness models of rural development (which reflect the hybridized vices of outdated developmentalist paradigms and neoliberal myopia); and 3) highlight the need for more inclusive forms of policy-making for more sustainable and equitable territorial development.

Colonial Heritage and Contemporary Socio-environmental Politics

Like most colonial countries, Brazil has retained the pernicious legacies of unequal social and geographical territorial organization that have been fostered by government policies and institutions, and extra-legal land and resource use practices that have thrived in the interstices of formal institutions

(Moraes, 2000). European colonization of Brazil was effectively initiated with the implementation of the *sesmarias* system, through which massive land grants were distributed to Portuguese noblemen, establishing a persistent, yet evolving, system of deeply differentiated access to land and marked land concentration. Land policies, economic priorities, the importance of transatlantic trade, the challenges of inland transportation stemming from Brazil´s geography, and other factors led to the concentration of people and anthropogenic transformations of nature throughout the coastal portions of Brazil.

These historical patterns of territorial occupation have resulted in the Amazon region remaining sparsely populated and having significant areas of forest cover. Starting in the latter half of the twentieth century, there was an intensification of population growth and deforestation as the region was treated as a blank slate for economic development and resource exploitation. These efforts were animated by misconceptions of the region as a pristine wilderness, which rendered the population of indigenous and traditional peoples largely invisible and ignored the anthropogenic nature of forests and other habitats.

Around the 1970s, the frictions caused by dispossession and environmental degradation from intensified development efforts in the Amazon resulted in increasing resistance from indigenous and, traditional peoples, and environmental activists who sought to redefine paradigms of territorial occupation, resource use, and economic development, and to increase local decision-making and autonomy in these matters. These efforts resulted in substantial changes in the 1980s, 1990s, and early 2000s in patterns of territorial occupation, and the ways in which people in rural peripheries were integrated as political and economic subjects. For example, there was a proliferation of conservation units, indigenous territories, and traditional territories, and indigenous peoples and the rural poor came to exert some influence on environmental decision-making around conservation units, watersheds, and municipalities (as discussed previously).

Before discussing persistent challenges and contradictions in the Brazilian Amazon today, it is worth recounting some important developments and trends in the recent Brazilian history of socio-environmental politics. The 1980s and the 1990s were formative years in socio-environmental politics in Brazil. A time period is marked by Brazilian democratization, the establishment of a new Constitution, a strengthening of socio-environmental movements, policies, and institutions, and the consolidation of a more neoliberal political and economic paradigm. A strong and active civil society emerged, linking environmentalists, environmental scientists, traditional peoples, and anthropologists around shared goals of promoting more sustainable and inclusive forms of development. Conservation of biodiversity, poverty alleviation, and land rights of the poor, previously seen to be at odds in some respects, became increasingly viewed and addressed as connected matters.[2] This period saw the establishment of a decidedly more robust institutional and policy framework to address socio-environmental concerns. An increasing political prioritization of this issue within the federal government, and a greater space for civil society influence in decision-making processes (Banerjee et al., 2009; Conklin and Grahm, 1995; de Castro, 2014; Drummond and Barros-Platiau, 2006; Hochstetler and Keck, 2007; Turner, 1993, 1995, 1999).

In the early 2000s, the PT party (*Partido dos Trabalhadores* or Worker's Party) took over the government with the election of Luiz Inácio da Silva (Lula). Under Lula there was an increased role for diverse actors in environmental decision-making processes. Environmental politics took on an even greater openness to social agendas in some respects, with social inclusion and sustainable development figuring prominently at the start of his administration. In practical terms, this included the greater inclusion of social movement actors in government positions, the further expansion of formal channels for decision-making (e.g., conservation unit councils, watershed councils, and federal, state, and municipal environmental councils), and a deepening of established trends toward the decentralization of environmental governance and oversight. A strong political commitment to address deforestation and social inclusion by the president and the minister of the Environment (Marina Silva, who began her political career as an activist in rubber tapper and socio-environmental movements) led to increases in law enforcement. In addition, effective NGO advocacy and strategic alliances with corporate organizations contributed to important gains in conservation efforts (Viola, 2013: 111–12).

Under Lula, despite greater openness to social justice agendas in some respects compared to previous administrations, there were also some "puzzling contradictions" and a persistence of tensions between socio-environmental and economic development goals and projects (de Castro, 2014). Notwithstanding the various developments and trends noted above that fortified the socio-environmental agenda, a closer look reveals many persistent issues and contradictions, some of which appear to have been intensified over the course of Lula's administration and that of his successor, Dilma Rousseff.

The socioenvironmental agenda was selectively integrated into government policy-making and planning, with many government actions countering and undermining the positive changes noted. During the course of the Lula administration, the initial transversal (i.e., effective collaboration across different Ministries) and participatory approach to environmental policy design and implementation was supplanted by a more top-down approach with a developmentalist, pragmatic perspective characterized by a national discourse of sustainability under deepened socio-environmental conflicts (de Castro, 2014). Economic development in the 2000s increasingly focused on export-oriented commodities from agribusiness and mining, a pattern that intensified in the latter half of the decade. The percentage of export value from commodities in Brazil rose from 23 percent to 46.7 percent from 2000 to 2012 (de Castro, 2014). Export-oriented economic development continued to largely trump environmental concerns as well as the needs and interests of traditional peoples and the rural poor.

These development efforts have often been animated by, and framed within, sustainable development discourses, but on closer inspection, they have involved conflicting and ambiguous implications for environmental sustainability and socioeconomic inclusion. The model of agriculture and rural development pursued has given heightened attention to agricultural sustainability and improvements in farm management, but this occurs alongside agribusiness efforts that continue to exacerbate a variety of long-established issues including environmental degradation, the heavy use of chemical inputs, political and socioeconomic inequalities, land concentration and insecure land tenure, and rural violence.

Important to understanding the contradictions and regressions of socio-environmental policies in the last five to ten years is the wider political and economic scenario. The increasing reversion to a developmentalist model of growth and the stagnation and rolling back of socio-environmental concerns coincided with the global economic crisis and a slowing down of Brazilian economic growth after an unprecedented economic boom. A slowing down of demand for Brazilian commodities due to economic crises and an attenuation of growth in other countries (e.g., China) diminished growth in Brazil. In this context, the ambitious PAC project was seen as an important means to promote growth in Brazil. Furthermore, political and economic pressures intensified to cut costs and increase exports, profits, productivity, and growth. With private and state enterprises intensifying pressure to rid themselves of the barriers of environmental licensing, facilitate access to lands and natural resources, transform more undeveloped areas into economically productive territories, and grant themselves greater autonomy over land and resource use decisions on their landholdings. With the deepening of Brazil's economic (and political) crisis starting in around 2014, these pressures have become even more acutely pronounced, with the marginalization of socio-environmental issues rationalized through the rhetoric of austerity and the primacy of growth. This rhetoric became even more salient under the administration of Michel Temer (who took office following the impeachment of Rousseff in 2016). This emphasis on austerity and growth has been reiterated by the president elect Jair Bolsonaro, who has made his allegiance to the neodevelopmentalist agenda of agribusiness and mining interests clear, and has repeatedly expressed overt hostility toward environmental concerns and the rights of traditional peoples and landless peasants.

It is important to note that dam construction, agribusiness development, the revision of the Forest Code, and other development efforts have been and continue to be animated and legitimated by sustainable development discourses. Environmental sustainability is mobilized by champions of agribusiness, hydropower, and other development in very circumscribed ways. The social and equity concerns that are central to conceptions of sustainable development among civil society actors and local resource uses, are partially or entirely misunderstood or ignored, consistently subordinated to a narrow vision of economic growth. Thus, key issues underlying the contradictions and inconsistencies above are contested conceptions of growth, development, sustainability, nature, and the question of how to define these conceptions, and who defines them, which brings us back to the matter of socio-environmental institutions and governance and the historical persistence of poverty in Brazil.

The Limitations of Civil Society Participation

Despite the many persistent issues and setbacks raised in the preceding discussion, there has been a substantial transformation of the policy and institutional matrix of socio-environmental governance in Brazil since the 1980s. Among the major advances of the last few decades have been the strengthening of socio-environmental governance institutions and mechanisms, in combination with a democratization of environmental decision-making in some respects.

Although there has been an overall trend in Brazil toward an increased valorization of indigenous and traditional peoples within conservation and sustainable development discourses, this valorization contains contradictory impulses that often reproduce historical asymmetries of power, simultaneously exalting traditional peoples while continuing to circumscribe their agency in the conception, development, and implementation of conservation efforts. Despite the widely publicized success of participatory conservation and resource management projects, the scale and impact of these programs in specific locales has been quite heterogeneous and inconsistent. Many challenges and contradictions impede their potentially wider impact on social, economic, and environmental issues. Poverty continues to be a persistent reality.

These deficiencies of socio-environmental decision-making provide the foundation for the myopic vision of growth that continues to prevail, where short-term political preservation or increased profits (or cost cutting) of selective sectors or companies trump collective needs and sustainable growth. As a result, the prevailing and highly deficient notion of sustainable development that is put into practice is one that is forged by and supports the interests of political actors and their elite allies in private and state enterprises. It is a notion in which many important components of sustainable development fall by the wayside (including, e.g., social inclusion, poverty alleviation, land reform, and the diversification of productive activities for economic and environmental resilience).

Despite a widely acknowledged need to think of economic development in more holistic terms (or to question the very notion of growth), in practice there has in many respects been a regression or reversion to narrowly conceived and outdated notions of growth, that continue to ignore important externalities and lack longer-term vision, leading to emphases on economic activities that have negative socio-environmental impacts and lack economic and environmental sustainability. While also failing to promote much-needed economic diversification, essential to ensure stability and resilience in the face of economic and environmental uncertainties and volatility.

Socio-environmental Transformation and Food Sovereignty

Uncertainties mark this new era in Brazil, but in order to appraise the possibilities of social-environmental transformation it is necessary to understand how society functions and reacts to its historic "contradictions." The preceding discussion serves to orient our subsequent discussion of recent patterns in spatial occupation, economic development, and food sovereignty in the Amazon. Following are our chief concerns:

1) How poverty alleviation programs (e.g., *Bolsa Família*, a conditional cash transfer program) are contributing to changing resource use practices, the ways in which people acquire food (i.e., increased dependence on markets and foods produced elsewhere), and increased urbanization;
2) How the complex and unique urban–rural dynamics of the Amazon—including its seasonality tied to environmental dynamics—are important for understanding development and resource use in the region;
3) How current government social, economic development, and environmental policies undermine the possibilities for socially and environmentally resilient resource use paradigms, due to the privileging of economic development and poverty alleviation over sustainable resource use, the participation of peripheral peoples in decision-making, and addressing systemic social, economic, and political asymmetries.

The patterns of urbanization in the Amazon have changed significantly during this century, especially in Brazil, after 2002, with the implementation of the social development policy during the first Lula presidency and then during his second term and continuing with Dilma Rousseff's social inclusion policy. This policy emphasized a conditional cash transfer program, *Programa Bolsa Família* (PBF), which provides poor families with monthly payments on the condition that their children are in school and participate in vaccination programs, as well as pensions, welfare benefits, and bank loans. These income transfer and inclusion programs have had a strong monetization impact on local economies and thus on urbanization, especially in localities where economies that were relatively nonmonetized (i.e., where barter and exchange were more central to economic relations). This process has brought challenges in terms of sustainability, economic development, and implications for poverty alleviation, especially in more socioeconomically vulnerable regions such as the Amazon. These programs also have complex implications and impacts for socio-environmental politics.

There is a long tradition of city studies and urbanization in the Brazilian Amazon (Corrêa, 1987; Godfrey, 1988; Becker, 1990, among others). Most of them have focused on three distinct axes. Historically, urbanization in the Amazon has been dealt with on a regional scale. The studies performed by IBGE (Faissol, 1971; Amazonas-SEPLAN, 1976, 1990; Carneiro, 1998; IPEA / IBGE / UNICAMP / NESUR, 1999; BRAZIL-IBGE, 2007) and by Bertha Becker (1982) who in the 1980s recognize the Amazon as a "urbanized forest" are good examples of how the regional scale determined the understanding of the process and dynamics of urbanization in the Amazon basin. From the earliest writings of Bertha Becker (1974), the question of the role and consequences of urbanization in the Amazon and its impact on the natural environment has ruled the political and academic debate. There is a myriad of possible interpretations and methods of analysis of the urbanization process. Key authors in English, such as Bunker (1984), Roberts (1992), Browder and Godfrey (1997), and Moran and Brondizio (2005); in Portuguese, Trindade Jr (2013a, b; 2011), Oliveira (2000), Becker (1974, 1978, 1999, 2012); and in Spanish, Botía (2012) have been an important effort to understand the complex role of urbanization in transforming and / or the conservation of natural resources and the economic dynamics in Amazon in each country. Despite these efforts, there is still a gap in theoretical and methodological terms to understand the relationship between the contemporary process of urbanization, the diversity of cities, and the conservation of natural resources. Society and nature are often still conceptually separated when proposing the analysis of cities in the Amazon.

An interesting way to look at the relations between society and nature is through diverse forms of urbanization and economic (market) structures. One place where the discourse and practice of sustainable development have been most persuasive is in the Amazon. Understanding how national politics impacts the Amazon can help explain the transformations and persistence of "puzzling contradictions" of sustainable development. It permits an appreciation for the incongruities between commitments to sustainability and inclusiveness expressed within a socio-environmental framework, and prevailing political and economic models of development that recurrently undermine a socio-environmental agenda. There is a variety of ways in which this analysis can be done; we choose to consider access to food, urban networks, and poverty in the western part of the Brazilian Amazon, Amazonas State.

In order to analyze the cities and the urban network in western Amazon, it is necessary to move beyond the dendritic image of the urban network where all fluxes (material and symbolic, human and nonhuman) flow down the river; traditional literature has mostly emphasized downriver dynamics (Corrêa, 1987). Recent research involving extensive fieldwork has shown that this is not necessarily the full picture, suggesting that there are more complicated and dynamic patterns of sociomaterial flows (Schor et al., 2016, 2014, 2018; Schor and Oliveira, 2011; Schor and Azenha, 2017). Different cities and locations have different roles in terms of organizing their territories, especially those that have an important economic interest such as specific resources from the forest that interact with global market structures, such as acai berry (Marinho and Schor, 2012; Pegler, 2015). Focusing studies on the cities along large rivers, in this case the Solimões and Amazonas rivers in Brazil's Amazonas State, enables novel ways of understanding the impacts of socio-environmental policies, which include socioecological changes due to urbanization, monetization of the local economy, and globalization. At the same time, the policies allow for continuity in local livelihoods, which is positive, while also perpetuating poverty, which is not. Modernity and tradition, development and poverty coexist in this vast contemporary habitat that is the Amazon forest and its river system (Oliveira and Schor, 2008, 2010).

In this context, it is necessary to understand that the city is not an isolated entity in itself but invariably belongs to a group of cities that in many cases can compose a network or even a region. The Italian geographer Dematteis (1999) defines two types of relations between cities. The first is the network of relationships which, for the author, implies belonging to the global network where interactions are influenced little or not at all by physical distance. These relationships reflect the division of labor at the supra-regional scale and indicate the degree of globalization of local urban systems. The second set of relationships that describe the interactions between the cities are what Dematteis calls territorial relations. In these, the friction of distance, physical forms (relief, hydrography, climate, etc.), and historical legacies have significant influence on the way cities are interrelated. For the author, these two groups of relationships are not mutually exclusive; on the contrary, they interact in their respective spheres. Network interactions transform the physical space and territorial relations structures in network connections. This is undoubtedly true in the Brazilian Amazon, especially for the state of

Amazonas, where territorial relations deal with frictions of physical forms in a unique way—the flood or ebb of the large rivers that determine transport flows and food production in the floodplain (Aguiar et al., 2014; Pinedo-Vasquez et al., 2011; Ronchail et al., 2016; Schor et al., 2014), extraction, hunting (Tavares-Pinto, 2015a), and the cost of living in cities (Moraes and Schor, 2010a, b)—and at the same time are re-organized by the network of relationships that insert certain locations within the global network.

Including the study of cities in the analysis of socio-environmental policies is essential in analytical terms. It breaks with the overall nation–state duality and focuses on the city in a variety of national sub-components, some deeply articulated with the global economy (Sassen, 2005), others less so. For the Amazon, with its low-density urban centers, to analyze the city is to understand the different forms of inclusion or exclusion that result from national policies and the development process, and how these affect conservation and poverty. The city as an economic and/or cultural event is an adequate scale to understand the diverse and contradictory social processes of modernity and its impact on poverty as an important component of human security (Gasper, 2005). The study of cities in Amazonas rescues the historical process of territorial formation. In doing so, at the same time it helps us to understand how a region considered a "hot spot" in terms of climate and biodiversity conservation can show seemingly contradictory outcomes in terms of sustainability policies and various networks of globalization markets implicated in deforestation and other unsustainable forms of economic development. Focusing on cities permits a specific geographic strategy that links places with each other and with the economic dynamics of globalization that, in turn, as Dematteis shows, modifies territorial relations and restructures urban dynamics. This perspective requires integrating the global scale with the local scale, understanding nanoscale as a possible way of considering the social and natural dynamics affecting the region from a local perspective. However, how should these scales be put in motion?

Multi-scalar Analysis

Cities in the Brazilian Amazon are privileged places for the analysis of contradictory processes of sustainable development policies, with the implication of inserting the local population in national and global markets. These contradictory processes can be seen in environmental debates, in the promises of obtaining wealth from biodiversity or the knowledge of traditional populations, or in international research agendas worldwide (Schor and Moraes, 2011) with a strong discourse of expanding cattle and monocultures such as soya beans. Understanding urban dynamics and cities in the Amazon is important to comprehend the mechanisms by which national policies, with their "puzzling contradictions," interact with local dynamics, including and excluding processes and people in this contradictory movement that is socio-environmental policies in Brazil. Rejecting the idea of a dendritic network is a first step to understanding the complex national-local dynamics in the region and at the same time embedded geohistorical and biogeographical dimensions important in understanding the network that streamlines the urban western Brazilian Amazon.

Studies conducted by Schor and Oliveira (Schor et al., 2009, 2011, 2014, 2015, 2016; Oliveira and Schor, 2009, 2010, 2011; Schor and Oliveira, 2011) created a multi-scalar methodology of analysis of urbanity in the Amazon and from the results obtained in extensive fieldwork in the period 2006–2008 proposed a typology that incorporates relational parameters. These parameters incorporate socioeconomics and demographics with hydrology, biogeography, access to natural resources, and other issues of interaction between society and nature. As classical studies in geography, based on descriptions and primary data collection, these studies aimed to understand the role of each city in the structure of the urban network and how these roles are related to national socio-environmental policies. This understanding was intended to create knowledge likely to support specific urban public policies for the region, without which the possibility of development—understood here according to Sen (2000) as freedom of the people who inhabit these places and sustainability of life within them—would be reduced to homogeneous patterns of national scale policies that do not change social structures locally and thus do not alleviate poverty, address inequalities, or maintain biodiversity.

To build the typology of the urban network of Amazonas State, some institutional frameworks were constructed that could, if taken together, establish an urban network. Nine sets of institutional frameworks were developed and organized under the following categories: historical variables, intra- and inter-urban relations, services and trade, tax collection, inputs for regionalized food staple basket, construction index, extractive products, urban infrastructure, and transport flows (Schor et al., 2014, 2016). Each of these frameworks included a set of variables to be analyzed. The results allowed the construction of a city typology for the urban network. This typology helped understand the internal and external dynamics of each city. These procedures were described in detail by Schor and Oliveira (2011) and Schor et al. (2016). The proposed typology has two entries, one related to size in demographic terms, classifying the cities as small or medium. The other entry related to more subjective aspects: Territorial Responsibility; External Economic Dynamics; Intermediate and Special Cities.

As a result, the following typology was developed: Medium Cities of Territorial Responsibility (Tabatinga, Tefé and Parintins); Medium Cities with External Economic Dynamics (Coari); Medium Intermediate Cities

(Manacapuru and Itacoatiara); Small Cities of Territorial Responsibility (Bejamin Constant, Santo Antonio do lça and Fonte Boa); Small Cities with External Economic Dynamics (Codajás, Iranduba and Carreiro da Várzea); Special Cities (São Paulo de Olivença, Amatura, Tonantins, Jutaí, Uarini, Alvarães, Anori, Anama, Silves, Urucurituba and Urucará) (mentioned in Oliveira and Schor, 2010; Schor and Oliveira, 2011; Schor et al., 2014, 2016).

In addition to the proposed typology, this new perspective yielded some interesting results. First, it is very important to recognize that it is not possible to consider the area along the river Solimões-Amazon as one urban network; rather, the two rivers have different urban networks that need to be analyzed separately (Schor and Costa, 2011; Schor and Oliveira, 2011). In order to better understand this typology, it is necessary to make some more specific comments regarding the differences between the types of cities. Special Cities are those that depend strongly on the urban network to function as cities. They cannot be considered individually due to the fact that some important functions happen in another city. An interesting example is Tonantins and Santo Antonio do Içá. All the judiciary of Tonantins is located in Santo Antonio do Içá, so if a citizen needs any judicial help they have to travel for approximately one hour up the Solimões River to Santo Antonio do Içá or wait for the monthly visit of the judge and prosecutors to town.

The cities classified as having External Economic Dynamics have an economic dynamic that does not aggregate value in the urban network, and does not rely on the urban network to function. An example is Coari, where the economy is based on natural gas extraction by the Brazilian Gas and Petroleum Enterprise (Petrobras). The royalties paid back to Coari increase official revenues and the municipal budget but do not affect the urban network in which it is inserted (Rodrigues and Santana, 2015; Santana, 2015; Tavares-Pinto, 2015b).

The cities of Territorial Responsibility carry out various urban functions and contain institutional arrangements that are important not only for the immediate municipality, but also for the cities and municipalities around them, and especially for what in the Amazon is configured as the interior of the municipality, the hinterlands. The importance of the city originates in the historical-geographical development that constitutes the region and urban network. The economic development of these cities tends to aggregate and/or disaggregate value in the region depending on what direction the development takes (Schor, 2013; Schor and Oliveira, 2016).

Cities of Territorial Responsibility strategically exercise functions in the network and region that go beyond their own characteristics, since they become loci of visibility for invisible acts, places of dispute and conflict, not only in terms of struggles for land and the use of resources but also, and mainly locally, in terms of disputes in the symbolic plane. It is in these cities of territorial responsibility that several territories overlap, both those that are established and those in the process of destruction, reconstruction, or constitution and reconstitution. The cities of Territorial Responsibility are often chosen as sites for the most diverse manifestations of these conflicts, from the murder of rural workers to suicides of young people (indigenous or otherwise). Because of their importance on the symbolic level, these cities are not simply "node cities" [*cidades polo*], as the literature conventionally treats them, because their importance goes beyond their economic dynamics, they do in fact exercise political, economic, cultural, and symbolic responsibility (Schor, 2013; Schor and Oliveira, 2016).

This typology permits a more substantial understanding of the role played by each individual city in the context of the urban network. It is understood that it is important to consider these relational differences strategically when planning and executing policies with the objectives of alleviating poverty and making ecosystems more sustainable. The city understood through the urban network is undoubtedly a useful scale for analyzing policy interactions in the Amazon.

Food as an Analytical Category

The urban network can be studied in a variety of ways, as shown in the institutional frameworks proposed above and in other studies (Schor et al., 2016). However, what are the variables that could help us relate urbanity with biodiversity and people? In our initial studies, one aspect caught our attention: the Food Staple Basket price analysis. The seasonal nature of relations within the urban network is especially noticeable when observing local food production. In dry periods, the production of floodplain crops increases, affecting the availability of local produce and prices. Lower prices for fresh food enter the local market, significantly diminishing Food Staple prices. When the plain floods, a lot of fresh produce comes by boat from Manaus. Food Staple prices increase, and so does the level of dependency in relation to Manaus and the monetary structure of the market. Most fresh vegetables come from Manaus during high-water season.

With this in mind, a research project was structured that aimed to calculate an indicator of the cost of the Food Staple basket for these cities in relation to the hydrological regime of the Solimões-Amazonas River. It was noted that the items in the National Food Staple Basket do not correspond with actual staple foods in the area studied, and therefore it was necessary to adapt the National Food Staple Basket into a Regional Staple Food Basket (RSFB). This idea was originally drafted by the Economic Development Commission for Amazonas State (CODEAMA), but after the dismantling of CODEAMA and the literal burning of archives in the 1990s, no further research was done on this theme. The main difference between the National Food Staple Basket and the one proposed by CODEAMA was that the latter was to include essential staple elements for the local population such as manioc flour, tomato, lemon, and fish, and exclude products not consumed locally, such as potatoes. Using this approach, data were collected for eleven cities along the Solimões-Amazon River. The results were impressive: the prices of the RSFB are weakly related to distance and transportation costs but strongly related to river seasonality. This is particularly the case for fresh vegetables that can be produced in the floodplains such as tomatoes (Moraes and Schor, 2010a, b).

In the low-water period, when food can be produced in the floodplains, the price of tomatoes, lemon, and manioc flour, as well as fish, decreased. This decrease was significant and reflected in the overall cost of living in cities. The expectation that transportation costs would be a determinant for the price of RSFB was fundamentally challenged, because as the seasonality of rivers is annual, traders anticipate the periods when access will be more difficult (dry) and manage to keep the prices of industrialized products stable. The major impact came from local agricultural production, with floodplain production supplying the cities and influencing prices (Aguiar et al., 2014). This demonstrated how a simple exercise to calculate the cost of the RSFB can illuminate issues so central to geography and generate knowledge about supply and eating habits in the cities of the Amazon in an interdisciplinary and innovative way. This was so much the case that the research methodology has since been used in many other contexts, such as classrooms and community groups, and was later transformed into a book focusing on teachers and showing how this simple practical exercise of monitoring food prices can act as an interdisciplinary project in schools (Moraes et al., 2016).

It was clear from the RSFB research, followed by a short questionnaire on eating habits among the local population, that diets are rapidly changing in the Amazon and the impact of the monetization of society is being clearly expressed in the structuring of food markets, where there have been significant changes in favor of processed and ultra-processed foods (Schor, 2015). The results showed how the people who live in riverine Amazonian cities are increasingly consuming industrialized chicken, egg, sausage, calabresa, etc., with bread, rather than fish and other regional products (Cost and Schor, 2015; Nardoto et al., 2011; Schor, 2015). These studies on the RSFB also showed the importance of analyzing fishing. The hydrological regime has a clear impact on fishing. High waters are difficult to fish because the fish "hide" in the flood forests, while in the low waters the fish are "trapped" in lakes. Fish is the main protein component of local diets in the Amazon (Dufour et al., 2016). Analyzing what people eat in the cities of the Solimões River in the ebb when there is plenty of fish confirms this pattern (Cost and Schor, 2015; Costa, 2015), and, even more strikingly, fish does not necessarily participate in market structures. The economy of affectivity, as Winklerprins (2005) suggests, is very active in riverine cities.

However, in the high-water period (usually from February to June) the RSFB increased in price due to the scarcity of locally produced fresh food. This fact, together with the greater access to processed food and the monetization of social relations due to federal income transfer policies, leads the population to low-cost canned foods, and chicken, eggs, and frozen chicken that come from Manaus. The hydrological regime establishes a connection between the cities, while river transport establishes physical links between them. The boats, locally called *recreios*, are responsible for transporting people and goods. The *recreios* buy and resell products, establishing themselves as another agent in the market structure and thus controlling the distribution and marketing of goods. Such differentiated functionality implies the high level of influence that transport has in the urban network of Amazonas. Regarding river seasonality, the main effects on flows are set by transport journey times, routes, and docking locations. The choices made by these companies regarding what to sell have an important impact on eating habits, as the boats function as mobile supermarkets and distributors. Processed and ultra-processed food, such as soft drinks, beer, and pre-cooked pasta, are the main products offered (figure 6.1).

Figure 6.1 Unloading a Boat in Tefé in 2011. *Source*: NEPECAB archives.

The study of eating habits introduces the approach of health geography. Public health is a topic of growing interest in the Latin American geographic community. The geography of health, since its inception, has been grounded in problem solving, allowing the identification of places and situations of risk, territorial planning of health actions and the development of prevention and health promotion activities. One of the main commitments of health geography in Brazil is to contribute to the strengthening of national health policy (SUS) and the reduction of social inequalities.

Geographers and epidemiologists have many characteristics in common. Both professions have long histories, often intertwined, of applying theoretical knowledge to practical problems. The development of medical geography was a result of the search for relationships between environmental conditions and diseases, while epidemiology developed as doctors, while still dedicated to the fundamentals formulated in antiquity by Hippocrates, became interested in the importance of the environment for quality of human life. It was this ecological perspective that laid the foundations of medical geography, especially for those inspired by Maximilien Sorre's theory of Pathogenic Complexes, according to which the ecological relationships between diseases and places would be mediated by biological agents, influenced by climate and other geographical factors acting as pathogens (Schor and Oliveira, 2016).

Sorre's ideas kept open this line of study and are now being incorporated in the most recent works of Latin American geographers studying health. However, there is a need for a more global view of the problem, since the focus on understanding disease is not sufficient for a comprehensive approach to public health; other concepts such as prevention, quality of life, and welfare of the population also need to be considered. Thus, there is a trend toward integration between eminently physical and biological approaches and the approaches of social geography. In this context, Latin American researchers have been greatly interested in the development of research projects involving broader scales. After all, in addition to being aware of the many health problems common to the American continent, the understanding of such problems and the capacity to intervene requires more analysis of the processes of landscape transformation. Analyzing how eating habits are changing rapidly, especially among the urban and rural poor, is an important perspective in a multi-scalar health geography approach.

In the Amazon this discussion gains other elements such as how the relation between society and nature is being molded by socio-environmental politics in a contradictory construction of economic development discourses that navigate between sustainability, austerity, and economic growth, ignoring the need for biodiversity conservation and having very little impact on poverty alleviation.

CONCLUSIONS

Efforts to promote economic inclusion and poverty alleviation have brought important benefits to the rural poor and to sustainable development efforts, including *Bolsa Família*, increased access to education and health, facilitated access to credit, the promotion of family farming, and access to electricity. However, these fairly modest efforts at socioeconomic inclusion have had uneven impacts within Brazil, having been undermined by the broader thrust of rural development policies and investments—which have privileged exclusionary development and the perpetuation of inequalities in income, wealth, and access to resources—and have had a limited impact on entrenched political and economic inequalities.

Moreover, many of these gains are being eroded by the current economic crisis and inflation. With the current economic and political crisis, there are indications that social and environmental issues are becoming increasingly marginalized in the political sphere. The crisis is being mined as a justification for a plethora of sweeping social, economic, political, and environmental reforms. While there is an acute need to rethink approaches to sustainability and economic development, and to question the premises and limitations to the notion of sustainable development, there is a danger in ushering deleterious reforms based on specious and myopic reasoning. Arguably, many of the existing problems are related to the implementation of policies, which do not consider specificities such as the urban network described in this chapter.

The socio-environmental problems faced today are dynamic, complex, and not fully understood. The solutions are far from simple, as they invariably involve tradeoffs between different needs and interests, but in effectively defining the problems and solutions there are many important lessons that can be drawn from the successes and failures of the past to provide a general roadmap to guide thinking, planning, and action around these issues. In developing guiding principles to cultivate more resilient and inclusive forms of land and resource management and use, four of the most important and interrelated overarching matters are the continued need for increased resource mobilization for government agencies responsible for matters related to environmental and traditional peoples; more effective and strongly institutionalized transversality in decision-making; more inclusive and diverse participation in socio-environmental decision-making; and a more expansive notion of growth that more fully integrates concerns about environmental sustainability and social inclusion. All of these need to take into consideration local diversity and the impact on populations on the local scale.

The promotion of more inclusive and sustainable forms of development requires more democratic participation in environmental governance and development planning in terms of poverty alleviation, addressing the issue of land concentration, addressing educational and infrastructural inequalities, promoting diversification of agricultural and extractive enterprises, and the continuation of efforts to promote an increased valuation of the agricultural production of food crops. A different perspective is needed, one in which contemporary forms of socio-environmental processes, such as urbanization in the Amazon and food sovereignty are key to policies that aim to promote economic resilience, poverty alleviation, and environmental sustainability.

The scale of the city is an important aspect of understanding poverty in the Brazilian Amazon. Unlike small rural communities, the Amazonian cities have a growing poor and vulnerable population, which is reflected in precarious housing, lack of social, cultural, and economic opportunities, and transformation of livelihoods. Violence is a harsh reality in these cities. The monetization of social relations in the Brazilian Amazon, in particular in the western section of Amazonas State, has clearly affected one aspect of local lives: eating habits. The consequences of this transformation have yet to be analyzed. What can be said is that local populations are becoming more dependent on processed and ultra-processed foods that are bought in the markets, bringing them closer in terms of eating habits to the urban poor in Brazil. This dependence has a clear nutritional-health impact and transforms food security into an issue when poverty comes into the picture. In the Amazon, where historically local populations were able to extract their food products from the forest and river, the increased dependence on money to buy food will transform poverty, reconfiguring the nature of food security by increasing its dependence upon and sensitivity to global markets and government social policies. Access to monetary revenues has many benefits but in some cases can bring more insecurity, as might be the case in the Brazilian Amazon. These changes also introduce complex, diverse, and poorly understood transformations in the character, dynamics, and intensity of human interactions with forests, rivers, and other habitats.

Changing eating habits from a traditional diet based on locally produced or extracted food from the forest and rivers into a supermarket-based diet has an impact not only in the increase of poverty but also in the loss of knowledge of the uses of resources from the forest and rivers. Will future generations, with their transformed eating habits, be able to identify and prepare a biodiverse meal? Without this knowledge, will the forest have a role in alleviating poverty? Will local populations still value the forests and aim for sustainable development, which maintains the forest, or will they prefer deforestation to grow cash crops and raise cattle? Research and public policy agendas should take these questions into consideration in order to link sustainability of ecosystems to poverty alleviation in the Amazon. Food sovereignty understood as participatory decision-making in a sustainable development program is key to any socio-environmental policy in a country where environmental threats and historical inequalities persist.

NOTES

1. We are cognizant that the transformative potential and inclusiveness of various political innovations in Brazil and other parts of Latin America in recent decades that aim for more participatory forms of policy-making and implementation have often been overstated. Councils (e.g., federal, state, and municipal environmental councils; watershed managements councils; and conservation unit councils) often have limited influence and can often be themselves exclusionary (e.g., privileging technical experts and private sector representatives over local resource users), falling short of the promise of participatory democracy and inclusion that some advocates, political actors, and academics ascribe to them (see Wampler and McNulty, 2011 for a discussion of the limitations of participatory governance initiatives). Despite their shortcomings, in many cases they have been successful mechanisms to broaden participation and to fortify the influence and monitoring capacity of civil society actors, as in the case of Conama. Similarly, there are also instances where local participation in environmental licensing for large infrastructure projects in Brazil has also effectively enabled increased input from and responsiveness to local stakeholders.

2. This relationship, of course, has not been entirely harmonious and tensions between local resource users, natural scientists, social scientists, and environmentalists persisted, but a broadly shared political agenda was created, which sought to valorize local livelihoods and environmental protections, while attempting to reign in the socio-environmental excesses of more predatory and extensive extractive and agricultural enterprises. Despite often widely divergent understandings of sustainability, development, and growth, these heterogeneous groups converged around a (contested and dynamic) project of sustainable development.

BIBLIOGRAPHY

Aguiar, D. G., Sampaio, F. P. R., Filizola, N. P., and Schor, T. 2014. Cesta Básica Regionalizada e Variabilidade Hidrológica: O Caso das Cidades de Manacapuru (AM) e Óbidos (PA). In: *Dinâmica Urbana na Amazônia*, Schor, T. (ed.). Valer: Manaus, pp. 151–65.

AMAZONAS—SEPLAN. 1976. *A Rede Funcional Urbana do Amazonas*. SEPLAN: Manaus.

AMAZONAS—SEPLAN. 1990. *Estudo de hierarquia urbana: Estado do Amazonas*. SEPLAN: Manaus.

Banerjee, O., MacPherson, A. J., and Alavalapati, J. 2009. Toward a Policy of Sustainable Forest Management in Brazil: A Historical Analysis. *Journal of Environment and Development*, 18(2), 130–53.

Becker, B. K. 1974. A Amazônia na Estrutura Espacial do Brasil. *Revista Brasileira De Geografia*, 36(2), 3–33.

Becker, B. K. 1978. Uma Hipótese sobre a Origem do Fenômeno Urbano numa Fronteira de Recursos no Brasil. *Revista Brasileira de Geografia*, 40(1), 160–84.

Becker, B. K. 1990. *Amazônia*. Ática: São Paulo.

Becker, B. K. 1999. Os Eixos de Integração e Desenvolvimento e a Amazônia. *Território*, 6, 29–42.

Becker, B. K. 2013. *A Urbe Amazônida: A Floresta e a Cidade*. Garamond Universitária: Rio de Janeiro.

Botía, C. G. Z. (ed.). 2012. *Espacios Urbanos y Sociedades Transfronterizas en la Amazonia*. Universidad Nacional de Colombia: Bogotá.

BRASIL—IBGE. 2007. *Regiões de Influênciasdas Cidades*. Instituto Brasileiro de Geografia e Estatística: Rio de Janeiro.

Browder, J. and Godfrey, B. J. 1990. Frontier Urbanization in the Brazilian Amazon: A Theoretical Framework for Urban Transition. *Yearbook of the Conference of Latin Americanist Geographers*, 16, 56–66.

Bunker, S. G. 1984. Modes of Extraction, Unequal Exchange, and the Progressive Underdevelopment of an Extreme Periphery: The Brazilian Amazon, 1600–1980. *American Journal of Sociology*, 89(5), 1017–64.

Carlos, A. F. 1996. *O Lugar no/do Mundo*. Hucitec: São Paulo.

Carneiro, M. C. 1998. Regiões de Influência das Cidades: Rede de Lugares Centrais e Áreas de Atuação das Cidades Brasileiras. IBGE/DEGEO: Rio de Janeiro.

de Castro, F. 2014. Environmental Policies in the Lula Era: Accomplishments and Contradictions. In: *Brazil under the Workers' Party*. Palgrave Macmillan: London, pp. 229–55.

Conklin, B. A. and Graham, L. R. 1995. The Shifting Middle Ground: Amazonian Indians and Eco-Politics. *American Anthropologist*, 97(4), 695–710.

Corrêa, R. L. 1987. A Periodização da Rede Urbana da Amazônia. *Revista Brasileira de Geografia*, 49(3), 39–69.

Costa, E. A. C. and Schor, T. 2016. Transformações e Permanências de Hábitos Alimentares em Idosas nas Cidades de Tefé, Alvarães e Uarini. In: *Geografia da Saúde: Ambientes e Sujeitos Sociais no Mundo Globalizado*, Oliveira, J.A. and Souza, G.A. (eds.). EDUA: Manaus, pp. 199–214.

Dematteis, G. and Sforzi, F. 1999. *The Italian Urban System: Towards European Integration*. Ashgate: Burlington.

Drummond, J. and Barros-Platiau, A. F. 2006. Brazilian Environmental Laws and Policies, 1934–2002: A Critical Overview. *Law and Policy*, 28(1), 83–108.

Faissol, S. 1974. Tipologia das Cidades e Regionalização do Desenvolvimento Econômico: Um Modelo de Organização Espacial do Brasil. *Boletim Geográfico*, 30, 25–57.

Gasper, D. 2005. Securing Humanity: Situating "Human Security" as Concept and Discourse. *Journal of Human Development*, 6(2), 221–45.

Godfrey, B. J. 1988. Frentes de Expansão na Amazônia: Uma Perspectiva Geograficohistórica. *Geosul*, 6, 7–20.

Hochstetler, K. and Keck, M. E. 2007. *Greening Brazil: Environmental Activism in State and Society*. Duke University Press: Durham.

IPEA/IBGE/UNICAMP/NESUR. 1999. *Caracterização e Tendências da Rede Urbana do Brasil*. Unicamp/Nesur/ Ipea/ IBGE: Campinas.

Marinho, T. P. and Schor, T. 2012. Nos Interflúvios do Rural e do Urbano na Amazônia: O Caso de Codajás-Amazonas, Brasil. *Acta Geográfica (UFRR)*, 6, 69–81.

Moraes, A. C. R. 2000. *Bases da Formação Territorial do Brasil: O Território Colonial Brasileiro no 'Longo' Século XVI*. Hucitec: São Paulo.

Moraes, A. O. and Schor, T. 2010a. Mercados, Tabernas e Feiras: Custo de Vida nas Cidades na Calha do Rio Solimões. *Mercator*, 9, 101–15.

Moraes, A. O. and Schor, T. 2010b. Redes, Rios e a Cesta Básica Regionalizada no Amazonas, Brasil. *Acta Geográfica (UFRR)*, 7, 79–89.

Moran, E. F. and Brondizio, E. S. 2000. *Human Ecology from Space: Ecological Anthropology Engages the Study of Global Environmental Change (draft)*. Dartmouth College: Hanover.

Oliveira, J. A. 2000. *Cidades na Selva*. Valer: Manaus.

Oliveira, J. A. and Schor, T. 2008. Das Cidades da Natureza à Natureza das Cidades. In: *Cidades Ribeirinhas na Amazônia: Mudanças e Permanências*, Trindade Jr, S.-C.C. and Tavares, M.G.C. (eds.). UFPA: Belém, pp. 15–26.

Oliveira, J. A. and Schor, T. 2010. Urbanização na Amazônia: O Local e o Global. In: *GEEA—Grupo de estudos estratégicos Amazônicos—Caderno de Debates*, Val, A.L. and Santos, G.M. (eds.). INPA: Manaus, pp. 147–89.

Pegler, L. 2015. Peasant Inclusion in Global Value Chains: Economic Upgrading but Social Downgrading in Labour Processes? *The Journal of Peasant Studies*, 42(5), 929–56.

Pinedo-Vasquez, M., Ruffino, M. L., Padoch, C., and Brondízio, E. S. (eds.). 2011. *The Amazon Várzea: The Decade Past and the Decade Ahead*. Springer Science+Business Media.

Rappaport, R. A. 2001. *Ecology and the Sacred: Engaging Anthropology*. University of Michigan Press: Michigan.

Roberts, J. T. 1992. Squatters and Urban Growth in Amazonia. *Geographical Review*, 82(4), 441–57.

Rodrigues, P. K. A. and Santana, P. V. 2015. Finanças Públicas Municipais e Saneamento Básico em Coari/AM. In: *Dinâmica Urbana na Amazônia Brasileira*, volume 2, Schor, T. and Santana, P. V. (eds.). Valer: Manaus, pp. 79–98.

Ronchail, J., Schor, T., Sabot, M., Pinheiro, H., Filizola, N., Gomez, P., Drapeau, G., Michot, V., Guyot, J. -L., Martinez, J. -M., and Sultan, B. 2016. Hydrologie et Production Agricole dans le Nord-ouest de l'Amazonie. *Bulletin de l'Association de Géographes Français*, 3, 270–86.

Santana, P. V. 2015. As Corporações e o Espaço Urbano Amazonense. In: *Dinâmica Urbana na Amazônia Brasileira*, volume 2, Schor, T. and Santana, P. V. (eds.). Valer: Manaus, pp. 13–38.

Sassen, S. 2005. The Many Scales of the Global: Implications for Theory and Politics. In: *Critical Globalization Studies*, Appelbaum, R.P. and Robinson, W. I. (eds.). Routledge: New York, pp. 82–93.

Schor, T. 2016. Urbanização e Modernização do Território: A Geografia da Saúde e da Alimentação no Coração da Floresta Amazônica. In: *Geografia da Saúde: Ambientes e Sujeitos Sociais no Mundo Globalizado*, Oliveira, J. A. and Souza, G. A. (eds.). EDUA: Manaus, pp. 179–98.

Schor, T. and Azenha, S. G. 2017. Ribeirinho Food Regimes, Socioeconomic Inclusion and Unsustainable Development of the Amazonian Floodplain. *EchoGéo*, 41. Available at: http://echogeo.revues.org.

Schor, T. and Costa, D. P. 2011. Rede Urbana na Amazônia dos Grandes Rios: Uma Tipologia para as Cidades na Calha do rio Solimões—AM. In: *As Cidades e a Urbanização no Brasil: Passado, Presente e Futuro*, Pereira, E. M. and Dias, L. C. D. (eds.). Insular: Florianópolis, pp. 129–46.

Schor, T. and Moraes, A. O. 2011. Programas de Pesquisa em Meio Ambiente e o Urbano: Um Ensaio sobre a Ausência. *Revista Geonorte*, 2, 1–24.

Schor, T. and Oliveira, J. A. 2011. Reflexões Metodológicas sobre o Estudo da Rede Urbana no Amazonas e Perspectivas para a Análise das Cidades na Amazônia Brasileira. *Acta Geográfica (UFRR)*, sp. vol., 15–30.

Schor, T. and Oliveira, J. A. 2016. Parintins: A Geografia da Saúde na Formação da Cidade Média de Responsabilidade Territorial no Amazonas. In: *Parintins: Sociedade, Territórios e Linguagens*, Bartoli, E., Muniz, C., and Albuuerque, R. (eds.). UFAM: Manaus, pp. 35–58.

Schor, T., Azenha S. G., and Bartoli, E. 2018. Contemporary Urbanization in the Brazilian Amazon: Food Markets, Multisited Households and Ribeirinho Livelihoods. *Confins*, 37. Available at: http://journals.openedition.org/confins/15682; doi:10.4000/confins.15682.

Schor, T., Costa, D. P., and Oliveira, J. A. 2009. Cidades, Rede Urbana e Desenvolvimento na Amazônia dos Grandes Rios. In: *Pequenas e Médias Cidades na Amazônia*, Trindade Jr., S.-C.C., Carvalho, G., Moura, A. and Gomes Neto, J. (eds.). FASE/UFPA: Manaus, pp. 35–58.

Schor, T., Marinho, R., Costa, D. P., and Oliveira, J. A. 2014. Cities, Rivers and Urban Network in the Brazilian Amazon. *Brazilian Geographical Journal*, 5(1), 258–76.

Schor, T., Oliveira, J. A., Moraes, A. O., and Santana, P. V. 2016. Apontamentos Metodológicos sobre o Estudo de Cidades e de Rede Urbana no Estado do Amazonas, Brasil. *PRACS: Revista Eletrônica de Humanidades do Curso de Ciências Sociais da UNIFAP*, 9(1), 9–35.

Tavares-Pinto, M. A. 2015a. A Caça e a Pesca na Beira de Tabatinga: Um Estudo do Mercado de Recursos Naturais na Tríplice Fronteira Brasil-Colômbia-Peru. MSc dissertation. UFAM: Manaus.

Tavares-Pinto, M. A. 2015b. Vulnerabilidade Fiscal e Rede Urbana no Amazonas: Um Estudo para as Cidades Médias na Calha do Rio Solimões-Amazonas, Amazonas Brasil (2000–2010). In: *Dinâmica Urbana na Amazônia Brasileira*, volume 2, Schor, T. and Santana, P.V. (eds.). Valer: Manaus, pp. 59–77.

Trindade JR., S.-C. C. 2011. Diferenciação Territorial e Urbanodiversidade: Elementos para Pensar uma Agenda Urbana em Nível Nacional. *Cidades*, 7, 227–55.

Trindade JR., S.-C. C. 2013a. Das "Cidades na Floresta" às "Cidades da Floresta": Espaço, Ambiente e Urbanodiversidade na Amazônia Brasileira. *Papers do NAEA*, 321, 1–22.

Trindade JR., S.-C. C. 2013b. Uma Floresta Urbanizada? Legado e Desdobramentos de uma Teoria sobre o Significado da Cidade e do Urbano na Amazônia. *Espaço Aberto*, 3, 81–100.

Turner, T. 1995. An Indigenous People's Struggle for Socially Equitable and Ecologically Sustainable Production: The Kayapó Revolt against Extractivism. *Journal of Latin American Anthropology*, 1(1), 98–121.

Turner, T. 1999. Indigenous Rights, Environmental Protection, and the Struggle over Forest Resources in the Amazon: The Case of the Brazilian Kayapo. In: *Earth, Air, Fire, Water: Humanistic Studies of the Environment*, Conway, J. K., Keniston, K., and Marx, L. (eds.). University of Massachusetts Press: Boston, pp. 145–69.

Turner, T. 2013. The Role of Indigenous People in the Environmental Crisis: The Example of the Kayapo of the Brazilian Amazon. *Perspectives in Biology and Medicine*, 36(3), 526–45.

Viola, E. 2013. Transformations in Brazilian Deforestation and Climate Policy since 2005. *Theoretical Inquiries in Law*, 14(1), 109–24.

Wampler, B. and McNulty, S. 2011. *Does Participatory Governance Matter. Exploring the Nature and Impact of Participatory Reforms.* Woodrow Wilson International Center for Scholars.

Winklerprins, A. M. G. A. and Souza, P. S. 2005. Surviving the City: Urban Home Gardens and the Economy of Affection in the Brazilian Amazon. *Journal of Latin American Geography*, 4(1), 107–26.

Chapter 7

Rethinking Fluid, Complex, and Uncertain Poverty in Amazonian Ecosystems in Bolivia and Brazil

Sergei Shubin

INTRODUCTION

The recent reformulation of poverty in social studies reflects a broader understanding of this concept as a complex set of practices and experiences, unfolding through interrelationships between local cultures, opportunities, and lifestyles (Alcock, 1997; Asen, 2002; Danziger and Haveman, 2002; Levitas, 2001; Room, 1995). In particular, different commentators stress the need to understand sociocultural mechanisms in the production of poverty and to develop a holistic approach to understanding disadvantage that connects material conditions of living with the lack of specific forms of cultural knowledge, skills, and social contacts valued by marginalized groups (Appadurai, 2004; Lin and Harris, 2008; Robinson and Oppenheim, 1998; Sen, 1999, 2006). Despite broader changes in thinking about poverty that reflect its diversity and complexity, anti-poverty policies still tend to overemphasize structuring of opportunity in the form of employment, services, housing and income, and to prioritize responses associated with addressing these standard measures of poverty. As Rao and Walton (2004) note, implementation of anti-poverty programs has often been undermined by the perceived tradeoff between poverty alleviation and preservation of cultural or sociocultural values. As a result, environmental marginality, cultural materials, and social practices creating poverty still remain on the margins of poverty-reduction strategies (World Bank, 2002) and are seen as a "minor issue," often overlooked in favor of tackling material and economic deprivation (Alkire, 2004: 207). This chapter contributes to the broader discussions in this book by unsettling the center/margins binaries and re-examining the mechanisms producing dominant representations, priorities, and frontiers in development discourse in the Amazon.

Interpretations of different poverties experienced by the disadvantaged people become even more complex when they are considered in relation to environmental change. It is widely acknowledged that degradation of ecosystem services, which are seen as the benefits that people derive from ecosystems (MEA, 2005), exacerbates poverty, mainly because the well-being of those in need tends to be dependent on the provision of these services (Shackleton and Shackleton, 2012). Literatures exploring links between environmental degradation and well-being have often focused on understanding poverty as a consequence of uneven processes of transformation of natural capital (forests, water, and soils) into other forms of capital (housing, education), and "trade-off" between economic and ecological components of well-being (Borner et al., 2007; Carpenter et al., 2009; Sjostedt, 2012). In this context, dependence on the environment has often been conceptualized in negative terms, with both environmental change and poverty discursively constructed as a set of problems associated with rural living (Dasgupta et al., 2005). As Suich et al. (2015) state, many of these studies concentrate mainly on material income at the expense of other aspects of well-being and do not develop a multidimensional approach to poverty. As a corollary, poor people often find themselves in a situation of double disadvantage when the loss of ecosystems services resulting from deforestation is compounded by misunderstandings of poverty within institutional structures dealing with sustainable ecosystem development and improvement of well-being (Rao and Walton, 2004). In support of the broader arguments in this book, the chapter re-imagines existing policy frontiers, ethical boundaries, and explores complex relationship between the poor people and the environment at the limit of development discourse.

In the Latin American context, these discussions are echoed by a stream of literatures that are often described as "pobretología" or "povertology" (Booth et al., 2006). Over the last forty years, conceptualizations of poverty have developed from earlier approaches exploring the marginality of rural people in the process of modernization to a "new rurality" perspective stressing the importance of external factors (globalization, market liberalization, migrations and agri-industrial change) in shaping the well-being of rural households (Kay, 2006). The recent turn toward a rural livelihoods approach reflects growing interest in developing a broader understanding of poverty beyond political economic definitions, focusing attention on the agency of the poor in developing different forms of capital (including natural capital), and the historical dynamics of poverty (Bebbington, 2004; de Haan and Zoomers, 2005). While attempts to model complex deprivation formulas like indicators of multidimensional poverty, such as those carried out by Paes de

Barros et al. (2006) in Brazil and Ballon and Krishnakumar (2008) in Bolivia, help in highlighting the scope and magnitude of poverty, they often overlook "messy" cultural and social problems experienced by rural people that cannot be categorized in a measurable form. Despite acknowledging the complexity of poverty, these approaches tend to concentrate on the structural causes of rural disadvantage (such as uneven land distribution and power relations, cf. Kay, 2006) and less on the nontangible aspects of poverty reflected in people's "capabilities" and potentialities in dealing with deprivation (Battiston et al., 2013). As a result, poverty emanating from social and environmental marginality, which does not conform to the obvious policy-related headings or deprivation indices, tends to be overlooked and poor people with "unusual" or conflicting experiences of disadvantage can drop out of policy networks.

This chapter addresses this gap by drawing attention to disconnections between anti-poverty policy-making and the everyday experiences of disadvantage expressed by rural people in the Amazon. It builds on the results of the ESPA-funded project (NE/I004467/1) conducted in Bolivia and Brazil in 2010–2011, which linked ecosystem services and poverty, focusing on the experiences of the forest-dependent poor at the agricultural frontier. It starts with a brief account of the changing formulations of poverty in policy-making controlling equitable management of forest ecosystems in the Amazon. It then considers the mechanisms of channeling poverty into policy discourses and discusses alternative ways that challenge categories, oppositions, and frontiers of development. By engaging with the poverty experiences of people in the Amazon, it then offers critical analysis of the links between everyday poverties in Bolivia and Brazil, their articulation in policy-making, and their translation into anti-poverty mechanisms.

POVERTY AND POLICY-MAKING IN BOLIVIA AND BRAZIL

Reflecting the changing interpretations of poverty in academic literatures, the meanings attached to poverty in policy-making in Bolivia and Brazil have also been changing. Traditional discursive treatments of poverty within policy-making, which emphasize normative translation of disadvantage, have dominated various modernization programs in the Amazon for the last forty years (see the excellent discussion in Kay, 1989). In early development policies, poverty was couched in forms of marginality, where the poor were seen as unable or unwilling to adequately participate in the initiatives governing forest access and use (Bicalho and Hoefle, 2010). From this perspective, policy-makers determined poverty and appropriated definitions of disadvantage by prescribing the norms and boundaries of "poor" identities, so that "marginality and poverty were largely reduced to certain attributes of individuals" (Kay, 2006: 460). These political constructions attempted to name a messy set of experiences of disadvantage and create what Deleuze and Parnet (2007: 96) describe as constructions of "rigid segmentarity." In this situation, mechanisms of naming disadvantage in policy-making divide all senses and practices of poverty into recognizable segments of moneyed/poor, dependent/independent, un/employed, and reproduce the sameness of poverty inherent in the structuring of opportunities. As Deleuze and Parnet (2007, original emphasis) would suggest, such policy-making deploys "the *devices of power* which code the diverse segments . . . overcodes them and regulates their relationships" with the environment, while separating cultural, social, and natural dimensions of disadvantage. In Brazil, conflicting modernization agendas provided by the federal and state authorities regarding land use and development further compounded definitions of poverty (Schmink and Wood, 1992). In Bolivia, contested development initiatives and the failure of formal government mechanisms to distribute resources equally diluted understandings of poverty (Redo et al., 2011). As a result, poverty has often been inadequately translated within policy-making, reducing it to a set of coherent appearances (stereotypes) and linear regularities to the exclusion of diverse populations with varied experiences of disadvantage.

In response to these treatments of poverty, more recently policy-makers in Bolivia and Brazil have increasingly tended to conceptualize rural disadvantage in terms of "social exclusion." These policy approaches have helped "to break with the economistic and individualistic parameters of traditional concepts of poverty," focusing on the relational character of rural problems and, in particular, closer connections between society and nature in environmental decision-making (Munck, 2005: 26). Reformulating poverty as social exclusion has also addressed issues of diversity in rural life, the complexity of socio-natural transformations in the Amazon and the heterogeneity of problems experienced by rural people (Gomes et al., 2012). These policy mechanisms attempted to focus on "differential incorporation" of the poor into economic and political processes by tackling three dimensions of disadvantage: economic, political, and cultural (Altamirano et al., 2003: 21). However, an interest in addressing the economic exclusion of disadvantaged people and their marginalization from the productive system still dominates policy discussions in Latin America (Kay, 2006). In this context, "peasants [are] depicted as being economically backward because they chose to be so, and for non-economic reasons to do with culture and the nature of peasant society itself" (Brass, 2002: 3). Unsurprisingly, social welfare and environmental policy-making continue to prioritize development as a centralized mechanism of rational power distribution, so that the multiple nature of poverty processes (including nonrational emotions and affective movements) are often ignored. Inter-organizational approaches to regional development in Latin America often rely on technological structures that reduce the difference and heterogeneity of different groups of actors to orderly, predictable relationships between specific organizations, thus risking oversimplification and privileging certain forms of poverty over others (Batiston et al., 2013). As a result, in Brazil both the federal and regional governments prioritize "traditional" poor rural groups at the expense of other disadvantaged people, while failing to consider the significance of cultural and emotional attachments to the environment in the development of rural problems often creates misunderstandings around gendered access to land and forest resources (Barsted, 2005). In Bolivia, the rural poor tend to be defined as "landless," with environmental policies developed in the "best interests" of landowners but not vulnerable people (Rudel, 2007).

Critics of the "social exclusion" approach in development have advocated using the concept of social capital as a way of understanding rural disadvantage. One of the arguments for using this approach in the Amazonian context was the important opportunity to divert attention from economic capital to other forms of capital, particularly capital embodied in natural resources (Kay, 2006). Natural capital has been described in terms of benefits and losses of living using land and water resources and their biodiversity, in the form of food production, ensuring livable conditions and security and other means of supporting human well-being (Kareiva et al., 2011). Social capital is typically described as "resources that come from the structure of social relationships," through measures increasing civic participation, support, trust, and mutually beneficial collective action (Anderson and Bell, 2004: 235). Development professionals have argued that natural capital can also be turned into financial resource, or helped to develop social links to address social ills such as poverty. "It is argued that poor people are often able to obtain some material benefits from ecosystems" (MEA, 2005: v), harnessing ecosystem services to provide specific contributions (material and cultural) to the poor. The ideas behind social-ecological systems research are used to develop analysis of the political economy of access to ecosystem services and their use, with the anthropocentric premise defining poor people as both threatening to and excluded from using environmental resources (Fischer et al., 2014). From this perspective, ecosystem services are evaluated in terms of their usefulness as a source of either dependency or potential wealth for the poor and treated as a product (rather than a process) subject to specific mechanisms of control and exchange. Within environmental conservation strategies poverty is constructed mainly in terms of objective and measurable structures (e.g., linked to the Payment for Ecosystem Services framework), ignoring experiences, histories, and changeable living relations between poor people and the environment (Robertson, 2012). The logic of calculation and abstraction used in determining the utility of ecosystem services leads not only to the segregation of nature in different typologies of productivity, but also to attempts to define the value of poverty and poor people themselves.

In Brazil, development policies often determine rural poverty in relation to people's ability to demonstrate "useful" and "necessary" improvements to the land, so that forest-covered lands and traditional people living on them are often deemed "unproductive" and subject to development (Araujo et al., 2009). Persistent ideology linking deforestation in the Brazilian Amazon with progress presents development as a calculative and rationalistic process, where the poor are also portrayed as coherent agents making rational choices about their engagement with the environment (Hirakuri, 2003). More recently, "new developmentalism" agenda, first defined by Brazilian economist Luiz Carlos Bresser-Pereira (2007), focused on state activism and support of domestic companies in technological innovation, investment opportunities at the global scale, while also commitment to maximizing labor resources and addressing existing inequalities (Ban, 2013). This approach, implemented with political support from the Brazil's Workers' Party (PT), has been criticized for destruction of the environment, limiting scope for overcoming poverty, and deepening social conflict (dos Santos, 2019). The task of balancing human needs, such as poverty alleviation, and supporting nature (the nonhuman) has been undermined by conflicting interests, divergent values, and contradictory political interventions (Bratman, 2019).

Similarly, in Bolivia both land and people are expected to contribute to economic development and comply with their assumed "socio-economic function" (Redo et al., 2011). Development initiatives in Bolivia have not only promoted forest clearing, but also destroyed the traditional relationships poor people had with the environment and discredited nontimber users of the forests, thus increasing deforestation and aggravating rural poverty (McDaniel et al., 2005). Recent retrogressive legal reforms in Bolivia undermined the earlier programs for environmental protection, which affected rural poverty (Schilling-Vacaflor, 2016). While the new Bolivian government of Movement Towards Socialism (MAS) since 2006 set to tackle political inclusion and redistribution of resources in favor of marginalized communities, these measures often presented a "repetition" of the previous development programs (Cusicanqui, 2012: 12). Overall levels of moderate and extreme poverty in Bolivia between 2007 and 2015 have been reduced due to the economic boom and extractivist-based (oil, gas, monocrop agribusiness) growth (Beverinotti, 2018). However, overreliance on large-scale projects in strategic sectors and limited sustainability of social policies relegated the environment to an inferior position compared to the "productive" public economy (Wanderley, 2018). Unsurprisingly, the translation of nature and social relations into capital forming part of enumerated models have tended to overlook complex relations between society and nature, reduced different experiences of poverty to "unproductive" stereotypes and assumed predetermined identities for the poor. To address these issues, this chapter attempts to go beyond the language of investment, exchange, and measurement (such as those used in PES approach) in describing rural people's experiences of poverty in the Brazilian and Bolivian Amazon. Unlike existing approaches to rural poverty, it highlights the role of spatiotemporal uncertainties in relationships between needy people and the environment in producing disadvantage. Building on poststructuralist ideas about different understandings of space and time, it offers alternative ways of thinking about poverty in the Amazon.

RETHINKING RURAL POVERTY IN THE AMAZON

First, the chapter argues for a *fluid* articulation of rural poverty. It builds on an understanding of time and space as always developing and becoming, reflecting the "unpredictability and precariousness of lives" (Horschelmann, 2011: 379). In this context, poverty should no longer be seen as an entity or a product but rather a process co-constructed by different actors (human and nonhuman) and expressing a sense of lack in both material and nonmaterial forms. On the one hand, lack here relates to both incompleteness of the "poor" subject, who is always situated with others in the world and is always incomplete without recognition from others, unstable and internally fractured (Shubin et al., 2014). This approach to poverty challenges the separation of the poor from the environment and takes into account "unpredictable connections among materials and processes, forces and events" creating socionatural disadvantage in the Amazon (Lee and Motzkau, 2011). On the other hand, this lack relates to the impossibility of spatiotemporal completion of the subject in the system, which is not complete and closed, but continuously evolving. Relationships between the poor and the environment are not limited to rational systems of exchange, but rather create excessive energies and affective connections irreducible to calculation or totalization. This approach undermines assumptions about ownership and mastery of the environment and challenges utilitarian logic and binaries of poor/nonpoor, productive/unproductive, useful/not-useful evident in existing development policies. In expressing fluid poverty and attending to its dynamism, the chapter addresses the call by Skeggs (2004) for "thinking beyond exchange-value" and challenging the restricted economy of meaning (negative/positive) in determining both poverty and the environment.

Second, the chapter expresses *complex* poverty assembled through the often-unexpected contributions of multiple (human and nonhuman) actors. It draws on the conceptualization of "assemblage" by Deleuze and Guattari

(1987: 88) referring to the arrangements of "bodies, actions and passions" that combine and recombine (assemble) to create new connections and ranges of flows. This approach attempts to articulate the multiplicity of relations between always emerging elements of poverty, including feelings, artefacts, and symbols, which tend to fall into the "grey" area in traditional development policies.[1] Drawing on Deleuze and Guattari's ideas, the analysis of "socionatural assemblages" of meanings, materials, social orders has already been successfully used to explore the production of vulnerabilities in the Global South (McFarlane, 2009; Doshi, 2013; Ranganathan and Balazs, 2015). Although not applied specifically to poverty studies, assemblage-thinking is particularly fitting for the analysis of deprivation as it develops alternative formulations of injustice and helps "to recover the agency of non-humans" in capital-nature relations (Ranganathan, 2015: 1305). In this context, difference-producing poverty should not be seen as a given (in the form of indices and monetary values), but instead perceived as continuously recreated, blurring the established boundaries and frontiers of disadvantage. Deleuze (1986: 67) challenges the power of negation (as a "difference from") and advocates affirmation (as "difference in itself"), which helps to re-articulate poverty. Instead of reducing poverty to binary oppositions and difference from accepted measurements such as the poverty line, this can help to focus on the creation of complex poverty, the process of becoming different, the very dynamism of change. Poverty in this context develops as neither subject nor object and can be best described in terms of determinations of magnitude. In this context, an assemblage of poverty can be said to emerge as "a new means of expression, a new territorial/spatial organisation . . . a new reality, by making numerous, often unexpected, connections" (Parr, 2010: 19). The chapter opens space for the interrelation of forces producing different poverties to be expressed, rather than subordinating them to representational logic as a matter of opposition or analogy. It works in and through complex poverty by means of recreating combinations of its elements (both material and non-material) and re-articulating the links between them.

Third, the chapter attends to *relational* and *uncertain* poverty, which is not defined solely from an anthropocentric position. Traditionally, poverty is described as a lassitude or weakness to be overcome, a failed condition calling for restorative action, or resulting from withdrawal from the world (Harrison, 2008). In this context, poverty is constructed in relation to the "outside" world, which is unified, orderly, and "simply an experience of our perceived environment" (Linglis, 1996: 13). Drawing on the work of Blanchot (1995), this chapter questions the existence of the "outside" world that is made known through the application of reason and reduced to the conceptually structured experience incorporated into human knowledge. On the one hand, it argues that the complex and heterogeneous conceptualizations of poverty advocated for earlier cannot simply be brought into knowledge. Poverty involves nonintentional and confusing relations with others (people and things), interruptions, and uncertainties that cannot be simply incorporated into rational discourse (Shubin and Sowgat, 2019). In its openness to the unforeseen it exceeds the ability of knowledge to contain and make it reasonable—it involves what Bataille (1988: 51) calls "non-knowledge." On the other hand, poverty in the form of intuition, openness, and susceptibility is not something that can be owned by the (anthropocentric) actor: "it is vulnerability, enjoyment, suffering, whose status is not reducible to the fact of being put before a spectator subject" (Levinas, 1991: 63). This approach undermines assumptions about mastery of the relationship between the poor and the environment and challenges the development logic of self-enclosed agents making autonomous, rational, and reasonable choices in relation to nature. Drawing on these theoretical assumptions, the chapter explores relationships of poverty that develop in relation to exposure to others, uncertain connections and practices that create possibilities for different ways of knowing at the limits of reason.

CONTEXT AND METHODS

This analysis of poverty draws on the findings from the international project conducted in 2010–2011 in the Brazilian and Bolivian Amazon, which focused on understanding the links between ecosystems and change and poverty. In Brazil, research activities focused on rural areas around the city of Santarém in the State of Pará, an area exposed to different development initiatives with devastating environmental consequences, particularly in the form of large-scale deforestation (Stone, 2006). This area has been the focus of several policies aiming to rebuild forest-based ecosystem services and increase environmental benefits for poverty alleviation (Börner et al., 2007). In particular, subsidized credits for production services to enhance ecosystem services as part of the federal *Proambiente* program, combined with compensatory cash transfer mechanisms, contributed to a reduction in recorded levels of poverty (Soares et al., 2006). On the state level, until recently, expanding soybean production was the main development objective, while environmental programs (such as the Tropical Forests Pilot Program) tended to focus on decentralization of forest management and attempts to involve the poor in decision-making (Stone, 2006). However, lack of co-coordination of environmental and anti-poverty efforts, rural violence, and conflicts over land distribution often meant that small farmers received limited benefits from the development programs. Despite expectations that economic growth would contribute to a reduction in poverty in the area, Pará's forest-based populations have been negatively affected by expanding agricultural production, mining, logging, and rapid urban development. As a result, in rural Pará poverty measured as the proportion of the population living below the poverty line was reported as 55 percent in 2014 (IPEADATA), only a slight reduction from the level of 58.6 percent reported in 2000 (Verner, 2004).

In Bolivia, research focused on rural areas around the city of Cobija in the Pando Department, where a poverty line defining household consumption at a level of US$ 2 per day was used to classify 72.4 percent of people as poor in 2001 (O'Hare and Rivas, 2007). Since 1996, important changes in the country's forest regime involving the decentralization of forestry management (using Forest Management Plans), accompanied by extensive land tenure reform, have significantly affected rural poverty. These changes resulted in increased support for local co-operatives, wider opportunities for nontimber extraction and support for sustainable forms of subsistence agriculture that positively affected the forest-dependent poor (Zenteno et al., 2013). In particular, poverty indices using income–expenditure data suggest that extreme rural poverty (defined in terms of minimal nutritional requirements for adults and local eating habits in relation to income distribution) in Pando dropped from 59 percent in 1994 to around 37 percent in 2002 (Spatz, 2006), and then remained largely unchanged until 2010 (Vargas and Garriga, 2015). Gains in service provisions (such as education and health facilities) have been significant, but they did not always affect smaller towns and villages, putting the forest-dependent poor at a relative disadvantage. Furthermore, development initiatives in Pando have been accompanied by significant in-migration of poor peasants from the Bolivian mountains, leading to increased pressure on available resources, forest clearances, and exacerbation of poverty for traditional residents of this region. Due to the persistence of structural inequalities, poor people in rural Pando, particularly in remote municipalities like Cobija, continue to experience extreme poverty and there has been little progress in terms of the numbers of people whose basic needs are not being met (O'Hare and Rivas, 2007).

To understand the poverty experiences of rural people in Pando and Pará, this analysis builds on 119 semi-structured interviews developed in several rural locations to encourage in-situ reflections on disadvantage in "small-scale, isolated, rural communities" (Jackson, 1983: 40). Rural people were interviewed in their native languages (Spanish and Portuguese), with the help of interpreters, in their villages (including remote communities of the extractive reserve, RESEX near Santarém, where subsistence and extractive practices are allowed and encouraged). Conversations were conducted in respondents' homes to maintain confidentiality and avoid any stigmatization that might have resulted from being poor. Criteria for selecting "poor" people were based on social disengagement from the village (defined by respondents themselves), closeness of relationships with the environment and other members of the community (see Shubin, 2010), and qualitative information about earnings and expenses shared by rural people. Interviews focused on themes identified through preliminary policy analysis and included discussions about life trajectory, understandings of poverty, relationships with the forest, impacts of government and anti-poverty programs, political activism, and coping strategies. The project also worked closely with NGOs, policymakers, and activist groups in the areas to continuously question definitions of poverty emerging from the fieldwork and develop a broader understanding of the ramifications of poverty. Findings from the interviews were complemented by participant observation to reflect on complex poverty, the emergent nature of disadvantage, and relationships that respondents developed with the environment. All interviews were fully transcribed, translated into English, and then analyzed using NVivo 10 software. All names used are pseudonyms.

FLUID AND DYNAMIC POVERTY

Rejecting common descriptions of rural disadvantage in Bolivia and Brazil as a product, this chapter offers a fluid definition of poverty as a process, often shaped in relation to lack. As the earlier literature review showed, dominant visions of poverty in development discourse tend to be individualistic, blaming the "poor" for their problems and separating them from the world they live in. On the one hand, this happens through the construction of the "poor" person as a more or less stable entity, who is allocated a particular place as the object of development:

> The authorities want us to organize a group of people to write a plan, and then take this plan to the government, [. . .] so that the government can see our needs and say: "They really do need help there, they are poor, they need this." In a group . . . everyone has a number so their need can be evaluated and a price put on it. (Matilde Castelo, Curí village, Brazil, 35 years old)

Naming ('they are poor') in this case creates poverty as a product, it fixes the order by defining the poor person as someone who needs help. From the outset, the poor person is seen as a universalizable subject, a problem that requires resolution. As Deleuze and Guattari (1987: 16) would suggest, language is deployed to stabilize a developing poor person within the "system that pre-exists the individual, who is integrated into it at an allotted place." Importantly, to place the individual within the linguistic system of coding the state uses "units of measurement" that regulate individuals' relationships with the development apparatus. In the above quote, it is the evaluation of one's needs that creates the object of development, which is expected to transform itself into a subject of lack. The number attached to poverty (the "price put on it") is used as "a universal concept measuring elements according to their emplacement in a given dimension" (p. 8), where individual progression is expected along the scale of assets or resources that form part of deprivation indices.

However, poverty is not stable and it creates changeable relationships due to its different temporalities (seasonality of income, portfolio of activities) and the developing nature of coping arrangements. Many poor people in the Amazon experience insecurity due to changing government regulations (such as definitions of what constitutes "illegal" logging), and ebbs and flows of "extreme" poverty due to changes in the accessibility of support (Stone, 2006). Similarly, two of our respondents expressed poverty as a changeable process:

> This family benefit, education benefit and so on. We tried to get that. But no-one got anything so we gave up, we left it. Because we were spending more money than we would have received. So this means we are no longer poor? (Guilherme Carneiro, Curí, Brazil, 40)

> I think that poverty comes in different forms. If someone doesn't work, he'll cause problems for himself because if he is on the land and doesn't work, he won't progress, he's not doing anything worthwhile. (José López, Prainha, Brazil, 45)

As these examples illustrate, there are continuous changes in poverty trajectories. They challenge illusions about consistent constructions of the poor subject—instead this subject is presented as evolving, incomplete (not/progressing), and potentially no longer poor or not worthy of support. A poor person in this case is not seen as a self-enclosed and enduring subject, but rather as internally fractured, exposed, and open to self-destruction. In Bataille's (1988: 23) terms, "a being that isn't cracked isn't possible"; poor people challenge the powers of totalization in the linguistic order and always exceed the bounds of representation as the "poor." Furthermore, poverty creates multiple subjectivities that do not fit within existing policy definitions, are unclear and unidentified. As one respondent noted, "We, river dwellers, are forgotten by the government, because . . . we have no resources, no structure, no support. We are different" (Carlos Acosta, Mentai, Brazil, 30). A fluid understanding of poverty challenges the dominant language establishing the boundaries drawn around poor identities.

On the other hand, a fluid approach to poverty undermines the separation of the poor from the world and the dominant utilitarian logic applied to their interactions with the environment. As one of our respondents stressed:

The forest brings everything; unfortunately it can be destroyed by money. Nature offers everything to you without asking for a money bag. (Javier Garcia, Trinchera, Bolivia, 52)

Although poor people inevitably frame their engagement with the environment in terms of available and useable entities (suggesting a variety of uses for wood and fruits produced by the forest), they are also involved in broader process of being with the world that exceed the use of ecosystems as resources within frameworks of technical calculation and manipulation. When elements of ecosystems are ordered into networks of use, these arrangements undermine the importance of relations of nearness (expressed in terms of care) and distance familiar to rural dwellers and which contribute to poverty (Joronen, 2013). A rubber tapper from Bolivia explains:

State policy to me is what comes to destroy and destroy, because it has major impacts on our way of life. The state could define a policy for proper use of a good forest, with the powers that are not just economic, not production. (Marcelo Martínez, Trinchera, Bolivia, 52)

The economic drive for ecosystems to fulfil "socio-economic function" and productivity denigrates nontimber uses of the forest (Redo et al., 2011). As this quote suggests, the economics-driven policies destroy traditional ways of life and transform ecosystems into calculable reserves. As a result, elements of the forest and people's relations with them are subjugated to efficient ordering within a closed system of exchange. This creates a sense of loss and alienation that aggravates poverty:

People were taking about selling oxygen produced by the forest. They were saying that Europeans were going to pay for that, but it was just a fantasy. We are getting poorer, and they do not pay us. (Geraldo Flores, Bioceánica, Bolivia, 46)

Utilitarian logic applied to ecosystems produces a sense of lack that is seen as a problem, which needs to be addressed. This lack refers to elements that have not yet been ordered and turned into calculable reserves, such as oxygen. However, the introduction of new techniques of use (such as the Payment for Ecosystem Services mentioned by the respondent) intrinsic to economic rationalities can further contribute to poverty. Many rural dwellers from our study value their relationships with the forest precisely because it offers them something beyond useable and tradeable resources, and creates possibilities for not feeling poor:

We feel for the forests. It is a part of us. When these big companies come into the forest we feel that, it may not be causing problems now, but we already

> feel that it could do later on [. . .]. We have that kind of evolving relationship. (Manoel Guardado, Coipiranga, Brazil, 26)

> Even in these challenging conditions that we live in, the forest gives us lots of things. A lot of the riches that are there we can use. [. . .] We suffer, but not because of the forest, more because of our health, education, which we need here. We are not poor, we just have a low income. (Isabel Villanova, São Pedro, Bolivia, 42)

As these quotes suggest, poor people's involvement with the forest always exceeds the calculative picture of poverty and accounts for the inexhaustible emergence of people-ecosystem links that go beyond the isolated selves. A fluid approach to poverty highlights the openness of people's relationships with the forest and reveals "a lot of the riches" that can emerge from such a connection. By highlighting the excess that exists beyond a system built on exchange, deconstruction of economic logic reconfigures poverty as dynamically changing and irreducible to "low income." Feelings of hope, joy, and suffering along with the forest challenge boundaries between inside and outside, and express openness to the world in all its strangeness and irreducible alterity. This acceptance of co-existence with the forest when the latter is not characterized as an entity or as occupying a certain role challenge expectation about the mastery of ecosystems. The fusion between an individual and the ecosystem, where the poor person is no longer seen as an isolated entity, undermines the binary logic of human versus environmental knowledge and productive versus unproductive definitions of forest-related poverty. It offers the opportunity to unravel more-than-human aspects of poverty and the unexpected connections that produce rural disadvantage, which the next section explores.

COMPLEX POVERTY

An approach focusing on complex rural poverty helps us to understand its emergence through a combination of different elements and transformations. It further rejects negativity (lack) in expressing poverty and instead focuses on internal differentiation. From an assemblage perspective, poverty can be considered as an evolving combination of objects, actions, bodies, and expressions that create new ways of functioning (Parr, 2010). Poverty includes memories of failure, distrust, and dependency on development programs:

> I live on the road, go from one community to another. I know CIPCA, they are promoting reforestation [. . .]. I don't have confidence in these programs

> because when land restructuring happened, land distribution created poverty and restricted rural communities. They have not fulfilled many things they promised. (Ernesto Moreno, Cobija, Bolivia, 35)

Poverty here emerges as an assemblage of perceptions of reforestation, attitudes to development programs and disappointments about "things promised." It is not limited to calculative positioning on the poverty measurement grid, but expresses a variety of unexpected connections. The itinerant lifestyle of the poor person ("I live on the road") demonstrates the nomadic composition of poverty as an assemblage, which is not pre-given (distributed according to an economic law of distance or labor market access), but developing and opening new differences and paths, desires and relations. Deleuze and Guattari (1987) stress that assemblages are created through the interplay of active and reactive forces that coalesce together. They view active forces as affirmative, life-enhancing, and able to transmute into (become) something else in a creative manner, while reactive forces restrict, capture, and decompose. Deleuze (1986) focuses on the dynamism of change through the interplay of active and reactive forces, which affirms difference as productive (as a "difference in itself") at the expense of difference as negation (as a "difference from").

The emphasis here is not on replicating the same and comparing it with the existing model ("indexed" poverty), but on processes with the greatest potential for creating difference and metamorphosis. A fragment from the interview with a Brazilian small farmer illustrates this point:

> For me, poverty, well there is poor and there is hard up. Today we are hard up, but we are not poor. [. . .] I might not have anything today, but I will have tomorrow. Here at home I might have nothing in the morning, but by mid-day I'll have it. If I don't have what I need in the afternoon, I'll find a way get it by the evening. So we are not poor, the poor are those who don't have warm clothes, who don't have a source of income, who go around begging. People look at me and say: "that poor rich man." (Tiago Teixeira, Arapiuns, Brazil, 48)

The respondent speaks about poverty as a developing process, changing from morning to evening, and about continuously emergent difference. This difference moves beyond economics-based definitions of poverty ("source of income") or dominant stereotypes ("begging"). Instead, this quote highlights the existence of a difference in kind, which is described as a mutation into an in-between creation of a "poor rich man." The process of becoming poor here does not refer to a position within the grid of poverty and wealth (an identity), but rather marks degrees and levels of intensity (being very hungry, not having anything). This approach challenges the logic of negation, which is often used to define poverty as a polar opposite to wealth within the dualistic framework underpinning development discourse.

Instead of belittling poor lives, a complex and transformative vision of poverty expresses them as a play of difference creating metamorphous arrangements of people and things (the forest). In this respect, Deleuze and Guattari (1987: 237) point toward "the existence of a very special *becoming-animal* traversing human beings and sweeping them away." This reflects a movement from major (the identity, the constant) to minor (the variable), from unity to complexity that escapes identity (expressed through the category "the poor") and determination (its goal is unclear). Quotes from Brazilian and Bolivian peasants speak to this point:

> In poverty we become like animals and depend on the forest so we are nothing without it. (Ana Guarin, Los Mandarinos, Bolivia, 22)

> If you dedicate yourself to work in the forest you will definitely change and gain some kind of freedom. [. . .] I had nothing, I ate and drank depending on my father. After having lived in the forest, I decided that I wouldn't work for an employer because I wanted to work on something that was my own. You kind of leave something within the forest. (Cristiano Soares, Arapiuns, Brazil, 34)

In the process of "becoming-animal" familiar forms of poverty become undone; they are situated on the borderline between the person and the forest. As López-Rivera (2013) claims, sensitivity to the materiality of nature, in particular the fusion of biophysical and social forms of being and their valorizations, directly affects the reproduction of inequalities. Freedom of being with the forest, to which one respondent refers, implies that poverty changes and passes into a different state that reflects a different degree of intensity. This kind of poverty escapes "human classifications" (Deleuze and Guattari, 1987: 244) and is beyond the possibility of naming due to its complexity and foreignness. A river dweller explains:

> A while ago, when I was younger, I wasn't thinking of planting anything! For me if I got some small change working for a day, that would do me. I didn't care. [. . .] Poverty became like pain, I became like a mule working for others. (Silvio Fontes, Curí, Brazil, 42)

Unlike the process of becoming-human, which is predicated upon the negation of nature, becoming-animal ("became like a mule") implies a transformation of poverty to include often marginalized expressive behaviors and experiences of desire, hunger, pain, and fear. These elements of poverty are often spiritual, and include "enchanted" animals, bush spirits (such as

Caipora), other-worldly creations (like witchcraft), as well as elements of conventional religion, which cannot be reduced to wealth or traditional stereotypes of disadvantage (Hoefle, 1999). Respondents from Brazil explain:

> I think that for people here it's the devil that's poor! Something like that, well, I don't really know. (Guilherme Carneiro, Curí, Brazil, 40)

> We are haunted by memories, forest spirits, scared of starting again—so people stay away, even if they can benefit from [anti-poverty] projects. (Tiago Teixeira, Arapiuns, Brazil, 48)

As Hoefle (2009) stresses, in the Amazon popular beliefs that land and water are alive contribute to the development of poverty as a symbiosis, an "enchantment" open to change and experimentation. The above narratives witness the emergence of uncertain or "larval" subjects in a poverty-tinged relationship with the forest, which is formed by continuously fluctuating forces "under conditions yet to be determined" (Deleuze, 1994: 215). The resultant poverty produced by different expressions and things emerges as more-than-human, elusive and unsettling. It appears as an "in-between poverty," where the poor emerge as neither subjects nor objects, but multiple (unlabeled) beings best described in terms of determinations and magnitudes. The next section explores this uncertain and relational poverty that emerges from such multiple and complex connections.

RELATIONAL AND UNCERTAIN POVERTY

A complex "becoming" approach to poverty destabilizes the limits of the "poor" self and the ways in which it forms human knowledge based on the principles of rationality and reason. Poverty is unpredictable and contradictory, as a carpenter from Brazil suggests:

> I have worked to improve my situation, but it hasn't improved much because we work, work, work but we don't know what is going to happen. We do these crafts here but we just get paid in dribs and drabs because the person who orders the work doesn't always have all the money to pay on the spot. (Alejandro de la Pena, Arapiuns, Brazil, 32)

This passage describes the uncertain nature of poverty that emerges outside existing expectations and categories of knowledge. In development discourse, poverty is often seen as a weakness to be overcome by work, but the above quote reveals the confusing relation between work and disadvantage in contradiction to the rational logic. This understanding of poverty as uncertain and something that cannot be fully grasped in knowledge also questions assumptions about how it can be framed as a rural problematic. When poverty exceeds the logic of reason, it reveals insufficiency and disconnections between development practices and rural disadvantage:

> The government tackles poverty through technical assistance, with supplies, cupuaçu saplings, and plantains. But peasants are not interested in those things. Saplings dry out at their doors. They only harvest Brazil nuts, sometimes rubber and wood. [. . .] They are expected to manage resources rationally. (Saúl Gutierrez, Bolpebra, Bolivia, 27)

Nonrational expressions of poverty that appear unaccounted and unjustified not only undermine the stability of development discourse, but also reflect alienation with regards to the human being. As we have seen earlier, in the process of emergence of heterogeneous poverty there is no stable "poor" subject that connects different expressions, actions and things and makes them intelligible. Perdigon (2015: 90) stresses that such "poverty is not effectuated in the recognizable form of a "political subject of poverty"—an "I, poor"—who would emerge in the interpretative act of establishing a semiotic relationship of some kind between her malnourishment [and] ill health." The resultant poverty developed within this changing context cannot be seen as a matter of human knowledge as it is commonly understood. Due to its unpredictable emergence, poverty eschews appropriation in the certainty and regularity of knowledge based on reason. Poverty expressions that develop from unintentional relations with the forest exceed the demands of rational recognition. Attending to this uncertain poverty requires going beyond what rural dwellers call "expert" knowledge:

> [Our] education interprets the knowledge of experts so that our knowledge can be made better. Currently, our education is really a European copy so we impose it instead of teaching [about the forest], we interpret experts' wisdom and do not move forward. [. . .] The forest has natural richness. What is missing for people is to discover how to have development and respect nature. (Quique Hernandes, itinerant, Bolivia, 35)

This quote refers to "natural richness" and "respect for nature," which involve constantly developing relationships between people and the forest and, due to their complexity, cannot be known in advance. They cannot be contained within the dominant Latin American development discourse, which treats "nature as a service provider of ecosystem services . . . and local populations . . . as ecosystem managers" (Dietz, 2017: 46). As Bataille (1988:

51) notes, such an open relationship is a matter of "non-knowledge," lying beyond the world of presuppositions and conscious thoughts based on belief in the possibility of defining and resolving problems. Poverty that involves opening up to the unfamiliar experiences of struggle, sudden exhaustion, and indeterminate conditions lying outside consciousness describes movement beyond individual experience toward an abyss of un-knowing. In this case, poverty "reconfigures the partition of the sensible" (Rancière, 1999: 53) and creates moments of suspense, radical doubt, and disconnection from discursive registers. It relates to the expenditure of energies, anguish, fear rather than sense, the rationality of usefulness. A rural dweller speaks about this kind of poverty:

> I am not exactly abandoned, but I don't care about anything. I don't always have nutritious food. [. . .] It is the pain of being on my own because my son-in-law who lives with me is away at the plantation. I am fearful, hurt and lost, I haven't been hunting any more, I forgot how walk around in the forest. (Danilo de Oliveira, Arapiuns, Brazil, 65)

Although such expressions of poverty (lack of food) fit within the existing stereotypes of disadvantage, they articulate the exposure of an individual in a moment of risk and pain, and movement beyond oneself. When an individual is lost and exposed to unbearably foreign forces in the emergence of poverty, the self-enclosure required for the possession of rational knowledge is no longer possible. As Bataille (1988: 52) asserts, "non-knowledge lays bare," it reflects the moments that cannot be grasped and cannot be reached without the withdrawal of the individual. In the above quote, this is evident in a passivity in relating to the foreignness that poverty brings and the movement of the exception in affliction: "I don't care about anything." This affliction (hurt) does not belong to the economy of time, it forgets itself as something present or meaningful (Derrida, 1991: 17).

The withdrawal of the individual ("I, poor") as a self-enclosed and rational agent acting upon nature opens up spaces for alternative interpretations for anti-poverty politics and its impacts in the Amazon. In many cases, this leads to increased dependence on government support, submission in weariness to fate or the order of things:

> Here today most people live off government support. [. . .] There is not much extractivism. The [government] says "no, you can't do that anymore', so the person just sits around waiting for the gift, the pittance that the government is giving out." (Tiago Teixeira, Arapiuns, Brazil, 48)

> We borrowed this money, but didn't actually see any of it, not actual money. Just rubbish, them bringing the amount in saplings. [. . .] They left it up to us.

> "Get on with it." We were left with just a debt. And we still don't know if the debt has been paid or not. [. . .] If we had received this money, we wouldn't have spent it working with our own hands, doing what we know best with the maniva [manioc leaves]. (Emilio Carvalho, Pedreira, Brazil, 37)

In these quotes, poverty appears as a process of growing dependence and uncertain (un/paid) accumulation of debt because of government support. The success of the *Bolsa Família* (Family Grant) in addressing poverty, portrayed as the Brazilian president's personal "gift" to the deprived (Hall, 2008), relies on clear identification of the "poor" and "very poor" on the basis of household income. As Derrida (1991: 11) explains, providing such a "gift" of government support requires that some "one" has to give some "thing" to some "one else." This understanding of a gift speaks to the values of subject, self, consciousness, and intentional meaning in the situation of uncertainty where poverty and subjectivity cannot be reduced to clear and recognizable forms. In Derrida's (1991: 23) terms, poverty in this case is interpreted as a possible outcome of an exchange, with "donor and donee constituted as identical, identifiable subjects, capable of identifying themselves." When the state gratifies itself (as a donor), it sets the conditions for reciprocity and exchange and annuls the "gift" because it amounts to doing harm. Effectively, such a form of support aggravates poverty since it presents the "poor" subject as the figure of circulation and expects her to obligate and further indebt herself. To escape poverty re-created within such a system of exchange, forest dwellers refuse the government's "gift":

> I have done one important thing to escape from poverty, I did not ask for a loan to harvest Brazil nuts. (Ignacio Montejo, Planchon, Bolivia, 40)

This example points toward the harm of receiving support as a part of rational development initiatives. To eschew the mechanisms of symbolic exchange and accumulation, Derrida (1991: 27) offers an alternative of "a gift without wanting, an insignificant gift" that he describes as "counterfeit money." To become unbound from the obligation of debt and exchange would require forgetfulness, reformulation of the gift (reviewing its implied generosity, restitution of the time spent waiting for the gift), and nonappearance of the subject. In the context of ecosystem-related disadvantage in the Amazon, poverty would need to be considered outside of the logic of circulation to make development interventions sympathetic to the suffering of disordered "poor" selves, irreducible to the "gift" economy of exchange and open to the uncertain expressions of rural malaise.

CONCLUSIONS

Drawing on the examples from fieldwork in the Brazilian and Bolivian Amazon, this chapter has questioned the subordination of the poor within clearly defined "problem" areas and structured "poor" social groups. It highlighted the disconnection between development discourses, the state's conceptualizations of poverty and local understandings of poverty and development, which aggravate rural poverty in the Amazon. The chapter argued that existing development interventions (such as Payment for Ecosystem Services) tend to draw on the logic of capital to prioritize economic dimensions of poverty and focus mainly on material practices of the poor, which separates them from nature and offers limited anti-poverty solutions. Alternative understandings of poverty as fluid, complex and uncertain developed in this chapter problematized traditional visions of developmental frontiers in the Amazon central to the debates in this book.

First, the chapter questioned the construction of poverty as a fixed entity and argued for its *fluid* articulation. It demonstrated that attempts to measure one's poverty contribute to creation of boundaries/frontiers around seemingly stable poor subjects separated from changing temporalities of disadvantage to the detriment of their living conditions. The examples from fieldwork demonstrate that the lives of poor people cannot be evaluated in terms of their emplacement and units of measurement in proportion to developmental progress. Attending to dynamic poverty expressions and practices calls for acceptance of the unstable and incomplete nature of poor "subjects" as "between things, interbeings" (Deleuze and Guattari, 1987). Furthermore, the chapter undermined the ordering of ecosystems in terms of their usefulness and moved beyond the "rigid segmentarity" (Deleuze and Parnet, 2007: 96) of existing policy approaches to highlight the impossibility of creating a uniform view of forest-related poverty. It questioned utilitarian approaches age in dominant anti-poverty policies and unsettled the existing binary oppositions between people and ecosystems, and the frontiers they create. The stories from disadvantaged people in the Brazilian and Bolivian Amazon suggested that attempts to reduce lives in the forest to orderly frameworks of exchange and calculation (i.e., translated into different forms of capital) oversimplify disadvantage and aggravate poverty.

Second, by engaging with the stories from the Amazon the chapter unraveled *more-than-human* and *assembled* dwelling poverties. In exploring the process of becoming poor, the chapter highlighted different intensities, potentialities, and complexities generating poverties assembled from combinations of objects, actions, bodies, and expressions. It used fieldwork examples to destabilize clear distinctions between people and the forest and to redefine the categories of minor and major in development discourse, thus contributing to the broader discussions about changeable frontiers and positioning central to this book. The discussion explored the "minor" processes that destabilize the notion of poverty, dynamically shift centers and margins, and rethink seemingly marginalized minor positions in the emergence of disadvantage. This complex approach to poverty goes beyond the vision of a "constitutive interconnectedness" (Therborn, 2015: 17) between nature and society, which often allocates the place to nature as an explanatory variable producing inequalities. By focusing on expressions of pain and fear and combinations of beliefs and memories in the production of forest-related disadvantage the chapter developed a vision of poverty as an "enchantment" (Hoefle, 2009), an always-developing symbiosis between different elements that redefines frontiers of change in the Amazon beyond material terms.

Third, the *relational* and *uncertain* approach to poverty unsettled the visions of disadvantage as an object of rational knowledge and further destabilized the oft-assumed frontiers of the poor "selves." It attempted to respond to the limited success of cash transfer programs (such as *Bolsa Família* in Brazil and *Bono Juana Azurduy* in Bolivia) in providing poverty relief and attending to the specific needs and concerns of the poor, which are constantly evolving and go beyond economic and self-advancement targets (Hall, 2006; Molyneux and Thomson, 2011). While in development policies poverty is often constructed as a failed condition requiring interventions in the form of work and income generation, this negative view assumes the ability of knowledge to contain poverty and make it reasonable. Several examples from the rural Amazon go beyond rational knowledge and outside demands of development discourse, thus challenging the predictability of poverty trajectories, the possibility of rational resource management and the intelligibility of poverty-related outcomes within the ordered force of (expert) representation. Moreover, these examples articulate the "non-knowledge" of the uncertain emergence of poverty, which expresses an opening up to the shared strangeness of pain, risk, and loss that exceed determinate conditions of consciousness and go beyond an individual experience (in the sense of *Ekstatikon*, cf. Shubin, 2015). Such experiences of poverty escape assimilation by the external authority of knowledge and challenge the positioning of the poor beyond the development frontiers, outside political rationalities and economic calculations, often condemned for their apparent inaction.

Importantly, the dynamic, complex and uncertain understanding of poverty developed in this chapter challenges the dualistic frontiers of knowledge used in development discourse. It presents the distinction between interiority (felt expressions) and exteriority of knowledge (expert knowledge) as false. As the research findings demonstrate, support for the poor based on clear identifications of subjects involved in an exchange (donor and donee), calculable interests, and intentional meaning can aggravate poverty. The chapter suggests a reconfiguration of the frontiers of exchange and subjectivity through passivity and withdrawal, where rejection of the government's "gift" to the deprived implies forgetfulness and nonappearance of the subject. This complex and uncertain approach contributes to recent discussions about socio-ecological disadvantage in Latin America by challenging the understanding of ecosystem services as producing "socially-relevant good" and as part-objects of compensatory policies addressing socioeconomic inequalities in terms of income and wealth (Görg et al., 2017). It supports recent criticisms of the social abstraction of nature, stressing that it is not possible to understand nature solely within the system of exchange and define "all of life as bearers of value" (Robertson, 2012: 388); this executes real violence on poor people. By calling for affirmation of difference and rejection of utilitarian logic in expressing frontiers of development it encourages increased attention to changing temporalities, unstable positions, "assembled" becoming and potentialities of poverty. In so doing, it adds to the debates on exploring alternative articulations of poverty by moving beyond inequalities of opportunity and positions to study dynamic and unpredictable frontiers in "entangled inequalities" (Costa, 2018).

NOTE

1. Thrift (2004) describes these emergent sensations, active outcomes of encounters and unactualized knowledge as affective thinking—a different kind of intelligence that makes sense of transient and metamorphic things even though they cannot always be named.

BIBLIOGRAPHY

Alcock, P. 1997. *Understanding Poverty*. Palgrave Macmillan: London.

Alkire, S. 2004. Culture, Poverty and External Intervention. In: *Culture and Public Action*, Rao, V. and Walton, M. (eds.). The World Bank: Stanford, pp. 185–209.

Altamirano, T., Copestake, J., Figueroa, A., and Wright, K. 2003. *Poverty Studies in Peru: Towards a More Inclusive Study of Exclusion*. ESRC Research Group on Wellbeing in Developing Countries, WeD Working Paper 05. University of Bath.

Anderson, S. and Bell, M. 2004. The Devil of Social Capital: A Dilemma of American Rural Sociology. In: *Rural Visions*, Cloke, P. (ed.). Pearson Education: Harlow, pp. 232–44.

Appadurai, A. 2004. The Capacity to Aspire: Culture and the Terms of Recognition. In: *Culture and Public Action*, Rao, V. and Walton, M. (eds.). The World Bank: Stanford, pp. 59–84.

Araujo, C., Bonjean, C., Combes, J. L., Motel, P., and Reis, E. 2009. Property Rights and Deforestation in the Brazilian Amazon. *Ecological Economics*, 8, 2461–68.

Asen, R. 2002. *Visions of Poverty: Welfare Policy and Political Imagination*. Michigan State University Press: East Lansing.

Ballon, P. and Krishnakumar, J. 2008. Estimating Basic Capabilities: A Structural Equation Model Applied to Bolivia. *World Development*, 36(6), 992–1010.

Ban, C. 2013. Brazil's Liberal Neo-developmentalism: New Paradigm or Edited Orthodoxy? *Review of International Political Economy*, 20(2), 298–331.

Barsted, L. 2005. Brazil: The Legal Status of Women in the Context of Agrarian Reform. In: *Gender and Land Compendium of Country Studies*, García-Frías, Z. (ed.). FAO: Rome, pp. 37–81.

Bataille, G. 1988. *Inner Experience*. Trans. L. Boldt. SUNY Press: Albany.

Battiston, D., Cruces, G., Lopez-Calva, L., Lugo, M., and Santos, M. 2013. Income and Beyond: Multidimensional Poverty in Six Latin American Countries. *Social Indicators Research*, 112, 291–314.

Bebbington, A. 2004. Livelihood Transitions, Place Transformations: Grounding Globalization and Modernity. In: *Latin America Transformed: Globalization and Modernity*, Gwynne, R. N. and Kay, C. (eds.). 2nd ed. Arnold: London, pp. 173–92.

Beverinotti, J. 2018. *Development Challenges in Bolivia*. IDB Policy Brief 289. Inter-American Development Bank: Washington, D.C.

Bicalho, A. M. S. M. and Hoefle, S. W. 2010. Economic Development, Social Identity and Community Empowerment in the Central and Western Amazon. *Geographical Research*, 48, 281–96.

Blanchot, M. 1995. *The Writing of the Disaster*. University of Nebraska Press: Lincoln.

Booth, D., Grigsby, A., and Toranzo, C. 2006. *Politics and Poverty Reduction Strategies: Lessons from Latin American HIPCs*. Overseas Development Institute: London.

Börner, J., Mendoza, A., and Vosti, S.A. 2007. Ecosystem Services, Agriculture, and Rural Poverty in the Eastern Brazilian Amazon: Interrelationships and Policy Prescriptions. *Ecological Economics*, 64(2), 356–73.

Brass, T. 2002. Latin American Peasants: New Paradigms for Old? *Journal of Peasant Studies*, 29(3–4), 1–40.

Bratman, E. 2019. *Governing the Rainforest: Sustainable Development Politics in the Brazilian Amazon*. Oxford University Press: Oxford.

Bresser-Pereira, L. C. 2007. The New Developmentalism and Conventional Orthodoxy. *Iberoamericana: Nordic Journal of Latin American and Caribbean Studies*, 37(1), 31–58.

Carpenter, S. R., Mooney, H. A., Agard, J., Capistrano, D., DeFries, R. S., Díaz, S., Dietz, T., Duraiappah, A.K., Oteng-Yeboah, A., Pereira, H. M., Perrings, C., Reid, W. V., Sarukhan, J., Scholes, R. J., and Whyte, A. 2009. Science for Managing Ecosystem Services: Beyond the Millennium Ecosystem Assessment. *Proceedings of National Academy of Sciences*, 106, 1305–12.

Costa, S. 2018. Entangled Inequalities, State, and Social Policies in Contemporary Brazil. In: *The Social Life of Economic Inequalities in Contemporary Latin*

America, Approaches to Social Inequality and Difference, Ystanes, M. and Strønen, I.Å. (eds.). Palgrave Macmillan: London, pp. 59–80.

Cusicanqui, S. 2012. *Violencias (re)Encubiertas en Bolivia*. Otramérica: Santaner.

Danziger, S. and Haveman, R. 2002. *Understanding Poverty*. Harvard University Press: Cambridge.

Dasgupta, S., Deichmann, U., Meisner, C., and Wheeler, D. 2005. Where is the Poverty-Environment Nexus? Evidence from Cambodia, Lao PDR, and Vietnam. *World Development*, 33(4), 617–38.

de Haan, L. and Zoomers, A. 2005. Exploring the Frontier of Livelihoods Research. *Development and Change*, 36(1), 27–47.

Deleuze, G. 1986. *Nietzsche and Philosophy*. Continuum: London.

Deleuze, G. 1994. *Difference and Repetition*. Trans. P. Patton. Columbia University Press: New York.

Deleuze, G. and Guattari, F. 1987. *A Thousand Plateaus: Capitalism and Schizophrenia*. University of Minneapolis Press: London.

Deleuze, G. and Parnet, C. 2007. *Dialogues II*. Columbia University Press: New York.

Derrida, J. 1991. *Given Time: I. Counterfeit Money*. University of Chicago Press: London.

Dietz, K. 2017. Researching Inequalities from a Socio-Ecological Perspective. In: *Global Entangled Inequalities: Conceptual Debates and Evidence from Latin America*, Jelin, E., Motta, R. and Costa, S. (eds.). Routledge: New York, pp. 76–92.

Doshi, S. 2013. The Politics of the Evicted: Redevelopment, Subjectivity, and Difference in Mumbai's Slum Frontier. *Antipode*, 45(4), 844–65.

Dos Santos, F. 2019. Neo-development of Underdevelopment: Brazil and the Political Economy of South American Integration under the Workers' Party. *Globalizations*, 16, 216–31.

Fisher, J. A., Patenaude, G., Giri, K., Lewis, K., Meir, P., Pinho, P., Rounsevell, M. D. A., and Williams, M. 2014. Understanding the Relationships between Ecosystem Services and Poverty Alleviation: A Conceptual Framework. *Ecosystem Services*, 7, 34–45.

Gomes, C. V. A., Vadjunec, J. M., and Perz, S. G. 2012. Rubber Tapper Identities: Political-Economic Dynamics, Livelihood Shifts, and Environmental Implications in a Changing Amazon. *Geoforum*, 43, 260–71.

Görg, C., Brand, U., Haber, H., Hummel, D., Jahn, T., and Liehr, S. 2017. Challenges for Social-Ecological Transformations: Contributions from Social and Political Ecology. *Sustainability*, 9, 1045; doi:10.3390/su9071045.

Hall, A. 2008. Brazil's Bolsa Família: A Double-edged Sword? *Development and Change*, 39(5), 799–822.

Harrison, P. 2008. Corporeal Remains: Vulnerability, Proximity, and Living on after the End of the World. *Environment and Planning A*, 40, 423–45.

Heidegger, M. 1996. *Being and Time*. SUNY Press: Albany.

Hirakuri, S. R. 2003. *Can Law Save the Forest? Lessons from Finland and Brazil*. Forest Law Enforcement, Governance and Trade Series. CIFOR: Bogor.

Hoefle, S. 1999. Religious Worldview and Environment in Sertão of North-East Brazil. *Ethics, Place and Environment*, 2, 55–79.

Hoefle, S. 2009. Enchanted (and Disenchanted) Amazonia: Environmental Ethics and Cultural Identity in Northern Brazil. *Ethics, Place and Environment*, 12, 107–30.

Horschelmann, K. 2011. Theorising Life Transitions: Geographical Perspectives. *Area*, 43, 378–83.

IPEADATA. 2018. Economic Database of the Instituto de Pesquisa Economica Aplicada. Available at: http://www.ipeadata.gov.br.

Jackson, P. 1983. Principles and Problems of Participant Observation. *Geografiska Annaler B*, 65(1), 39–46.

Joronen, M. 2013. Heidegger, Event and the Ontological Politics of the Site. *Transactions of the Institute of British Geographers*, 38, 627–38.

Kareiva, P., Tallis, H., Ricketts, T., Daily, G., and Polasky, S. 2011. *Natural Capital: Theory and Practice of Mapping Ecosystem Services*. Oxford: Oxford University Press.

Kay, C. 1989. *Latin American Theories of Development and Underdevelopment*. Routledge: London.

Kay, C. 2006. Rural Poverty and Development Strategies in Latin America. *Journal of Agrarian Change*, 6(4), 455–508.

Lee, N. and Motzkau, J. 2011. Navigating the Biopolitics of Childhood. *Childhood*, 18, 7–19.

Levinas, E. 1991. *Otherwise Than Being or Beyond Essence*. Kluwer: Dordrecht.

Levitas, R. 2001. What is Social Exclusion? In: *Breadline Europe: The Measurement of Poverty*, Gordon, D. and Townsend, P. (eds.). Policy Press: Bristol, pp. 357–85.

Lin, A. and Harris, D. 2008. *The Colors of Poverty: Why Racial and Ethnic Disparities Exist*. Russell Sage Foundation: New York.

Lingis, A. 1996. *Sensation: Intelligibility in Sensibility*. Humanities Press: Atlantic Highlands, NJ.

López, R. 2013. *Flows of Water, Flows of Capital: Neoliberalization and Inequality in Medellín's Urban Waterscape*. Working Paper Series 38. desiguALdades.net: Berlin.

McDaniel, J., Kennard, D., and Fuentes, A. 2005. Smokey the Tapir: Traditional Fire Knowledge and Fire Prevention Campaigns in Lowland Bolivia. *Society and Natural Resources*, 18(10), 921–31.

McFarlane, C. 2009. Translocal Assemblages: Space, Power and Social Movements. *Geoforum*, 40, 561–67.

MEA. 2005. *Ecosystems and Human Well-Being: Policy Responses*. Island Press: Washington, D.C.

Molyneux, M. and Thomson, M. 2011. Cash Transfers, Gender Equity and Women's Empowerment in Peru, Ecuador and Bolivia. *Gender and Development*, 19(2), 195–212.

Munck, R. 2005. *Globalization and Social Exclusion: A Transformative Perspective*. Kumarian Press: Bloomfield.

O'Hare, G. and Rivas, S. 2007. Changing Poverty Distribution in Bolivia: The Role of Rural-Urban Migration and Urban Services. *Geojournal*, 68(4), 307–26.

Paes de Barros, R., De Carvalho, M., and Franco, S. 2006. *Pobreza Multidimensional no Brasil*. Texto para Discussão No. 1227. IPEA: Rio de Janeiro.

Parr, A. 2010. *The Deleuze Dictionary*. Edinburgh University Press: Edinburgh.

Perdigon, S. 2015. "For Us It Is Otherwise": Three Sketches on Making Poverty Sensible in the Palestinian Refugee Camps of Lebanon. *Current Anthropology*, 56, 88–96.

Ranciere, J. 1999. *Disagreement: Politics and Philosophy*. University of Minnesota Press: London.

Ranganathan, M. 2015. Storm Drains as Assemblages: The Political Ecology of Flood Risk in Post-colonial Bangalore. *Antipode*, 47(5), 1300–20.

Ranganathan, M. and Balazs, C. 2015. Water Marginalization at the Urban Fringe: Environmental Justice and Urban Political Ecology Across the North–South Divide. *Urban Geography*, 6(3), 403–23.

Redo, D., Millington, A., and Hindery, D. 2011. Deforestation Dynamics and Policy Changes in Bolivia's Post-neoliberal Era. *Land Use Policy*, 28, 227–41.

Robertson, M. 2012. Measurement and Alienation: Making a World of Ecosystem Services. *Transactions of the Institute of British Geographers*, 37, 386–401.

Robinson, P. and Oppenheim, C. 1998. *Social Exclusion Indicators: A Submission to the Social Exclusion Unit*. IPPR: London.

Room, G. 1995. *Beyond the Threshold: The Measurement and Analysis of Social Exclusion*. Policy Press: Bristol.

Rudel, T. 2007. Changing Agents of Deforestation: From State-Initiated to Enterprise-Driven Processes, 1970–2000. *Land Use Policy*, 24, 35–41.

Schmink, M. and Wood, C. (eds.). 1992. *Contested Frontiers in Amazonia*. Columbia University Press: New York.

Sen, A. 1999. *Development as Freedom*. Knopf: New York.

Sen, A. 2006. Conceptualising and Measuring Poverty. In: *Poverty and Inequality*, Grusky, D., Ravi Kanbur, S., and Sen, S. (eds.). Stanford University Press: Stanford, pp. 30–40.

Shackleton, S. E. and Shackleton, C. M. 2012. Linking Poverty, HIV/AIDS and Climate Change to Human and Ecosystem Vulnerability in Southern Africa: Consequences for Livelihoods and Sustainable Ecosystem Management. *International Journal of Sustainable Development*, 19, 275–86.

Shubin, S. 2010. Cultural Exclusion and Rural Poverty in Ireland and Russia. *Transactions of the Institute of British Geographers*, 35(4), 555–70.

Shubin, S. 2015. Migration Timespaces: A Heideggerian Approach to Understanding the Mobile being of Eastern Europeans in Scotland. *Transactions of the Institute of British Geographers*, 40(3), 350–61.

Shubin, S. and Sowgat, T. 2019. Contested Experiences and Potential Justice at the Limit of the Law for Poor Urban Children in Bangladesh. *Children's Geographies*. doi:10.1080/14733285.2019.1630711.

Shubin, S., Findlay, A., and McCollum, D. 2014. Imaginaries of the Ideal Migrant Worker: A Lacanian Interpretation. *Environment and Planning D*, 32, 466–83.

Sjostedt, M. 2012. Ecosystem Services and Poverty Reduction: How Do Development Practitioners Conceptualize the Linkages? *European Journal of Development Research*, 24, 777–87.

Skeggs, B. 2004. Exchange, Value and Affect: Bourdieu and 'the Self'. *The Sociological Review*, 52(2), 75–95.

Soares, F., Soares, S., Medeiros, M., and Osório, R. 2006. *Cash Transfer Programmes in Brazil: Impacts on Inequality and Poverty*. Working Papers 21. International Policy Centre for Inclusive Growth: Brasília.

Spatz, J. 2006. *Poverty and Inequality in the Era of Structural Reforms: The Case of Bolivia*. Springer: New York.

Stone, S. 2008. *Country Case Study: Forest Tenure and Poverty in Brazil: With an Emphasis on the States of Acre, Amazonas, and Pará*. Rights and Resources: Rio de Janeiro.

Suich, H., Howe, C., and Mace, G. 2015. Ecosystem Services and Poverty Alleviation: A Review of the Empirical Links. *Ecosystem Services*, 12, 137–47.

Therborn, G. 2015. Movements of Equality: Today's Latin America in a Global Historical Context. In: *A Moment of Equality for Latin America?* Fritz, B. and Lavinas L. (eds.). Ashgate: Farnham Surrey, pp. 13–28.

Thrift, N. 2004. Intensities of Feeling: Towards a Spatial Politics of Affect. *Geografiska Annaler: Series B*, 86, 57–78.

Vargas, J. and Gariga, S. 2015. *Explaining Inequality and Poverty Reduction in Bolivia*. IMF Working Paper 15/265. IMF: Washington, D.C.

Verner, D. 2004. *Poverty in the Brazilian Amazon: An Assessment of Poverty Focused on the State of Pará*. World Bank Policy Research Working Paper 3357. World Bank: Washington, D.C.

Wanderley, F. 2018. Between Extractivism and Living Well: Experiences and Challenges in Bolivia. In: *Reframing Latin American Development*, Munck, R. and Delgado Wise, R. (eds.). Routledge: London, pp. 122–42.

World Bank. 2002. *Annual Report: Poverty*. World Bank: New York.

Zenteno, M., Zuidema, P., de Jong, W. and Boot, R. 2013. Livelihood Strategies and Forest Dependence: New Insights from Bolivian Forest Communities. *Forest Policy and Economics*, 26, 12–21.

Chapter 8

Illegal Gold Mining and the Struggle to Save the Amazon in Peru

Lynn Holland

On a recent mission dubbed "Operation Mercury," the Peruvian military descended into the jungles of the Amazon in Peru's latest effort to stem illegal gold mining in the area. Led by the ministers of Defense, Interior, and Environment, the mission involved the installation of three military bases with the "alpha" base located at Tambopata National Reserve. The reserve, a thousand square miles of protected area in southeastern Peru, lies at the heart of illegal gold mining activity in the Peruvian Amazon (Manifiesto, 2019). In addition to 300 soldiers, the mission included 1,272 police agents and 70 representatives from the Public Prosecutor's Office, all dispatched to an area where an estimated 30,000 to 40,000 unlicensed miners make their living (Ramshaw, 2019).

The idea of Operation Mercury was derived from the sporadic raids that have been conducted in the area for the last decade. In these raids, squads of soldiers and police officers have chased down hundreds, even thousands, of illegal miners living in transient mining towns. And while relatively few have actually been arrested, hundreds of thousands of dollars in mining equipment have been destroyed and fuel reserves have been confiscated along with the personal possessions of the miners. Despite all of this, miners usually return to the disputed areas within weeks (Post, 2016). The difference with Operation Mercury, says the Defense Minister, is that the military will remain until the area has been "completely reforested" (Peru Opens, 2019).

The aim of this chapter is to explore the recent rise in illegal gold mining in Peru by focusing on the Madre de Dios region, a center of gold rush activity and the inevitable damage it brings to the Amazon. We consider the history of illegal as well as "informal" gold mining, the work conditions of small-scale artisanal miners and their peculiar relationship to formal gold mining companies, the impact of neoliberal development policies, and current efforts to remedy environmental damage in Peru. Along these lines, new norms of environmental protection are emerging through "transnational environmental advocacy," a multitude of environmentally oriented networks arising from the intensification of the relationships between the industrialized North and resource-rich South in recent decades. Transnational advocacy involves both the commitment of international actors within government and international organizations, and the mobilization of civil society including indigenous, environmental, religious, and human rights groups, university experts, and NGOs.

At the same time, the steady drumbeat of neoliberal policy in Peru has produced an ever-widening gap between rich and poor with the latter seeking survival through dangerous work in informal metal mining. Because these workers are more often the target of environmental restrictions than large corporate firms, they typically resist this narrow sort of state-led effort at environmental protection. For this reason, any strategy that focuses on environmental protection without considering the needs of these workers will face vigorous opposition.

ENVIRONMENTALISM: JUSTICE AND PROTECTION

A look at the literature on environmentalism reveals a divide between localized movements for justice in poor communities and those of color, and broader movements for protection of the environment on a global scale. In both perspectives, neoliberal policies including the reduction of environmental protections, privatization of communally held lands, and removal of support for local agriculture while favoring of the extractive sectors, are seen as a major cause of environmental damage.

For environmental justice advocates, it is the poor and people of color who experience the worst effects of environmental contamination which is, in turn, perpetrated by the most privileged groups. Urkidi and Walter trace the origins of the movement for environmental justice to the mobilization in 1982 of a community of poor rural African Americans in North Carolina, against the dumping of PCBs into a landfill near their homes. Community members brought national attention to the issue by marching, protesting, and lying down in the road to prevent trucks laden with PCBs from proceeding to the landfill. In this, a civil rights orientation came together with the movement for environmental protection. In the years that followed, similar actions against landfills, dirty industrial plants, and other toxic projects were taken by Latinx, indigenous, and poor white communities (Urkidi and Walter, 2011).

For Newell, "race and class are key mediating structures in global environmental politics." As such, they are essential to understanding the process by which decisions are made about development and the distribution of harms and benefits that result from environmentally destructive projects. The author continues:

> While attention is drawn to the sorts of state-capital alliance that are responsible for environmental inequalities, either through blindness to the concerns of poorer groups or active exclusion of their voices within decision-making processes that privilege the interests of organized capital, systematic attention to racial and class dimensions is often lacking.

While academic accounts are guilty of this oversight, "activist oriented work" has been more attentive to the damaging effects that extractive projects impose especially with respect to indigenous peoples (Newell, 2005: 71–72).

Environmental justice movements go beyond resistance to specific environmentally harmful projects to challenge development policy and the decision-making processes that produce it. By pushing for greater inclusion in decision-making, the movement takes on a politically transformative character (Urkidi and Walter, 2011). Along these lines, Jacobs finds a strong and enduring commitment to environmental protection on the part of the urban poor in Brazilian favelas and a mutually reinforcing relationship between environmental politics and the struggle for democracy in that country (Guimarães, 1991).

Carruthers (2008: 4) distinguishes environmental politics in Latin America and from its counterparts in the United States and Europe. As he points out, "Latin American environmental groups are less able to advance agendas on their own terms" than those in the United States and Europe where large philanthropic foundations and government entities provide support. Latin American groups, by contrast, rely more on "denunciation and defense" incorporating broad-based issues such as land rights into the struggle.

In taking a constructivist perspective, Svampa (2015) cites David Harvey on "extractivism" as development strategy. This strategy, as he argues, is predicated on the idea that there is a consensus about extractivism that emerges from the modern global demand for raw materials. The dispossession of native peoples and others of land and resources in the form of strip mining, oil, energy, and gas exploitation, construction of large hydroelectric dams, expansion of commercial fishing, logging, and agribusiness is simply the price of progress. Any criticism of these activities is deemed antimodern or "infantile ecologism" promoted by nongovernmental organizations (NGOs) and foreign agents.

This development narrative is directly challenged by the Andean philosophy of *Buen Vivir* or "living well." *Buen Vivir* draws from ancestral conceptions of harmonious living conveyed in the Quechua and Aymara languages, the most commonly spoken indigenous languages in the Andean region (Salazar, 2015). This perspective recognizes the intrinsic value and "rights" of nature and conveys a demand for the restoration of the environment and ecological justice. Through a powerful movement comprising a coalition of indigenous groups, proponents succeeded in incorporating the idea of *Buen Vivir* into the Ecuadorian Constitution of 2008 as well as the Peoples' Climate Summit in Cochabamba, Bolivia, in 2010 (Svampa, 2015).

At times, leftist politicians in Latin America have allied with indigenous movements while, at others, the two have been at cross-purposes. Historically, the redistribution of wealth through taxation and social programs has been a hallmark of leftist political philosophy. During the recent era of rising commodity prices, leftist politicians who held sway throughout much of Latin America, often joined their redistributive policies to the extractivist model. Through taxation, the earnings from various extractive industries were directed into new programs and services for marginalized population. For indigenous and other rural groups, however, water shortages and environmental contamination continued leading to repeated clashes with government forces (Farthing and Riofrancos, 2017).

Marginalized populations including indigenous groups are not always at the forefront of the struggle to protect the environment, however. Questions have been raised about the level of unity and harmony that actually exist within indigenous communities with respect to environmental protection and heated debates have arisen over the suggestion that indigenous environmental knowledge has, in fact, been mythologized. Critics purport that indigenous environmental practices arise instead from the impact of historic events and exposure to Western notions of conservation (Dove, 2006).

The idea of "transnational environmental advocacy" arises from a pluralistic perspective and takes in the role of local individuals, groups, and associations along with various actors within government, international organizations, and NGOs. Baud et al. (2011: 80–81) see potential for "new approaches to natural resource management stress(ing) self-governance and higher levels of participation for civil society and private enterprises" as a consequence of the shift from state-centricity toward decentralized and privatized development. Over time, civil society groups including local, national, and international environmental groups, religious groups, university experts, human rights activists, political leaders at various levels of government, and NGOs, have "developed an alternative governance perspective for local users and communities."

Carruthers describes an internationalized framework arising from intensifying "social, economic, and political relationships between Latin America and the industrialized North" that are the consequences of global neoliberalism. These "solidarity links to the South present dense countervailing concentrations of the myriad 'transnational advocacy networks' that have arisen with various aspirations to humanize the workings of the global economy" (Carruthers, 2008: 11–12). Others describe the "glocalization" of environmentalism or the generation of new initiatives resulting both from international pressure and "bottom-up claims from civil society" (Baud et al., 2011: 80–81, 86).

Nonetheless, "persistent inequality, poverty, corruption, violence, limited institutional capacity and elite capture" conspire to deepen power inequalities and inhibit effective implementation of environmental initiatives. There is a need therefore for an integrated approach that considers the creation of structural, local development strategies while, at the same time, promoting environmental justice (Baud et al., 2011).

ALLUVIAL GOLD MINING IN THE PERUVIAN AMAZON

Peru is one of nine contiguous countries in South America containing an area of the Amazon forest. With more than 60 percent of its territory covered by Amazon forest, Peru contains the second-largest portion after Brazil (Orta-Martínez and Finer, 2010). About sixty different indigenous groups inhabit the Peruvian Amazon with around fifteen of these groups living in voluntary isolation to avoid exposure to deadly diseases. Indigenous groups hold land titles to 35.2 percent of the Peruvian Amazon with another 10 percent designated as national parks and sanctuaries (Orta-Martínez and Finer, 2010).

Small-scale mining has a long history in the Peruvian Amazon. Mining was first documented in 1887 by Antonio Raimundo, who explored the Madre de Dios area for its potential for alluvial gold mining. Between 1930 and 1960, and again in the 1980s, Madre de Dios received waves of migrant gold miners who joined local residents in the pursuit of gold. The Banco Minero, established in 1941, made credit available to "small mining producers" for the first time and remained the only effective source of institutional support for small-scale miners until it was closed in 1993. In 1971, the government passed a law formally recognizing small and artisanal mining. By the 1980s, however, formal mining companies were joining forces to stem the flow of incoming migrants and roll back the legal gains of small-scale miners (Valencia, 2015).

While Peru is one of the world's leading producers of cocaine, illegally mined gold has now surpassed cocaine as its largest illicit export. Estimates of its annual worth over the last fifteen years have ranged from US$ 500,000 to US$ 1 billion. The resulting rise in illicit gold mining in recent years has devastated the Peruvian Amazon, especially in the Madre de Dios region. Over 60,000 hectares of the region's forest has been destroyed, up from

10,000 hectares twenty years ago (Hill, 2016; US AID, 2018). Satellite photos of the region reveal large swaths of sand-colored land stripped of foliage and marked by pits that are twenty to forty feet deep.

This destruction owes largely to illegal deforestation along with the use of mercury to mine gold by unlicensed miners. A highly toxic metal, mercury is used to separate gold from sediment. While imports to Peru were officially stopped in 2015, its use by unlicensed miners does not appear to have slowed. In fact, miners in Madre de Dios dump an astounding forty tons of mercury into the rivers, streams, and wetland areas every year (Reuters, 2016). In addition, they fell trees, destroy vegetation and riverbanks, and turn water sources into lifeless discolored pools causing untold damage to the Amazonian ecosystem. Heather Williams' analysis of water samples from in Jila Central revealed a "suite of heavy metals" including lead, copper, zinc, and arsenic as well as mercury, all released through unlicensed mining in the area with devastating consequences. In addition,

> [T]oxic waste rock sediment from the informal mines began pouring onto fields and pastures through irrigation canals. Potato plants and barley withered. Pastures of hardy alfalfa grasses thinned and many sheep and cows who grazed in them died from intestinal maladies; lambs and calves were stillborn, some with horrendous deformities such as organs on the outside or distended bowels. (Williams, 2016)

Human health has also been seriously affected by the ingestion of mercury. Even small amounts in the body can cause significant damage including visual and hearing impairment, loss of coordination, muscle weakness, chronic headaches, and brain damage including in unborn children who suffer as a result of ingestion by the mother (Global Initiative, 2016). According to one Peruvian official, the health of an estimated 50,000 residents of Madre de Dios or 41 percent of the population, has been affected by mercury contamination (Jamasmie, 2016). The Carnegie Institution found that nearly four in five adults in the region's capital tested positive for dangerous levels of mercury (Weaver, 2018). In addition to direct contact during the gold mining process, mercury can enter the body by consuming contaminated fish and water, bathing in contaminated water, and breathing in mercury vapor (Global Initiative, 2016). For indigenous people, polluted rivers are often the only source of drinking water making them particularly vulnerable.

THE IMPACT OF NEOLIBERALISM

The implementation of neoliberal policies in Peru has also played a role in the surge in small-scale mining. From the 1990s on, government support for small farms and cooperatives were reduced while large-scale investments in mining and commercial agriculture were promoted.

In his study of development in Peru during this period, Tatsuya Shimizu found significant neglect of small farmers and rural businesses throughout the 1990s (Llosa et al., 2001). In 1991, the Agricultural Investment Promotion Law ended over two decades of agrarian reform that favored small farmers, communal agriculture, and small rural businesses. In its place, a system of individual titling of land holdings was instituted. Two years later, the new Peruvian Constitution legalized the division and sale of communally held lands (Deere and Leal, 2014). Limits on landownership were lifted and price controls and other obstacles to the movement of agricultural exports were reduced or eliminated altogether (Llosa et al., 2001).

Other measures cut back on subsidies for seeds, training, and agricultural inputs that small farmers had relied on (Llosa et al., 2001). The Agrarian Bank, which had made low-cost credit available to small farmers was closed as financial support shifted to large agricultural firms. Further, the Ministry of Agriculture was cut back to a fifth of its former size all but eliminating government mediation of land rights issues (Deere and Leal, 2014: 148–49).

The World Bank provides data for Peru from 1991 to 2017. The average value for Peru during that period was 32.82 percent with a *minimum of 27.41 percent in 2013* and a *maximum of 38.55 percent in 1993* (see figure 8.1) (Peru: Employment, 2019).

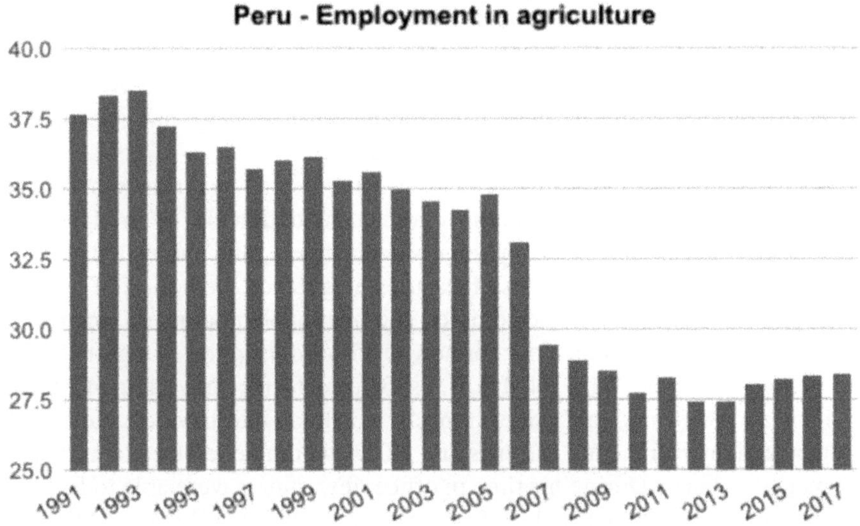

Figure 8.1 Employment in Agriculture, Percentage of Total Employment. *Source*: Graph created by author, data from theGlobalEconomy.com and The World Bank.

The resulting loss of farm labor in the countryside is reflected in the chart above where agricultural employment is shown as a percent of total employment in Peru between 1991 and 2017. From a high of around 38 percent in 1993, we see a steady decline to a low of about 27 percent in 2010. It then rises slightly and stabilizes at around 28 percent between 2015 and 2017.

The mining code was also rewritten at this time with the aim of liberalizing the mining sector. A 1991 decree removed restrictions and fees on the purchase, sale, possession, export, and import of raw and semi-finished gold. A law passed in 1995 granted mining companies the right to use land for exploration and drilling in return for financial compensation to existing owners. Mining companies also benefited from new exemptions on taxes and royalties and the liberalization of laws pertaining to the repatriation of profits. Tariffs on imports and exports were reduced and licensing procedures were simplified. As a result of these and other measures, Peruvian exports, of which minerals comprised about half, increased by over 85 percent between 1990 and 1998 (Bury, 2002).

For unlicensed miners, the "Law for the Promotion and Formalization of Small-scale Mining and Artisanal Mining," passed in 2002, set out a pathway for legalization. The process was stymied, however, under President Alan García, in office from 2006 to 2011, by the implementation of the *shock descentralista*. This "shock" consisted of a series of drastic reforms, which turned many central government duties over to regional authorities. Among these were the formalization and regulation of small-scale artisanal mining operations specified in the 2002 reform law. Amid the enormity and complexity of these processes, García's administration "failed to grant regional governments the necessary support to ease the transition" and regional governments received little in the way of the funding, training, and specialized staff needed for implementation (Law, 2012: 5).

The 2010 decree added further restrictions on small-scale miners including the suspension of new mining claims, the outlawing of dredges and other means commonly used by small-scale miners, and the establishment of mining exclusion zones. These measures heightened the notion of "illegal mining" and, together with the complex and overlapping nature of land rights in remote areas, opened the door for new levels of corruption in regional governments (Valencia, 2015: 14–15). Luis Otsuka, former governor of Madre de Dios and one-time gold miner himself, was known to have countered the influence of environmental activists and NGOs whom he referred to as "environmental mafias." During his time in office, few miners were able to formalize their operations and after he left, several officials at the Regional Ministry of Energy and Mines were imprisoned for accepting bribes in exchange for easing the process (Schwartz, 2019).

The completion of the Interoceanic Highway in Peru in 2011 further facilitated the flow of unlicensed miners into the region. A transcontinental highway linking the Pacific coast of Peru to the Atlantic coast of Brazil, the Interoceanic Highway wends its way through the region of Madre de Dios. A number of NGOs had warned early on that the highway could have a significant impact on the Amazonian environment including deforestation, soil erosion, and loss of biodiversity. As indicated on the map, the loss of forest and land cover is especially prevalent along this corridor as mining settlements have been established (Amazon Conservation, 2013).

As labor has exited the shrinking agricultural sector then, it has flowed into an expanding mining sector. Within this sector, informal mining is far more labor-intensive than formal mining. Where informal mining has accounted for about 15 percent of gold production in Peru, it is believed to employ between two-thirds and three-quarters of all who work in the mining sector (Williams, 2015). In Madre de Dios, the number of migrant workers flowing in over the last thirty years has more than doubled. With some 30,000 to 40,000 unlicensed miners, their production accounts for nearly 45 percent of the local economy of Madre de Dios and about 90 percent of the population depends directly or indirectly on this activity (Fisher et al., 2018).

As Fisher, Arora, and Rhee point out, these miners do not have the same ecological ties to the land as do indigenous populations in the area. The result is "a mosaic of economic land uses, and the introduction and expansion of a cash-based economy" (Fisher et al., 2018). Satellite images of the Peruvian Amazon taken between 1999 and 2016 reveal a steady process of deforestation only temporarily interrupted in 2012. By 2013, the high rate of deforestation had resumed as new mines opened up and, by 2016, some 1,287 acres of forest within the protected area had been destroyed (Rodríguez Mega, 2017).

THE SUPPLY CHAIN

The price of gold is affected by a number of factors including, it seems, public fear about the strength and stability of the economy. The chart below (figure 8.2) shows the price of gold per ounce between 1985 and 2015. On September 11, 2001, the day of multiple attacks on U.S. targets, the price of gold rose from US$ 273 to US$ 290 per ounce, a 6.2 percent gain (Aaron, 2017). Over the next decade, the price increased by an average of 18 percent per year or a total of 360 percent (Global Initiative, 2016). During the economic recession of 2007 to 2011, the price of gold tripled, reaching an all-time high of over US$ 1800 an ounce in 2012 (Smith and McCormick, 2018). The tax authority in Peru estimates that 35 tons of illegal gold valued at over US$ 1 billion were exported from Lima to the United States and Switzerland during just an eight-month period in 2014. While gold prices have slipped since then, they have remained fairly high.

Unlike cocaine, illegal gold loses its clandestine character as it passes into the world of legal markets. Drug traffickers diversify into gold not only because it is more lucrative than illegal drugs, but because it allows traffickers to appear as legal operators as it moves through the supply chain. With cocaine profits as "seed money," traffickers pose as legal traders by selling the metal through front companies to refineries outside of Peru. Once the agreements are reached, illegal gold is laundered into "clean cash" and underworld figures are deftly transformed into seemingly legitimate business people (Weaver et al., 2018).

Once raw gold reaches the refinery, it is very difficult to determine its origins. Comprising nearly 28 percent of Peru's total gold exports, illicit gold travels from Peru to Switzerland, Canada, the United States, India, and the UAE to be refined, then turned into jewelry, electronics, and other industrial products, or purchased by central banks or private investors (Global Initiative, 2016; Verité, 2013).[1] While passing through a series of intermediaries and exporters, gold from different places in the world may be mixed with stronger metals to form alloys, then combined with scrap gold. The gold is then purchased by manufacturers, central banks, and investors. An investigation by Verité, a labor research group, revealed that about 90 percent of the seventy-two companies examined "had purchased gold from refineries that had demonstrated a pattern of purchasing illegally mined gold." A collaborative report found direct links between some of the largest exporters of illegally mined gold from Peru and Colombia and refineries in the United States that provide gold to manufacturers and the U.S. Federal Mint (Verité, 2013). Peruvian prosecutors have also been persistent in investigating refineries for receiving illegally sourced gold (Robbins, 2019).

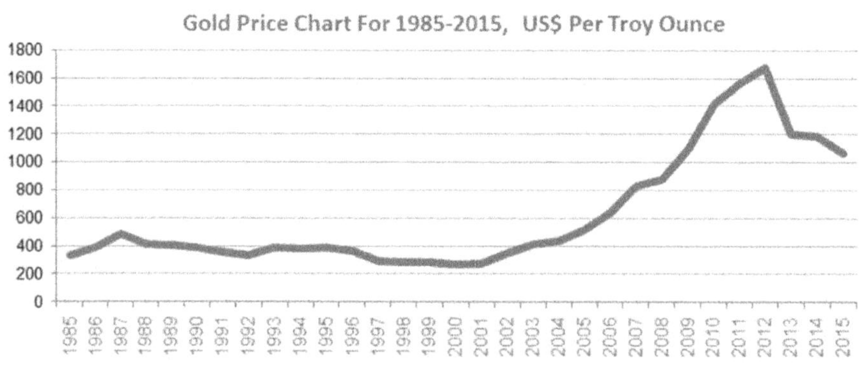

Figure 8.2 Gold Price Per Ounce, 1985–2015. *Source*: 30 Rates, 2016.

THE MINERS

Among unlicensed miners, the Peruvian government distinguishes between "illegal" and "informal" mining. On the one hand, "illegal mining" is carried out by those who have no intention of complying with laws regarding taxation, labor, or environmental protection. "Informal mining," on the other hand, refers to small-scale miners who have attained or applied for a concession but have not completed the various requirements for full legalization (Global Initiative, 2016). Failure to complete the many steps leading to full legalization is usually due to a lack of time, money, adequate information, and access to proper agencies. Some informal miners mistakenly believe that having a permit is enough. In other cases, the miners have filed on protected land or have conflicting claims to a land concession (Fisher et al., 2018). As one author notes,

> For mines to be considered legal and formal in Peru, they must have a title to the concession or an agreement with the concession owner, authorization to work the land, an approved environmental impact study, water rights, a certificate of non-existence of archaeological artifacts, authorization to carry out exploration and/or processing activities, and accreditation as a small-scale or artisanal miner, among other requirements. (Global Initiative, 2016)

Among informal miners, "small-scale artisanal miners" are defined by the size of their mining operations and the tools they use. Typically, these operations involve a few individuals or families, a cooperative, or moderately sized partnership and simple, often handmade, tools as opposed to the mechanized processes utilized by large formal firms. The "Tolba," for instance, a makeshift device layered with pieces of carpet used to capture gold deposits from river sediment, is commonly used by these miners (Bajakrodrigo, 2014). Small-scale artisanal mining is also more dangerous and lower-paid than work done by formalized mining companies (Global Initiative, 2016).

Informal mining operations can also be quite large in scale, involving millions of dollars of investment in heavy mining machinery and workforces of hundreds of migrant workers who arrive during gold rush periods. Because of the clandestine nature of these operations, the physical isolation of many mining sites, and the fact that migrants often lack identification papers, those who work in these larger operations are often vulnerable to exploitation by their employers. As a result, they may be denied adequate living conditions, forced to work against their will, or subject to extortion or wage theft (Verité, 2013). Webster describes the lives of informal miners in Madre de Dios in vivid detail.

> Standing waist-deep in muddy water, they chew coca leaves to stave off exhaustion and hunger. In the pit a minivan-size gasoline engine, set on a wooden cargo pallet, powers a pump, which siphons water from a nearby river. A man holding a flexible ribbed-plastic hose aims the water jet at the walls, tearing away chunks of earth and enlarging the pit every minute until it's now about the size of six football fields laid side by side. The engine also drives an industrial vacuum pump. Another hose suctions the gold-fleck-laced soil torn loose by the water cannon.

Webster learns that these miners earn US$ 70 most days but can make up to US$ 600. The work is dangerous and, among those working the thirty or so pits in the area, about four men die every week due to accidents such as the collapse of a pit wall. The miners' homes are made of wooden planks, many of them on stilts and covered with corrugated metal. Venders sell food, medicine, clothing, and auto parts along the streets where gold can be weighed, bought, and sold. Law enforcement agents generally leave this area alone having decided that it has already been sacrificed to mining (Webster, 2012).

The activities of formal and informal mining are not entirely distinct from each other. When a deposit is discovered, it triggers a gold rush bringing hundreds of miners into the area. These miners bring with them a wide range of extraction techniques from the rudimentary to the highly mechanized. Professional "middlemen" and transporters also arrive, bearing mercury, cyanide, drilling equipment, and explosives as well as food, water, and other necessities. Since hazardous substances such as cyanide and explosives are tightly regulated, only fully formalized companies have legal access to them. In fact, the middlemen who supply these items are often working for formal companies, whose accounting practices do not include these side activities (Kuramoto, 2012).

Formal firms also send their purchasing agents to collect shipments of minerals, whether raw or partially processed with mercury or cyanide, to be further processed by the firms. Legislative loopholes allow for underreporting and laundering of the resulting revenues (Kuramoto, 2012). Major firms also build roads to and from mining sites enabling them to control access to the sites. In Madre de Dios, many export companies

> buy gold from sources whose production has led to the destruction of forests, the contamination of rivers and sub soils with mercury, and the creation of multiple social problems. Furthermore, these export firms, often comprise criminal structures, involved in money laundering activities, which support the illegal and informal extraction of gold. (Friedli and Wiedmer, 2014)

In Peru, 60 of the 120 registered metal export companies have been investigated for possible financing of illegal mining, tax fraud, and money laundering. Among them, E&M has been investigated for exporting US$ 900 million worth of illegal gold to Switzerland while financially supporting illegal mining clans. Another exporter, AS Perú & CIA SAC, has been accused of money laundering related to gold sourced in Madre de Dios while UMT has been accused of money laundering, illegal gold mining, and selling gold to Russian Oleg Lipin, a major trader in illicit gold (Friedli and Wiedmer, 2014). As researchers from Verité note,

> Even the Peruvian government reports that its own legally constituted companies are being investigated for exporting illegally mined gold worth hundreds of millions of dollars. The corruption of government officials plays an important role in facilitating the link between major exporters and illegally mined gold. One investigation revealed that a company owned by the former Minister of Hydrocarbons within the Ministry of Energy and Mines had exported US$ 901 million in gold tied to the largest illegal mining clan in Peru (Verité, 2016).

Formal mining firms are also known to force indigenous and other local communities off their land in order to facilitate the mining process. Even when indigenous lands are legally protected, mining companies may still receive official grants of indigenous-owned land from the government prompting resistance on the part of indigenous groups. Three-way conflicts over land rights among indigenous peoples, informal miners, and formal companies are also common (Kuramoto, 2012).

Criminal organizations, both domestic and foreign-based, are able to take advantage of the geographic isolation of gold miners and the lack of official mediation where claims are concerned. These groups are quick to deploy existing paramilitary forces to intimidate miners and take over small-scale operations (Global Initiative, 2016). Among them are Shining Path in Peru, the Armed Revolutionary Front in Colombia (FARC), National Liberation Army (ELN), and criminal bands known as BACRIM all based in Colombia, the "Ndrangheta" based in Italy, and several based in China and Russia. Also included are family-based mining clans originating in Peru (Verité, 2016).

Environmentalists warn that, as the stakes become higher, these criminal organizations become more violent (Reuters, 2019). Between 2015 and 2016, the number of crimes against law enforcement officials in Madre de Dios more than tripled, while thefts doubled, and homicide was on the rise (Clavel, 2016). Around that time, environmental activist, Alfredo Ernesto Vracko Neuenschwander, was murdered near his home in the Tambopata Reserve, the likely victim of criminal groups involved in illegal mining. Vracko had filed numerous complaints regarding the activities of illegal miners in the area, but received no response from the government (Global Initiative, 2016).

Criminal organizations are also known to run human trafficking operations near the mining camps. One human rights organization found that 2,000 sex workers, over half of them minors, were employed in a single mining area of Peru (Weaver et al., 2018). Women and girls from other areas of Peru, some as young as twelve years old, are lured to the brothels through false offers of legitimate work in restaurants with good pay. Once the victims have been transported to the mining site, their IDs are confiscated, and they are told that they must pay for their clothing, lodging, and transportation, which they had understood would be free.

> If unable to pay, they may have to accept "*el pase*" (selling their virginity) under threats that their families could be harmed. Expert interviews indicated that new arrivals at the camps, especially virgins, are auctioned off to the highest bidder the first Friday or Saturday after they arrived at the camps. (Hill, 2016)

Many mine workers also find themselves trapped in abusive conditions. As with the trafficked women, they may arrive under false pretenses only to find out that they must pay for their own lodging, clothing, and tools even before work has begun. From there, the workers are paid a wage calculated to prolong that debt. This, together with the threat of violence, may make it difficult or impossible to walk away from the job.

State-centered Action

With rising awareness of illegal gold mining, pressure to take action increased under President Ollanta Humala, who took office in 2011. Promising sweeping action, he issued a decree in February 2012 setting out a four-to-eight-year sentence plus fines for anyone engaging in hazardous mining with respect to mineral, metallic, and nonmetallic resources (De Echave, 2019). The Ministries of Defense, Public Works, Environment, and Interior joined forces to deploy 1,500 troops along the Inambari, Madre de Dios, Tambopata, and Malinowski rivers. In raids against several mining camps in the region, these forces destroyed seventy-five dredges and seized fifteen vehicles from informal gold miners (Butler, 2011).

These raids were forcefully opposed by small-scale miners who protested that they were unfairly targeted while fully licensed mining companies were getting away with pervasive levels of contamination and environmental damage. In addition, they protested, the costs of full licensing were prohibitive. As one author notes, the requirements involved six administrative steps in connection to five government agencies and "were thoroughly out of step with the realities of most miners living hand to mouth" (Williams, 2016). In retaliation, the National Confederation of Artisanal and Small-Scale Gold Miners of Peru carried out a strike throughout most of the country. While Humala was able to negotiate peace in most areas of the country, the protest was fierce and prolonged in Madre de Dios where three were killed and several more suffered serious injuries (Dube, 2012).

Two years later, the government announced a nationwide ban on illegal mining and undertook one of the toughest efforts in Latin America. In partnership with Colombian authorities, the government ordered customs agents to verify the origins of wholesale gold shipments leaving the country. Illegal gold was confiscated at airports and illegal refineries in the coastal cities of Chala and Nazca were dynamited (Bajakrodrigo, 2014). In Madre de Dios, military personnel and police units were deployed to cut the supply of fuel, chemical inputs, and heavy machinery into the region. The mining town of La Pampa was entirely destroyed affecting some 5,000 miners and numerous businesses (Emery, 2014). Unauthorized dredges and gold mills were destroyed, and the personal possessions of miners and their families confiscated (Bajakrodrigo, 2014). In the end, however, the effort had little impact on illicit mining. This was, in part, the result of the cutback in sanctioning powers of the Ministry of Environment at that time, preventing it from issuing fines in the period that followed. With this, regulatory authority shifted to the powerful and mine-friendly Ministry of Energy and Mines (Williams, 2016).

Unlicensed miners have continued to clash with government forces in the period since then. In 2015, violent protests including rock throwing and blockades, erupted again in Madre de Dios (Williams, 2016). In July 2016, a two-day government operation resulted in the burning down of more than fifty mining camps, detention of six individuals on suspicion of human trafficking and the blowing up of dozens of motors that powered makeshift dredges in the mining pits (Reuters, 2016b; AP, 2016). Along the way, soldiers helped themselves to television sets, soccer balls, and other personal possessions belonging to the miners (Daley, 2016). The most recent incursion in Madre de Dios has followed the script established by these earlier raids. In a two-week period, military troops and police agents made arrests and destroyed millions of dollars in equipment and assets (El Comercio, 2019). The central government also pledged US$ 150 million for the development of sustainable alternatives including fish farming, agriculture, coffee, and cacao (La Pampa, 2019). Yet, no effective legal framework has been established for cleaning up contamination and restoring the forests or for making it easier to sue or fine major perpetrators.

In MDD, another group of interests poses even greater opposition to government clean-up efforts than unlicensed miners. According to former

Deputy Minister of the Ministry of Environment José De Echave, this group controls considerable territory in the region and represents over half the region's economy, and is well connected to informal and illegal mining. This group has penetrated the powers of the regional government including the judiciary and municipalities, and even at times, the office of the governor itself. Through their control of various public offices, the group has succeeded in shaping an economy that has become dependent on informal and illegal mining and is quite capable of thwarting efforts to end illegal mining. Combatting these powerful interests, says De Echave, will require a massive commitment of government resources that goes well beyond policing and raids to the development of a fully sustainable economy (De Echave, 2019).

From the perspective of formal gold mining companies, the raids would seem to serve a purpose. Formalizing and otherwise "cleaning up" the industry can effectively create new barriers to entry thereby reducing competition from informal miners and bolstering the price of gold. According to the president of Inca One Gold, a Canadian-based mineral processing company, the company backs this campaign "100 percent" and, in fact, plays a "pivotal role" in the formalization process (International Mining, 2014). In its corporate presentation in 2016, Inca Gold One enthusiastically noted that, "A MAJOR TRANSFORMATION of the industry is underway." Further down, it continues, "Low regulation and formality means lower barriers to entry," while "[h]igh regulation and formality means higher barriers to entry (Inca One Gold, 2016: 10–11)."

TRANSNATIONAL ENVIRONMENTAL ADVOCACY

The idea of transnational environmental advocacy featuring the role of civil society and various other actors, provides an alternative to state-centered thinking about environmental protection. In light of this perspective, we can consider the role of Peru's diminished yet determined Ministry of Environment. Over the years, officials within this ministry have partnered with various NGOs including Forest Trends, Ecodecisión, and CONDESAN, as well as the Swiss Agency for Development and Cooperation, to promote ecosystem support and low carbon development. The Ministry has also responded to the call by World Wildlife Fund (WWF) to address the problem of mercury emissions. This has included the promotion of National Action Plans for the development of mercury-free mining practices along with responsible retail and consumer purchasing (Fikri et al., 2018). WWF has also urged compliance with the Minamata Convention on Mercury of 2013 calling for the elimination of mercury in the mining process. Peru has responded by ending imports of mercury and stepping up training and technical assistance for informal miners to improve efficiency and reduce environmental contamination (Gonzalez, 2018; Kuramoto, 2012).

Peru has also responded to calls from other international organizations as well. Among them is the eight-member Amazon Cooperation Treaty Organization (ACTO) established in 1985, whose members include Bolivia, Brazil, Colombia Ecuador, Guyana, Suriname, and Venezuela in addition to Peru. ACTO provides a forum for management decision-making and a clearing house for information about activity in the Amazon. Since the spike in illicit gold mining in 2012, ACTO has put greater pressure on member countries to eliminate illegal mining.

In a more proactive fashion, Lima has hosted a number of international conferences having to do with environmental protection. In 2008, the fifth Latin America and Caribbean-European Union Summit was held in Lima. Peruvians also hosted the Conference of the Parties or COP20 in conjunction with the UNEP Sustainable Innovation Forum in 2014. Resolutions arising from this conference were incorporated into the Paris Agreement on Climate Change that emerged the following year. Along the way, Peru has constructed a number of plans for addressing climate change and environmental damage including a national payments program for ecoservices law, a Sanitation Sector Reform Law, a national strategy for forest conservation (World Wildlife Fund, 2019).

From out of these international involvements has come a new level of awareness of environmental destruction in the Amazon and elsewhere. Climate change and environmental degradation formed a major part of the agenda of the Latin America and Caribbean-European Union Summit inspiring the observation of "Earth Hour," a single hour in March during which all lights are turned off (World Wildlife Fund, 2019). In March 2019, nearly a quarter of the country's population in all twenty-four departments of the country participated. Activities have steadily expanded to include fairs and other educational initiatives for raising awareness about climate and the need to protect the biodiversity of the Amazon.

At the grassroots, indigenous and other longstanding communities also play an important role in protecting as well as restoring the environment. Over the years, these groups have organized and participated in local and country-wide marches, demonstrations, vigils, and speaker forums to draw attention to environmental degradation caused by mining. In Madre de Dios, indigenous groups have come together with environmental and agrarian associations to form the Alliance to Protect the Tambopata National Reserve aimed at reducing gold mining and revitalizing the ecosystem. Local individuals have also played a role. Native born Victor Zambrana has trained a number of area organizations, including one consisting of 5,000 families, in sustainable agricultural practices and has personally planted 19,000 trees of over 120 different species to reforest his family's land (National Geographic, 2016).

LABOR, LIVELIHOOD, AND GUARDIANSHIP

In Peru, as elsewhere, neoliberal policies have had multiple impacts on society, among them new vulnerabilities and new struggles in response. Government deregulation of the environment designed to attract foreign investment in mining has prompted open protest from indigenous groups, environmentalists, human rights organizations, religious groups, and others. This awakening of civil society is facilitated by a complex network of environmentally oriented interests and partnerships between North and South aimed at protecting the Amazon. Another consequence of neoliberalism has been the significant reduction of government support to the agricultural sector. This has been especially harmful to small farmers and rural workers. On being thrown out of work, these individuals and their families have managed to migrate into the margins of the mining sector where they survive in a dangerous and environmentally destructive business. While these miners have formed their own advocacy organizations, their struggle to survive often runs counter to the goals of transnational environmental advocates.

In an effort to bring convergence to these two struggles, many are calling for "a labor and livelihood" orientation to development. This approach considers the needs of miners, farmers, and migrants along with those of indigenous communities and the protection of the Amazon. It includes a call for maintaining and increasing government support for small farmers and communal agriculture along with the cultivation of other sustainable, labor-intensive livelihoods as alternatives to mining. With respect to land use, environmentally oriented rural associations have a central place in decision-making at all levels, and the notion of the right to prior consultation with indigenous communities is vigorously defended. For informal miners, it includes education and support around the goal of environmental protection, and health and safety standards, along with training in taxation and labor law as a pathway to legalization. It also includes efforts to protect these miners against land invaders and exploitation by more powerful legal and illegal mining operations.

Also consistent with this approach is the denunciation of environmental destruction caused by large-scale formal firms. Firms like Yanacocha in Cajamarca, for instance, use dynamite on a much larger scale than informal miners, carving out massive craters in the earth and releasing arsenic into the air and surrounding environment. Furthermore, despite much-vaunted efforts to contain them, toxic metals used to process the ore are frequently leached into the water supply through runoffs, leaks, and spills. Better monitoring of large-scale mines would include strict limits on the amount of water used and vigorous enforcement of prohibitions against the sale of cyanide and other toxic substances. Among the benefits of these measures might well be an increase in government legitimacy among informal miners.

Finally, the "labor and livelihood" approach incorporates a role for consumers as guardians of the environment. We can begin by remembering that the Amazon region is an ecosystem that serves not only indigenous tribes and other forest-dwelling communities, but the rest of the world as well. Its forests are essential in filtering and reprocessing harmful carbon dioxide and releasing oxygen. These forests stabilize the Earth's climate and provide a home to a great many of the world's plant and animal species. Our efforts to reduce our dependence on specialized metal-laden products and unnecessary upgrades, and recycle the metals in these products is a cost-effective alternative to mining that reduces deforestation, contamination, and climatic effects.

For many of us, the Amazon remains a romanticized place filled with cloud forests and exotic wildlife, an adventure that awaits. Yet for anyone flying over the Madre de Dios region, one glance at the barren, cratered consequences of illegal mining brings a stark reminder of its fragility.

NOTE

1. By comparison about 30 percent of gold mined in Bolivia, 77 percent of gold mined in Ecuador, 80 percent of gold mined in Colombia, and 80–90 percent of Venezuelan gold is produced illegally.

BIBLIOGRAPHY

30 Rates. 2016. Gold Price Per Ounce and Gold Price Per Gram. *The Economy Forecast Agency*. Available at: http://30rates.com/gold-price-per-ounce-and-gold-price-per-gram.

Aaron, C. 2017. Bullion Exchanges. September 11 and Gold Prices (Is It a Good Time to Buy?) 11 September. Available at: https://bullionexchanges.com/blog/2017/09/11/september-11-gold-prices.

Achtenberg, E. 2011. A Mining Ban in El Salvador? *NACLA*, September/October 2011. Available at: https://nacla.org/sites/default/files/A04405005_1.pdf.

Amazon Conservation Association. 2013. Fact Sheet: Illegal Gold Mining in Madre de Dios, Peru, March. Available at: http://www.amazonconservation.org/pdf/gold_mining_fact_sheet.pdf.

Associated Press. 2016. Raids on Illegal Mining Camps Target Brothels. *Daily Mail*, February 26. Available at: https://www.dailymail.co.uk/wires/ap/article-3465086/PICTURED-Raids-illegal-mining-camps-target-brothels.html.

Bajakrodrigo, F. 2014. Peru Security Force Trash Illegal Mining Machines. *Washington Examiner*, April 29. Available at: https://www.washingtonexaminer.com/peru-security-forces-trash-illegal-mining-machines.

Baud, M., Castro, G., and Hogenboom, B. 2011. Environmental Governance in Latin America: Towards an Integrative Research Agenda. *European Review of Latin American and Caribbean Studies*, 90, 79–88.

Bury, J. T. 2002. Livelihoods, Mining and Peasant Protests in the Peruvian Andes. *Journal of Latin American Geography*, 1(1), 1–19.

Butler, R. 2011. Peruvian Authorities Raid Illegal Gold Mining Operations. *Mongabay*, 7 November. Available at: https://news.mongabay.com/2011/11/peruvian-authorities-raid-illegal-gold-mining-operations.

Carruthers, D. V. (ed.). 2008. *Environmental Justice in Latin America: Problems, Promise, and Practice*. MIT Press: Cambridge.

Clavel, T. 2016. Illegal Gold Mining Fueling Crime, Violence in Peru. *InSight Crime*, 10 October. Available at: https://www.insightcrime.org/news/brief/illegal-gold-mining-fuels-crime-violence-in-peru.

Daley, S. 2016. Peru Scrambles to Drive Out Illegal Gold Mining and Save Precious Land. *The New York Times*, July 25. Available at: https://www.nytimes.com/2016/07/26/world/americas/peru-illegal-gold-mining-latin-america.html?hp&action=click&pgtype=Homepage&clickSource=story-heading&module=second-column-region&region=top-news&WT.nav=top-news&_r=0.

De Echave, J. 2019. Los Operativos en Madre de Dios. *CooperAcción*, February 20. Available at: http://cooperaccion.org.pe/los-operativos-en-madre-de-dios.

Deere, C. D. and Leal, M. L. 2014. *Empower Women: Land and Property Rights in Latin America*. University of Pittsburgh Press: Pittsburgh.

Dove, M. R. 2006. Indigenous People and Environmental Politics. *Annual Review of Anthropology*, 35, 191–208.

Dube, R. 2012. Peru Suspends Protests. *Wall Street Journal*, March 15.

El Comercio. 2019. Madre de Dios: Instalan Tres Bases Militares y Policiales en La Pampa. 3 May. Available at: https://elcomercio.pe/peru/madre-de-dios/madre-dios-instalan-tres-bases-militares-policiales-pampa-fotos-noticia-613673.

Emery, A. 2014. Peru's Illegal Mining Crackdown Slashing Gold Exports. *BN Americas*, 11 June. Available at: http://web6.bnamericas.com/en/analysis/miningandmetals/trend-perus-illegal-mining-crackdown-slashing-gold-exports.

Farthing, L. and Riofrancos, T. N. 2017. The State of the Left in Latin America: Ecuador and Bolivia After the Pink Tide. *NACLA*, July 19. Available at: https://nacla.org/news/2017/07/20/state-left-latin-america-ecuador-and-bolivia-after-pink-tide.

Fikri, E. G., Koster, H., and de Wit, W. 2018. Healthy Rivers Healthy People, World Wildlife Fund. Available at: https://www.wwf.org.uk/sites/default/files/2018-11/WWF%20-%20Healthy%20Rivers%20Healthy%20People.pdf.

Fisher, J., Arora, P., and Rhee, S. 2018. Conserving Tropical Forests: Can Sustainable Livelihoods Outperform Artisanal or Informal Mining? *Sustainability*, 10(8), 2586. https://doi.org/10.3390/su10082586.

Friedli, A. and Wiedmer, C. 2014. The Responsibility of the Gold Refineries in Human Rights Violation and Illegal Activities in Peru, Society for Threatened Peoples, March. Available at: https://www.gfbv.ch/wp-content/uploads/gfbv_peru_engl_web-1.pdf.

Global Initiative Against Transnational Organized Crime. 2016. *Organized Crime and Illegally Mined Gold in Latin America*, April. Available at: https://arcominero.infoamazonia.org/GIATOC-OC_Illegally-Mined-Gold-in-Latin-America-3c3f978eef80083bdd8780d7c5a21f1e.pdf.

Gonzalez, D. J. X. 2018. Treaty Does Not Stop Illicit Mercury Trade in South America. *Yale University*, February 13. Available at: https://yaleglobal.yale.edu/content/treaty-does-not-stop-illicit-mercury-trade-south-america.

Guimarães, R. P. 1991. *The Ecopolitics of Development in the Third World: Politics and Environment in Brazil.* Lynne Rienner: Boulder.

Hill, D. 2016. Gold-mining in Peru: Forests Razed, Millions Lost, Virgins Auctioned. *The Guardian*, May 1. Available at: https://www.theguardian.com/environment/andes-to-the-amazon/2016/may/01/gold-mining-in-peru-forests-razed-millions-lost-virgins-auctioned.

Inca One Gold Corp. 2016. Corporate Presentation, May, pp. 10–11. https://www.incaone.com/_resources/presentations/IO-Corporate_Presentation_May_2016.pdf.

International Mining. 2014. Peru Tames a $2 Billion Rogue Gold Industry, October 2. Available at: https://im-mining.com/2014/10/02/peru-tames-a-2-billion-rogue-gold-industry.

Jamasmie, C. 2016. State of Emergency in Peru Over Mercury Poisoning from Illegal Gold Mining. *Mining.com*, May 24. Available at: http://www.mining.com/state-of-emergency-in-peru-over-mercury-poisoning-from-illegal-gold-mining.

Kuramoto, J. 2012. Small-scale and Informal Mining: A Big Problem for Latin American States, Evidence and Lessons from Latin America (ELLA). Available at: https://assets.publishing.service.gov.uk/media/57a08a60e5274a27b2000583/121129_ECO_ExtIndConMan_BRIEF4.pdf.

La Pampa: The Illegal Mining City Peru Wants Wiped Out. 2019. *The Guardian*, March 25. Available at: https://www.theguardian.com/cities/2019/mar/25/la-pampa-the-illegal-mining-city-peru-wants-wiped-out.

Law, P. 2012. Artisanal and Small-scale Mining in Peru: A Blessing or a Curse? Peru Support Group. September. Available at https://www.perusupportgroup.org.uk/files/fckUserFiles/file/Artisanal%20and%20Small-scale%20Gold%20Mining%20in%20Peru.pdf.

Llosa, J. A. S., Gómez, R., and Shimizu, T. 2001. Modernization of Agriculture in Peru in the 1990s, March. Available at: http://www.ide.go.jp/library/English/Publish/Download/Las/pdf/01.pdf.

Manifiesto. 2019. Gobierno Instala Bases Militares en la Amazonía para Frenar Minería Ilegal. March 10. Available at: https://manifiestoperu.com/2019/03/10/gobierno-instala-bases-militares-en-la-amazonia-para-frenar-mineria-ilegal.

National Geographic. 2016. Environmentalists from Peru and Tanzania Win, National Geographic/Buffett Award for Leadership in Conservation, June 14. Available at: https://www.nationalgeographic.org/newsroom/environmentalists-from-peru-and-tanzania-win-2016-national-geographicbuffett-award-for-leadership-in-conservation.

Newell, P. 2005. Race, Class, and the Environmental Politics of Environmental Inequality. *Global Environmental Politics*, 5(3), 70–94.

Orta-Martínez, M. and Finer, M. 2010. Oil Frontiers and Indigenous Resistance in the Peruvian Amazon. *Ecological Economics*, 70, 207–18.

Peru: Employment in Agriculture. 2019. TheGlobalEconomy.com. Available at: https://www.theglobaleconomy.com/Peru/Employment_in_agriculture.

Peru Opens Military Base in the Amazon to Tackle Illegal Mining and Deforestation. 2019. *MercoPress*, March 7. Available at: https://en.mercopress.com/2019/03/07/peru-opens-military-base-in-the-amazon-to-tackle-illegal-mining-and-deforestation.

Post, C. 2016. Peru Spends Millions in Losing Fight Against Illegal Mining. *Peru Reports*, March 10. Available at: https://perureports.com/peru-spends-millions-in-losing-fight-against-illegal-mining/3432.

Ramshaw, D. 2019. A Military Solution: Will This Be Enough to Rid Peru of Illegal Mining? *Grey Dynamics*, March 26. https://www.greydynamics.com/greyzone/organised-crime/1337.

Reuters. 2016a. Peru Declares Mining-related Emergency in Remote Part of Amazon, May 23. Available at: https://www.reuters.com/article/us-peru-environment/peru-declares-mining-related-emergency-in-remote-part-of-amazon-idUSKCN0YE2C5.

Reuters. 2016b. Peru Renews Raids on Illegal Gold Mines as Wildcatting Spreads, July 15. Available at: https://www.voanews.com/a/peru-renews-raids-on-illegal-gold-mines/2865535.html.

Reuters. 2019. Peru Cracks Down on Illegal Gold Mining to Save Deforested Amazon Area. *NBC News*, February 20. Available at: https://www.nbcnews.com/news/latino/peru-cracks-down-illegal-gold-mining-save-deforested-amazon-area-n973551.

Robbins, S. 2019. Peru Targets Multinationals Linked to Sale of Illegal Gold. *InSight Crime*, April 10. Available at: https://www.insightcrime.org/news/analysis/peru-targets-multinationals-illegal-gold.

Rodriguez Mega, E. 2017. Amazon Gold Rush Continues to Destroy Peru's Rain Forest. *Scientific American*, November. Available at: https://www.scientificamerican.com/article/amazon-gold-rush-continues-to-destroy-peru-rsquo-s-rain-forest.

Salazar, J. F. 2015. Buen Vivir: South America's Rethinking of the Future We Want. *The Conversation*, July 23. Available at: http://theconversation.com/buen-vivir-south-americas-rethinking-of-the-future-we-want-44507.

Schartz, L. 2019. *The Nation*, The Other Manmade Disaster Ravaging the Amazon. August. Available at: https://www.thenation.com/article/peru-la-pampa-illegal-gold-mining/.

Smith, N. C. and McCormick, E. 2018. Barrick Gold: A Perfect Storm at Pascua Lama. In: *Managing Sustainable Business: An Executive Education Case and Textbook*, Lenssen, G. G. and Smith, N. C. (eds.). Springer: Dordrecht, pp. 227–42.

Svampa, M. 2015. Commodities Consensus: Neoextractivism and Enclosure of the Commons in Latin America. *South Atlantic Quarterly*, 114(1), 65–82.

Urkidi, L. and Walter, M. 2011. Dimensions of Environmental Justice in Anti-Gold Mining Movements in Latin America. *Geoforum*, 42(6), 683–95.

USAID. 2018. Illegal Gold Mining, October 2, 2018. Available at: https://www.usaid.gov/peru/our-work/illegal-gold-mining.

Valencia, L. 2015. Las Rutas del Oro Ilegal: Estudios de Caso en Cinco Países Amazónicos. *Sociedad Peruana de Derecho Ambiental*. Available at: https://www.cedib.org/wp-content/uploads/2015/08/las-rutas-del-oro.pdf.

Verité. 2013. Behind the Scenes of the Gold Report, Amherst, Mass. Available at: https://www.verite.org/behind-the-scenes-of-the-gold-report/.

Verité. 2016. The Nexus of Illegal Gold Mining and Human Trafficking in Global Supply Chains: Lessons from Latin America, July. Available at: https://www.verite.org/wp-content/uploads/2016/11/Verite-Report-Illegal_Gold_Mining-2.pdf.

Weaver, J., Nehamas, N., and Gurney, K. 2018. How Drug Lords Make Billions Smuggling Gold to Miami for Your Jewelry and Phones, January 16. Available at: www.mcclatchydc.com/news/crime/article194860749.html.

Webster, D. 2012. The Devastating Costs of the Amazon Gold Rush. *Smithsonian Magazine*, February. Available at: https://www.smithsonianmag.com/travel/the-devastating-costs-of-the-amazon-gold-rush-19365506.

Williams, H. 2016. Peru's Media Friendly Mining Ban Conceals Toxic Inaction. *NACLA*, February 11. Available at: https://truthout.org/articles/peru-s-media-friendly-mining-ban-conceals-toxic-inaction.

World Wildlife Fund. 2019. Peru Turned Off Its Lights for Nature, March 31. http://www.wwf.org.pe/en/?uNewsID=345275.

Part III

IDENTITIES, CULTURES, AND SUBJECTIVITIES

Chapter 9

Hidden Histories

Frontier Situations and Indigenous Agency

João Pacheco de Oliveira

The unit of social analysis we call the nation, with all the cultural assets that glorify it, is ensured in the violent processes of the submission of differences and the systematic and routine eradication of heterogeneities and autonomies. The facts and characters involved in these processes are objects of strong social control, which present themselves to succeeding generations in a ritualized manner, invariably institutionalized in certain forms of perception and narrativities.

In parallel, constructed memories are consistently accompanied by forms of forgetfulness that penetrate the minds of citizens through the most varied and diaphanous forms, even though they have never consciously learned and used such accounts. The construction of national identities thus operates by characterizing certain regions as internal borders versus frontiers, instituting certain social groups as the "others," re-elaborating the past with the systematic exclusion and re-semantization of historical facts, processes by which distinctions are made, and protagonists are transformed into uniquely distinct characters.

In this text, I focus on indigenous peoples and different boundary situations in reference to Brazil, in light of data produced through my own investigations (Pacheco de Oliveira, 2016) and the work of fellow historians, geographers, and anthropologists. I believe that national self-representations constitute strategic access routes for indigenous peoples and the Amazon as they confront historical process and social practices active therein. This frame allows us to identify the form of forgetting that in a sense distinguishes the country, as well as the conflicts and forms of resistance emerging with great constancy.

The elaboration of the first systematic historical account of Brazil, written by Francisco Adolpho de Varnhagen in 1854, laid the foundation for reflecting on the relationship between the nation and its indigenous populations. At this time, Brazil was considered essentially a project and creation of the

Portuguese. Varnhagen's central concern was to highlight the elements of continuity with the institutions, doctrines, and meanings encountered in the life of the mother country.

The Portuguese heritage was what explained the political, geographical, and cultural unity of the country, together with the forms of governance that accompanied the process of independence and the construction of the principal national institutions. The violence of colonial action was omitted or minimized and persistently justified. Of special relevance, the *Indians*[1] were always described negatively—as cruel, primitive, bellicose, and unproductive—and their actions were narrated as devoid of rationality and morality.

Thus, in the historical reconstitution that Varnhagen undertook, relying on official documents and administrative sources, the Indians merely constituted an obstacle to the creation of the colony. In the accounts contained therein, invariably it was the Indians who attacked farms and cities, killing the settlers and destroying the goods and wealth that they had created. The destruction of villages, the extermination of entire collectivities, enslavement, and the compulsory displacement of others, were seen as integral to "just wars," which were presented as pedagogical and humanitarian activities (catechesis) that were part of a so-called civilizing project.

What is more, this ideologically biased reading of history was reflected not only in the relationships between Europeans and indigenous peoples, but also other aspects of Varnhagen's comprehension of socioeconomic formation. For example, the poorly investigated complexity of the social fabric of the colony, which besides Portuguese and Indians, also included different categorizations of mixed races (*mamelucos*, *cafuzos*, *caborés,* and *curibocas*). Likewise, the enormous differences in the potential of the regions resulted in very different forms of indigenous incorporation, not a homogenous treatment nor the establishment of a unilinear history. Conflicts with other European powers, like those that occurred with the French and Dutch invasions and in questions regarding the limits of Spanish America, meant the Indians were sought out as important allies. Likewise, the constant riots, rebellions, and attempts at secession, in which Indians and other segments of the population were implicated, were approached superficially and simplistically, as mere curiosities and exoticisms.

INDIGENOUS PEOPLES AND RECURRENT RACISM

In his massive documentary research and in the writing of his work, Varnhagen followed the directives outlined by the naturalist Carl Friedrich Philipp von Martius (1845), whose monograph won the contest promoted by the Historical and Geographical Institute of Brazil (IHGB), in 1840, entitled "how one should write the history of Brazil." Although the recommendation contained within the contest was that the nation should be described as the result of the mixture of three races—European, African, and indigenous—equality of treatment has never been afforded to each of them, whether regarding attention to the presentation and recording of data, or in the interpretation of the facts. In effect, though in very different forms, Indians and blacks were always treated along clear racist and prejudiced paths, the former as merely part of an expanding frontier whose inhabitants were deprived of rights as a matter of principle, that is, seen as clear enemies; the latter because they lived under the legal condition of slaves, were treated as simple commodities.

Thus, there is an outrageous negation, persistently repeated and naturalized in important interpretations on Brazil, regarding the participation of indigenous peoples.[2] From the first great synthesis elaborated by Varnhagen in the mid-nineteenth century to the twentieth century Marxist historians, the indigenous presence in the formation of nationality is treated as an exoticism, the outcome of interactions of chance, minor incidents, and picturesque report circumscribed by in large to its cultural dimension. From Catholic monarchism, slavery, and conservatism to investigations into the formation of the working class, what has remained secondary in such narratives and interpretive panels is the ethnic and racial diversity of the country.

Paradoxically, this logic continues to exist well into the twenty-first century. There is little point in adopting a liberal or socialist ideology, or in using sophisticated scientific categories, if the majority of the interpretations about Brazil continue to ignore the specificity of the process of national formation and persist in reading the history of the country within exclusively Eurocentric parameters.

To better clarify some of the many paradoxes that these national historical accounts continue to pose, we need to incorporate the autochthonous populations and their impacts on the configuration of young nations as the objects of investigation. This means we need to stop reproducing a naive and ethnocentric account of the expansion of Europe and its institutions and beliefs, which treat the colonial worlds as if they were merely a chapter of the "universal" history of the West. History thus written only reflects the perspective of those who affirmed themselves as the colonizers and their heirs, never the perspectives of the autochthones and those who were transformed into subalterns.

IMAGES OF THE COUNTRY AND ITS BORDERS

National self-representations express recurrent modes of thinking from the beginning of the young new country regarding the configuration of its regions and the social types that could be encountered therein, images that we should treat analytically and critically. These self-representations have a huge impact on the framing of patterns of citizenship and morality, as well as on the performance of cultural and artistic activities, including the configuration of the future métier of historians, geographers, and anthropologists.

The famous painting by Victor Meirelles, the *Primeira Missa no Brasil* [First Mass in Brazil], based on a letter from the fleet's scribe, Pero Vaz de Caminha, depicts the arrival of Pedro Álvares Cabral's fleet (see figure 9.1). It was painted in Paris and in 1861, and was received at Rio de Janeiro with great pomp and circumstance. Much celebrated in the Second Reign (1840–1889), it soon became the official image of the country's emergence, and was reproduced repeatedly on text-books and even on printed money. It is Brazil's strongest self-representation, inculcated by schools and naturalized by the entire literate population.

In it, the Indians neither are the protagonists of History, at most they are unconscious witnesses, akin to the trees and animals, nor express any concern about their own destiny. As suggested by the painting, they seem to merge with nature, slowly losing their contours, in a sort of vegetable death, slow,

Figure 9.1 ***Primeira Missa no Brasil*** **(First Mass in Brazil).** *Source*: Victor Meirelles, 1861.

Figure 9.2 ***A Conquista do Amazonas*** **(The Conquest of the Amazon).** *Source*: Antonio Parreira, 1906.

painless, and inexorable. For the public, the conclusion is obvious and molds deep certainties—the Brazilian nation is to be constructed essentially by Europeans, and by them alone, establishing farms, churches, roads, and cities. Years later, in the twentieth century and under the Republic, this canvas was hailed by Capistrano de Abreu, a prominent historian and influential intellectual, as the legitimate "birth certificate of Brazil."

Nothing similar exists about the imagined views of the constitution of the Amazon. Yet, an attempt to create an equally celebratory image of the conquest of the Amazon through painting led the Governor of Pará, at the turn of the twentieth century, to involve in soliciting a work of art to be produced by the artist Antonio Parreira, which would also become an emblematic pictorial element (figure 9.2).

However, in contrast to what transpired in the work referring to colonial Brazil, Parreira's painting does not express an idealized image of colonization. Light strikes the naked image of an Indian woman, to whom the gazes of whites turn devouringly. Far from celebrating a pact for the formation of a colony or a nation, what the painter insinuates to his public is an explanation of the fragility of colonization and the specific, predatory ends that drove its agents. The lofty ideals of religious conversion and the imperial political project seem to be submerged in an image of exploitation and prey.

The difference in the representation of how the two Portuguese colonies in America (that of costal Brazil and the Amazon) were integrated into the process of constructing a history and a national image was already rooted in the preceding colonial history itself. The Portuguese administration instituted them in different overseas colonies, with very different economic and sociocultural configurations and temporalities.

In the Brazilian colony, the sites occupied by Europeans began as fortifications and strongholds housing commercial enclaves expanding into the *sertões* [backlands] with the implantation of mills and plantations. Beyond these protected areas came the farms for extensive cattle breeding. Olinda, Recife, Salvador, and Rio de Janeiro, among others, were cities that followed this political and architectural strategy, and in some respects differed little from medieval cities.

In contrast, the exploration of the Amazon proceeded mainly by waterways and occasional expeditions that favored extractivist activities, which were temporary in nature, materialized into expeditions to collect *drogas do sertão* (extractive products, dried fish, turtles, etc.). Almost all of this production was focused on export so that it did not necessarily assume the establishment of forts and urban centers in the interior of the country. Occupation of the interior occurred only through the implantation of missionary settlements, productive units that were solely dependent on indigenous labor and, therefore, on their acceptance of religious conversion and the work of foreign religious agents.

The fundamental economic factor for the occupation of Amazonia, both for the extractive expeditions and the establishment of missionary settlements, was indigenous labor, called "red gold" by Father Antonio Vieira, upon which the entire wealth of the region was constructed (Hemming, 1978). It was the search for and control over this essential factor of production, for which no other successful alternative was created until the 1870s, that the settlers, the religious orders and official authorities, established fierce disputes to impose their respective and mutual interests.

Against the idealizing self-representations echoing the belief in the centrality and protagonism of the white man, the current researcher must rather reassemble the past in another fashion, one that seeks to understand the emergence of wealth-generating structures based on inequality and territorial expansion. Indigenous slavery, which preceded the importation of African slaves, is thus shown to be fundamental to the establishment of this Portuguese colony of exploitation in South America, and institution that intimately and inexorably associated the production of wealth with genocide, territorial expropriation, the destruction of environmental resources, and varying modalities of coerced labor (temporary slavery, patronage, and tutelage).

If we were to search the colonial archives for images closer to what this "birth" might have been, surely Jean-Baptiste Debret's engraving, entitled *Família Guarani Aprisionada por Soldados Mamelucos* [Guarani family enslaved by mameluco soldiers], is much more suggestive (figure 9.3). The transformation of the autochthonous population, once free and autonomous into subalterns, an intrinsic violent and arbitrary foundational act, was the answer to dominant economic interests. This was so with the appropriation

Figure 9.3 ***Família Guarani Aprisionada por Soldados Mamelucos*** **(Guarani Family Enslaved by Mestizo Soldiers).** *Source*: Jean-Baptiste Debret, 1830.

of land inhabited by the Indians and the compulsory acquisition of labor, articulated with the multi-ethnic formation of the Brazilian family, the consolidation of the ruling class, and the structures of government. This transformation never dispensed with a process of genocide—euphemistically called "pacification"—which corresponded to the fabrication of a permanent state of war that, in practice, justified the complete denial of any rights to the autochthonous population.

FRONTIER SITUATIONS

It is important to remember that, in Brazil, national formation was not achieved through the homogenizing structures of a republic, but rather the creation of an empire that articulated large regional heterogeneities, coexisting with a long period of legalized slavery and the strong and official presence of the clergy in politics and administration. Together this conferred a very conservative and authoritarian character to young nation, much like the ways in which Enlightenment itself was conceived in Portugal, something that in no way resembles republican Jacobinism.

Immense regions that are currently part of the political map of the country were far removed from the economic and sociocultural life of the nation even four centuries after the so-called "discovery." Isolated from each other, they constituted potential boundaries, internal colonies, which at different times were the object of distinct forms of exploitation.

Although historians and sociologists of the past followed the representations of the ruling elite, who thought of such regions as "demographic voids," the current research of anthropologists shows that extensive and distinct indigenous peoples inhabited these places, regularly and permanently, while maintaining control over these territories and continuing to develop their cultural practices in conditions of relative autonomy.

Such was the case in the southern provinces, with the Kaingang Indians; in northern Minas Gerais, with the Botocudos and Xacriabás; in southern Bahia, with the Pataxós and Tupinambás; in the central-west region, with peoples of the Ge tongue; on the borders of the Chaco region, with the Terenas, Kadiwéus, and Guaranis; in Mato Grosso, the Apiakás and Parecis stand out; and finally in the immense Amazon valley dozens of peoples formed a mosaic of languages and cultures. Certain areas where resistance was stronger were found in Amazonas, Mato Grosso, Maranhão, and Roraima, such as those inhabited by the Muras, Mundurukus, Crichanás, Xavantes, Teneteharas, and Makuxis, respectively.

The pace of occupation of the interior regions of Brazil was very slow even beyond the Amazon. In the Atlas of the Empire of Brazil (figure 9.4), produced in 1868 by Cândido Mendes de Almeida, the state of São Paulo consists of twenty-six counties, twenty-five of which are carefully identified as the municipal headquarters, towns, parishes, and the European colonies located within them. In the last, Itapeva, which extended over more than half of the state, there was no sign of Brazilian presence and a legend reads: "Lands occupied by ferocious Indians." This is in São Paulo, the province next to the Court and only 150 years ago!

A map (included here as figure 9.5) by the Brazilian Institute of Geography and Statistics (IBGE) shows that 100 years after Independence, in 1920, moving away from the coastal strip, the population density of Brazil remained extremely low; in the vast majority of the interior it was of 0.5 inhabitants per square kilometer.

By virtue of this broad spatial dispersion (south, east, central-west, northeast, and Amazon), its inordinate ecological and cultural diversity, as well as the continental dimension of the young nation, the processes of conquest occurred at different moments in history, which were not always undertaken for the same economic reasons and were subject to distinct political formats. For this reason, we feel it is not possible or indeed meaningful to write a history of the Indians in Brazil, but rather explore the multiplicity of forms and temporalities that the relationships between indigenous peoples and Brazilian society have come to assume.

The war declared on the Botocudos of Minas Gerais and Espírito Santo, as well as the Kaingang people and others who occupied the fields of Guarapuava, in Paraná, occurred in the first twenty years of the nineteenth

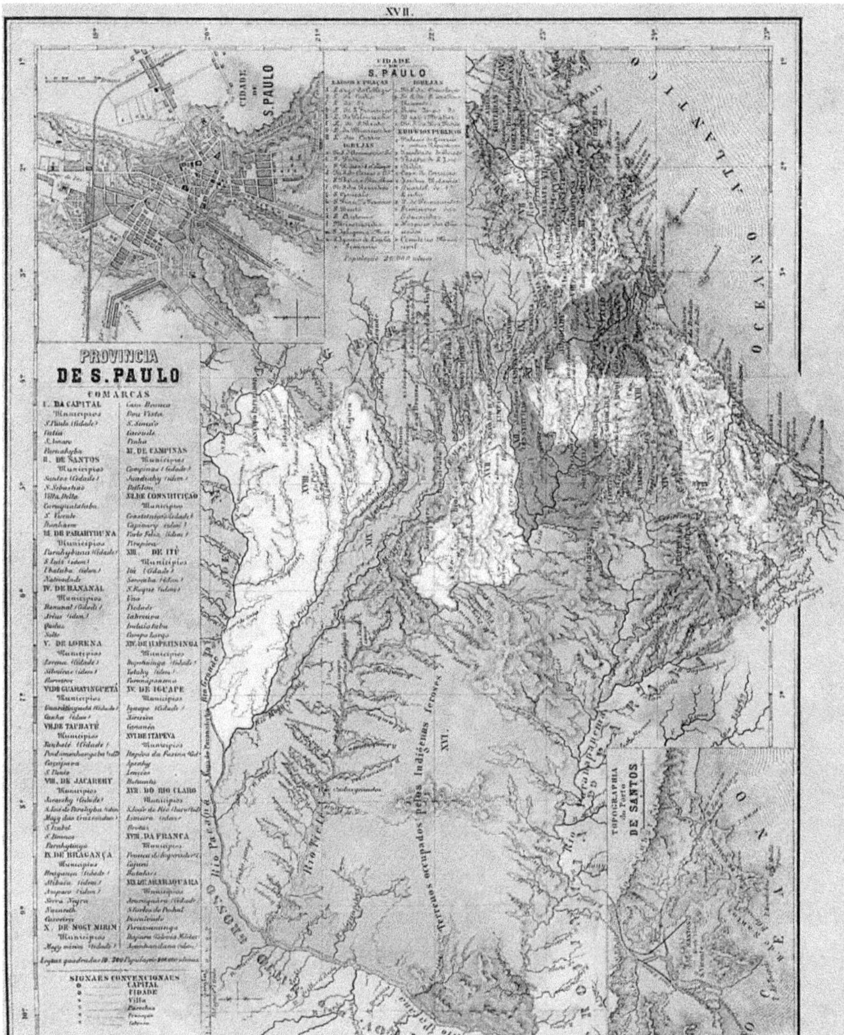

Figure 9.4 Atlas of the Empire of Brazil. *Source*: 1868.

century, beginning before independence in 1822. The religious missions deployed in the second half of the nineteenth century in various regions of the country (Amazonas, Maranhão, Bahia, Minas Gerais, Paraná States) were the spearheads of the colonists and *grileiros* [lit. land-grabbers], who stole indigenous territories. Extermination actions in the south against the Kaingang and Xokleng, carried out by *bugreiros* (Indian hunters) paid by

Figure 9.5 Population Density. *Source*: 1920.

farmers and settlers, extended until the first few decades of the twentieth century, as did similar actions in Mato Grosso against the Terenas and Guaranis (see table 9.1).

The conquest of the Amazon valley, reaching the most isolated areas, inhabited by "wild Indians," implied the extermination of indigenous peoples and communities in unknown numbers. This was carried out by rubber tappers during the expansion of rubber production in the late nineteenth and early twentieth centuries, especially between 1872 and 1911, and in the almost complete absence of the Brazilian state, a social process led by international markets and German, English, and North Americans banks.

The common fact, however, is that none of these conflicts were ever considered as important episodes in the life of the nation, but rather as strictly local events, supposedly devoid of major political or economic repercussions. Recurrent accounts of events involved the "surprise" of travelers and explorers who still encountered "wild Indians" in the nation's interior. In most

Table 9.1 Frontier Situations in Brazil

Frontier Areas / Brazilian States	*Indigenous Groups Involved*	*Period*
North of Minas Gerais and Espírito Santo (agricultural colonization projects)	Botocudos, Kakriabá, Krenak, Maxakali	1808–1815
Campos de Guarapuava (Paraná) (agricultural colonization projects)	Kaingangs	1808–1815
West of São Paulo (coffee plantation farms)	Kaingang, Terenas, Kadiwéus	1868–1920
Amazon Valley (rubber production) (Amazonas, Pará, Rondônia, Acre, Roraima)	(dozens of indigenous peoples)	1877–1920
Center-South of Maranhão (ranchos)	Guajajaras, Timbiras	1890–1940
South of Bahia (cocoa production)	Pataxós, Tupinambás	1910–1951
Southern Region (immigrant settlement areas) (Paraná, Santa Catarina, Rio Grande do Sul)	Kaingans, Xokleng	1910–1940
Central Region (ranchos) (Goiás, Mato Grosso, Pará)	Xavantes, Kayapós e Xinguanos	1945–1970
Legal Amazon Region (hydropower dams, roads, colonization projects) (Pará, Amazonas, Rondônia, Acre)	(numerous indigenous groups)	1970–1990

cases, the provincial authorities acted "pragmatically" in regards to the Indians who, echoing colonial chroniclers, were seen as an integral part of the "plagues" that "infested the *sertões*." Along these lines, considered as equivalent to snakes, mosquitoes, and numerous diseases that were endemic therein, the Indians were to be quickly and silently exterminated, thus removed from the path of "progress."

In the twentieth century, state intervention was conducted by a specific federal agency—the Indian Protection Service (SPI), followed by the National Indian Foundation (FUNAI)—which was driven by a protectionist and humanitarian ideology. During the "pacifications" of the Indians that the SPI promoted, the motto of its founder, Rondon, was "to die if need be, but never to kill." Although it represented official policy regarding indigenous peoples, the effective presence of the SPI was only a factor in very special cases, when conflicts with indigenous peoples were repeatedly reported by the press and gained a high profile in public opinion.

The pacifications conducted by the SPI, with great prominence in the press and the cultural and political environment of the Republic, occurred mainly in Mato Grosso and Rondônia in the first two decades of the twentieth century, affecting the territories occupied by Bororos, Parecis, and Nambiquaras (Souza Lima, 1994). In turn, the indigenous peoples and territories of the Upper Xingu region and parts of Central Brazil were only actually incorporated into the country's economic life in the 1940s and 1950s, together with several other regions of Mato Grosso, Pará, and Rondônia. Numerous indigenous peoples from Pará, Maranhão, Rondônia, Acre, and Amazonia only came to be contacted and settled by FUNAI in the 1970s and 1980s, in mitigating actions against large population displacements and environmental interventions resulting from the opening of roads (Belém-Brasília, Marabá-Santarém, Manaus-Porto Velho, Cuiabá- Rio Branco), the installation of hydroelectric and transmission lines, as well as legal and illegal mining activities and logging.

THE TUTELARY REGIME

Tutelage cannot be described on the basis of self-representations of indigenists and government officials alone, but must be approached ethnographically, as a sociological phenomenon, as we have proposed in several works (Pacheco de Oliveira, 1988a, 1998b, 2016). It is an exercise in coercive mediation that is supported by laws, but above all by uses and customs, and engenders and reproduces an "internal other," often in a biased and racist manner. The most varied forms of compulsory labor and modes of population confinement (ghettoization) are associated with the tutelary regime and the modes of construction of social group seen as the "internal others," that is, they have none of the rights generally attributed to citizens.

The position of tutor provides not only actions of protection and assistance, but also acts of restriction of rights and interests, repressive actions easily explainable by pedagogical intentions and ethical exemplarity. The tutelary condition establishes a social space legitimized for economic and political games in which the tutor's action oscillates between the interests attributed to the tutor and the counterpointing interests of circumstantial external agents.

Due to the diversity of forms of production and economic agents interested in indigenous territories, and the multiplicity of uses that could be attributed to indigenous labor, the exercise of tutelage assumes very distinct forms in the numerous regions that compose the country (Pacheco de Oliveira, 2016). The expansion of economic frontiers in the interior of the country did not involve disorderly and contradictory activities, but rather different adaptive experiences indirectly associated with political and ideological intermediations.

The radical polarity between Brazilian indigenist policy and that of other South American countries results largely from the intentional construction of a "showcase," which refers more to legal-formal, aesthetic, and philosophical self-representations and classifications than to concrete practices occurring in the different regions of the country. Exterminating or "pacifying" Indians, or placing them under the tutelage and protection of the state, while they may seem to be wholly contradictory actions and ideologies from the point of view of national self-representations, in the interior, they were often associated with the process of incorporation of indigenous territories.

When the occupation of the indigenous territories sought the implantation of highly profitable colonial enterprises, the extermination policy prevailed. This was the case with the arrival of Spaniards in search of precious metals in Mexico, the Caribbean islands, Colombia, and the Andes. In the 1880s, the newly created republics of Argentina and Chile, respectively, through the "Conquest of the Desert" and the "Battle of Temuco" (Briones, 2000; Lazzari, 2012; Pinto Rodrigues, 2003), used their armies to seize territories that were controlled by indigenous peoples and coveted by pastoral and agricultural enterprises.

In Brazil, given the important role that Indianism played in constructing a national identity, disseminating very positive representations of the Indians, in contrast to the Portuguese settlers, the expansion of frontiers could not happen in a similar manner. There were no wars and military campaigns against the Indians, and no official initiatives that directly involved extermination. However, the complete omission of government authorities allowed more than fifty indigenous peoples to be exterminated in the first half of the twentieth century.

And although the Indians were considered protected, the occupation of some frontiers was registered without any state presence, and extermination practices were freely conducted by private citizens. This was the case, for example, in the Amazon valley, during the "rushes" that occurred on rubber plantations (an activity inserted in an international financial network of native rubber production for North American industry), and in the south of the country with colonizing companies and European immigrants.

In other regions, where the potential for exploitation proved to be much lower or where other uses (above all geopolitical, in areas of international borders) were afforded the Indians, the policy of pacifying and protecting indigenous peoples through a state agency prevailed.

THE ORIGIN OF A FRONTIER

The generation of wealth in the colony, in the Empire, and the Republican eras, was never based on a closed economic system with limited resources and complementary (though antagonistic) roles in the production process. Surplus value is by no means the main factor of wealth generation in this society, where the holders of political power have never abandoned the so-called "primitive forms of accumulation,"[3] routinely transforming the appropriation of resources that comprise the basis of the way of life of indigenous populations. Decreed as "free" lands that are occupied by existing collectivities, considered "rudimentary" existing social practices, qualified as "criminals" who oppose you, these were recurrent arguments that justified the construction of an "internal other," to whom the rules that govern the coexistence between the other citizens do not apply.

A frontier is created whenever powerful groups of a nation associate to abolish the rights of another regarding the riches they possess and that they use for their reproduction and socio-cultural continuity. The present frontiers' situations in Brazil are not only located in the Amazon region, but can be found in the coastal strip of the northeast, where indigenous peoples and *quilombola* communities are being expelled from their former territories by large tourist and port enterprises, and where marginalized populations are being removed from valued areas in the peripheries of large cities.

As I have already observed in an article on the Amazon written several decades ago (1979), there are no natural frontiers, frontiers are political creations resulting from economic potentials activated by games of scale and structures of power. Approaching a nation from the point of view of frontiers is not to identify successive modes of production over time, but to seek to understand the complex process of the denial of rights, the construction of alterities and the exercise of tutelage that contributed to their formation and individuality in the most diverse manifestations of social life.

FINAL REMARKS

A large part of the university public and even Brazilian social scientists still consider the Indian to be a segment of the population with no demographic expression or major economic and political implications. They do not realize the importance that the Indian had and has in the country's recent history, as an inhabitant of the frontier situations in which the Brazilian economy came to exhibit its highest growth rates. This is the current situation of agribusiness and the production of riches for export (soybean, meat, and derivatives) that consider indigenous lands and areas pertaining to *quilombolas* and traditional populations as the main obstacle to national development.

Thus, the powerful ruralist faction established in 2015, in parallel with the impeachment of President Dilma Rousseff, a parliamentary commission of inquiry (PCI) to investigate the actions of INCRA (National Institute of Colonization and Agrarian Reform) and FUNAI in demarcations deemed wrong and fraudulent. In addition to these public agencies, the Brazilian Association of Anthropology (ABA) was violently attacked and investigated. In the final report, the PCI proposes the opening of criminal cases against important indigenous leaders and recommends the indictment of twenty-seven anthropologists who collaborated in conducting technical studies and producing expert reports.

What I want to propose here is another reading of the place of indigenous peoples and the frontier in the history of Brazil, seeking to write a kind of "history against the grain" (Benjamin) that helps to implode a vicious narrative structure unable to escape the self-representations of the elites and a merely legal-administrative history. There is an immense work to be done by the new generation of historians, geographers, and anthropologists. We must reread sources and historians of the past not as holograms, but as narratives engendered by social actors of their own time, since authors (as Pierre Bourdieu reminds us) have always written for their contemporaries. It is fundamental to investigate how only partial and fragmentary versions are consolidated in "historical truths," whose circulation and appreciation are carefully scrutinized by hierarchical, though sometimes contradictory, networks of institutions and powers.

The map of a country needs to be treated as a cognitive construction articulated with multiple interests, particularly from economically and politically dominant sectors, implying in parallel the exclusion of other views and projects (see Tomas Paoliello's recent doctoral thesis on the formation of the IBGE and geography itself as a university discipline). The traditional "colonial archives," full of lacunae and paradoxes, are susceptible to radically new readings, and need to be exposed to the memory records of the subalternized populations themselves, as well as to current political scrutiny.

We can no longer be just ethnographers in the "classical" sense, that is by limiting our investigative tools to the profile of an anthropology produced in a colonial context and with the deliberate erasure of its conditions of existence. This type of cognitive project engendered a peculiar form of sociocultural existence found in museums and in ethnological literature that is relativizing but also frozen and arbitrary and which could be conveniently grasped out of history, as criticized by Fabian (1982).

By refusing ethnography and ethnological theory to take as the object of the concrete forms in which indigenous collectivities have survived genocide and try to resist the multiple mechanisms of domination and subalternization—which aim to transform them into mere objects of administration—hegemonic and nonindigenous intellectuals decreed a kind of self-amnesty regarding the violent aspects of colonization, imposing an ethnographic invisibility of tutelage and transforming relativism into a unique tool of its ideological horizon.

Tacitly accepting the conditions of research conducted in a colonial context, ethnographers avoided investigating the processes of domination suffered by the Indians. They took sociocultural manifestations as if Indians proceeded from a permanent and immutable essence, completely immune to local relations and concrete political contexts, which in current idioms have been called "native ontologies" or even "cannibalistic metaphysics." Such anthropology is equally incapable of understanding how the indigenous peoples came to accept themselves as holders of rights, who are actively pursuing forms of empowerment and other forms of citizenship in the construction of nation-states.

NOTES

1. The appealing "Indian" is often used to refer to members of autochthonous communities, thus attributing a false and artificial unity to the native populations that preceded the arrival of Europeans in America. All persons thus classified are subject to special legislation (Law 6,001/1973 and Federal Constitution 1988). Indigenous leaders and intellectuals generally question this call and prefer to talk about themselves as "indigenous" by specifying to which people or ethnic group they belong.

2. The same could be said on Afro-Brazilians' studies until about sixty or seventy years ago. The work of sociologists, research into and studies on the African diaspora, and above all, the presence of intellectuals and black movements in the cultural and political life of the country has significantly changed this.

3. See Meillassoux (1973) and more recently Dorre (2010) and Mezzadra (2011) for thinking on primitive accumulation not as a circumscribed historical fact, but as a secondary form of exploitation, integrated with the economic and social reproduction of a social formation.

BIBLIOGRAPHY

Asad, T. 1973. Introduction. In: *Anthropology and The Colonial Encounter*. Humanities Press: New York, pp. 9–19.

Benjamin, W. 1986. Sobre o Conceito de História. In: *Obras Escolhidas: Magia e Técnica, Arte e Política*. 2nd ed. Editora Brasiliense: São Paulo, pp. 222–32.

Bensa, A. 2006. *Sur l'Exotisme: Essais d'Anthropologie Critique*. Anacharsis: Toulouse.

Biolsi, T. 1997. *Indians and Anthropologists. Vine Deloria, Jr., and the Critique of Anthropology*. University of Arizona Press: Tucson.

Bourdieu, P. 1982. *Ce que Parler Veut Dire*. Fayard: Paris.

Briones, C. 2000. *Pacta Sunt Servanda: Capitulaciones, Convenios y Tratados con Indígenas en Pampa y Patagonia (Argentina, 1742–1878)*. Vinciguerra Testimonios/IWGIA: Buenos Aires.

Buarque de Holanda, S. 1959. *Visões do Paraíso: Os Motivos Edênicos no Descobrimento e Colonização do Brasil*. Brasiliense: São Paulo.

Capistrano de Abreu, J. 1976. *Ensaios e Estudos: Crítica e História*. 2nd ed. Civilização Brasileira: Rio de Janeiro.

Carneiro da Cunha, M. (ed.). 1992. *História dos Índios no Brasil*. Companhia das Letras: São Paulo.

Castro, S. (ed.). 1985. *A Carta de Pero Vaz de Caminha. Descobrimento do Brasil*. L&PM: Porto Alegre.

Cezar, T. 2002. L'Écriture de l'Histoire au Brésil au XIXe Siècle. Essai sur une Rhétorique de la Nationalité. Le Cas Varnhagen. PhD thesis. École dês Hautes Études em Sciences Sociales: Paris.

Fabian, J. 1983. *Time and the Other: How Anthropology Makes its Objects*. Columbia University Press: New York.

Hemming, J. 1978. *Red Gold: The Conquest of the Brazilians Indians*. Macmillan: London.

Lazzari, A. 2012. ¡Vivan los Indios Argentinos! Etnización Discursiva de los Ranqueles en la Frontera de Guerra del Siglo XIX. *Archivos Virtuales de la Alteridad Americana*, 2(1). Available at: http://corpusarchivos.revues.org.

L'Etoile, B., Neiburg, F., and Sigaud, L. 2005. *Empires, Nations, and Natives. Anthropology and State-Making*. Duke University Press: Duham and London.

Lowenhaupt Tsing, A. 2005. *Friction: An Ethnography of Global Connections*. Princeton University Press: Princeton.

Mendes de Almeida, C. 1868. *Atlas do Império do Brazil: Comprehendendo as Respectivas Divisões Administrativas, Ecclesiasticas, Eleitoraes e Judiciárias* [sic]. Typographia do Instituto Philomatico: Rio de Janeiro.

Metcalf, A. C. 2005. *Go-Betweens and the Colonization of Brazil: 1500–1600*. University of Texas Press: Austin.

Monteiro, J. 2001. Tupis, Tapuias e Historiadores. Estudos de História Indígena e do Indigenismo. Thesis of 'Livre Docência'. Universidade Estadual de Campinas: Campinas.

Neiva, A.H. 1947. Aspectos Geográficos da Imigração e Colonização do Brasil. *Revista Brasileira de Geografia*, 2, 249–70.

Pacheco de Oliveira, J. 1988a. *O Nosso Governo: Os Ticuna e o Regime Rutelar*. Marco Zero/MCT-CNPq: São Paulo/Brasília.

Pacheco de Oliveira, J. (ed.). 1988b. *Indigenismo e Territorialização: Poderes, Rotinas e Saberes Coloniais no Brasil Contemporâneo*. Contra Capa: Rio de Janeiro.

Pacheco de Oliveira, J. 2016. *O Nascimento do Brasil e Outros Ensaios*. Contra Capa: Rio de Janeiro.

Pinto Rodríguez, J. 2003. *La Formación del Estado y la Nación, y el Pueblo Mapuche: De la Inclusión a la Exclusión*. 2nd ed. Dirección de Bibliotecas, Archivos y Museos: Santiago.

Raminelli, R. 1996. *Imagens da Colonização: A Representação do Índio de Caminha a Vieira*. Jorge Zahar: Rio de Janeiro.

Schwartz, S.B. 1988. *Segredos Internos: Engenhos e Escravos na Sociedade Colonial (1550/1835)*. Companhia das Letras: São Paulo.

Smith, A.D. 1986. *The Ethnic Origins of Nations*. Alan Sutton Publishing: Gloucester.

Sommer, D.W. 1993. *Foundational Fictions. The National Romances of Latin America*. University of California Press: Berkeley.

Souza Lima, A.C. 1994. *Um Grande Cerco de Paz*. Vozes: Petropolis.
Stocking Jr., G.W. 1991. *Colonial Situations. Knowledge Essays on the Contextualization of Ethnographic Knowledge*. University of Wisconsin: Madison.
Varnhagen, F. A. 1978 [1854]. *História Geral do Brasil Antes de sua Separação e Independência de Portugal*. Melhoramentos: São Paulo.
von Martius, K. F. P. 1845. Como se Deve Escrever a História do Brasil. *Revista Trimensal do Instituto Histórico e Geográfico Brasileiro*, 24(6), 381–403.

Chapter 10

Moving Beyond the Human-Nature Dichotomy

On the Cosmopolitics of the Amazon

Maria Fernanda Gebara

INTRODUCTION

Although a human-centered perspective has ruled humankind's approach to nature for millennia, we live in a time when human domination of nature is ever more the norm. Due to a series of profoundly transformative revolutions (e.g., cognitive, agricultural, scientific, industrial) our current geological epoch—known as the Anthropocene—has significantly intensified this type of domination (Harari, 2014; Manschot and Suransky, 2014; Noss, 2013). The concept of the Anthropocene, based on the notion that the Earth has moved into an epoch of unprecedented human domination of natural resources, is an increasingly prevalent framework for reflecting on contemporary environmental challenges, both in academia and in wider cultural and policy-making contexts (Malhi, 2017). Humans are an intrinsic part of the natural world, and the fact that human activity has a domineering influence over the natural environment gives us an opportunity to rethink how we view and interact with nature—and perceive our place in it. The concept of cosmopolitics (Latour, 2004a; Stengers, 2010, 2011) emerges as a useful frame for such a rethinking, and also for questioning the deeper political context behind the contemporary ecological crisis and its challenges.

In this chapter, I look at how cosmopolitics is being framed and provide a critique of the rationality in which contemporary structures of forest and natural resources governance are based. I argue that the failure of such structures to address central ecological problems has resulted from capitalist narratives (i.e., profit and growth orientation based on the commodification of natural resources). The cosmopolitics rhetoric, on the other hand, could represent an epistemic occasion to step outside of such rationality and think beyond the capitalist orientation to adopt an intellectual attitude that proposes, and thus creates, possibilities for new interpretations of natural resources and conservationism. In particular, the chapter asks how indigenous cosmologies—indigenous theories of the world (Lallemand, 1978)—might open up different practical ways of thinking and living, of conceiving the scope of our possibilities of being in the world, relating to the nonhuman and what we tend to perceive as "nature." I also look at Alexander von Humboldt's ideas of the cosmos and views of nature to complement such cosmologies from a Western perspective. Humboldt's angle comprehend nature in connection with humans and the movements of society, with language, economy, exchange, institutions, and architecture (Walls, 2009), and can therefore provide insights into the practice of cosmopolitics.

The importance of the Amazon forest for global climate regulation and as a host for a complexity of ecosystems is unquestionable (Lovejoy and Nobre, 2018; Nobre et al., 1991). However, from past hunter-gatherers to recent agriculturalists and urban settlements, the effects of human intervention in the Amazon (i.e., soil modification, climate alterations, biodiversity loss) are significant (Roberts et al., 2017). The largest remaining area of tropical forest in the world is currently threatened by increasing competition over natural resources (Fearnside, 2017; May et al., 2016). In recent years, the rate of deforestation has elevated sharply, and increased land use for agriculture and forestry activities is the second largest contributor to global climate change (INPE, 2018; IPCC, 2014). Moreover, in a globalized economy with its tremendous demand for forest and agriculture commodities, the Amazon has become the womb of the world's economy—the fertile grounds that provide much sought-after agricultural and extractivist commodities for countries like China, the United States, Russia, and Saudi Arabia, among others.

In the context of these facts, notions such as cosmopolitics demand that the broader political context is reframed in a way to allow a new approach to thinking about our interaction with nature (Latour, 2004a). In fact, the cosmos acts as an equalizing operator, meaning that all impacted parts of the cosmos, including what may be called "earth beings" (de la Cadena, 2015), should be considered in any political decision-making process. Cosmopolitics, then, is about decision-making in full awareness of the consequences for *all* beings on planet Earth. All beings are indispensable in the cosmos and thus in cosmopolitics. The prevalent political system, however, greatly influenced by capitalist ideas, displays little appreciation for the various ways we can relate to other "earth beings," including nonhuman beings (Sullivan, 2017). Capitalism is an economic system based on the private ownership of the means of production and their operation for profit. In such a system, natural resources are seen by different thinkers as scarce (Heyne et al., 2014;

Krautkraemer, 2005), and what prevails is the economic imperative of supply, demand, and competition. In contrast, the diverse and ancient practices of indigenous peoples in the Amazon have focused on care, respect, and sharing between humans and nature. The Amazon is a rich domain of bio and human diversity, and it is this diversity that creates ecosystem balance, which in turn perpetuates biological abundance, as opposed to scarcity.

Based on capitalist views of nature, the concept of "ecosystem services"—notably used in the Millennium Ecosystem Assessment and the Economics of Ecosystems and Biodiversity processes between 2005 and 2010—emerges as key to economic incentives for supporting environmentally positive actions (e.g., payment for environmental services; REDD+) that have been seen as innovative policy tools for forest conservation worldwide (Clot et al., 2015). Ecosystem services are defined as the tangible and intangible benefits provided by ecosystems that contribute to making human life both possible and worth living, such as water, climate regulation, biodiversity, and others. Such terminology exemplifies a contemporary shift in both ecological and economic spheres, whereby "nature" is consolidated as "natural capital" and the functions of natural systems as ecosystem services (Sullivan, 2017). Previously distinct domains (economics and finance vs. ecology and conservation) are coming together in an emerging "green economy" (Pretty, 2013) that connects actors, institutions, and calculative technologies through specific discourses of rational action and behavior. Here I question such discursive shifts that are part of the ecological modernization movement (Ewing, 2017), which underpins the creation of markets for ecosystem services, including for carbon, and are critical in constituting and normalizing the logic of forest governance in the Amazon. Markets then became a central political and epistemological object in the formation of a distinct neoliberal art—the governing of natural resources (Gebara and Agrawal, 2017; Robertson, 2011).

This chapter is structured as follows. The first section builds on Humboldt's ideas of the cosmos and views of nature, adding to Stenger's studies of cosmopolitics and Latour's analysis of the politics of nature in innovative ways. It questions how re-imagining current political narratives, mainly based on Western rationality, might inform the possibilities for a critical political reframing and response. The second section discusses the antagonism between contemporary structures for forest and natural resources governance and Amazonian indigenous cosmologies and views of nature. This is an original contribution to the debate on cosmopolitics, supplementing its focus on science and politics with attention to indigenous cosmologies in order to provide insights that will facilitate the practical implementation of cosmopolitics. The third section then links aspects of cosmopolitics with indigenous cosmologies, to discuss the phenomenological path toward thinking of humans and other-than-human beings as earthed and embedded with, not in, their environment (i.e., "earth beings"). Finally, the conclusion highlights how a practical engagement with indigenous cosmovisions may help us to move beyond current political narratives, and better integrate our choices with the different ways in which we can collaborate with our natural environment.

COSMOPOLITICS: RE-IMAGINING CURRENT POLITICAL NARRATIVES

Western Rationality and the Human-Nature Dichotomy

This sub-section provides insights on how the views of nature that form the basis of hegemonic political narratives were born, in order to then move on to the cosmopolitics debate as a response to such rationality. Deeply ingrained in Western thought since the Enlightenment, the human–nature dichotomy, and the assumption that natural entities are unalterable "givens," are still considered common sense. The distinction between humans and nature reflects a long-standing assumption that nature is external to, and different from, human beings. As Neil Smith (1984: 2) puts it, "external nature is pristine, God-given, autonomous; it is the raw material from which society is built." Thus, Western rationality has taken nature as an inherently nonsocial and nonhuman entity, captured in the term "the environment."

Hegemonic political narratives have historically been based on the antithetical separation of "Humanity" and "Nature" (de la Cadena, 2010). This, together with a second Western antithesis between supposedly superior and inferior humans, has led to the gradual extinction of many other-than-human beings and the worlds in which they existed. These processes, it is worth stressing, are tied to generalized processes of colonialism that have sought to erase diverse and pre-existing indigenous cosmologies and ways of life; to draw a line in time that leaves behind such practices in order to foster and establish new realities, new human-nonhuman and human-land relations, new collective and individual accounts of meaningful life from a Western perspective. Intersecting here are forms of technological rationality that have allowed Western societies to think of themselves, in ways often taken to be self-evident, as masters of the Earth (Marcuse, 1982). This prevailing technological rationality also works in processes of rationalization that have significantly reordered the terms of social life and worldly relations (Brown, 1993; Kalberg, 1980). The story of the contemporary ecological crisis is, in this sense, the story of the consequences and often unforeseen limits of these intersecting logics.

Of particular relevance here, and intimately tied in with the continued domination and exploitation of natural resources in the Americas, was the flow of newly available American resources to Europe from the late 1400s onwards, and the creation of "cheap natures" (labor, food, energy, and raw materials), which facilitated the subsequent rise of Europe to geopolitical dominance, and ultimately the Industrial Revolution (Malhi, 2017). This period also marked the beginning of the modern world, and it is seen by many as the beginning of the capitalist world economic system and of the Anthropocene, also known as the Capitalocene (Moore, 2014). It was, indeed, an upward inflection in global sociometabolism, where the resource consumption per unit of human population increased as energy and materials use intensified (Fischer-Kowalski et al., 2014). Regardless of how this new epoch is named (Anthropocene, Capitolecene, Chthulucene), the phenomena involved are complex and multifaceted. If we wish to make sense of some of the more relevant aspects of these transformative events, we should begin with the shift in patterns of subsistence and land use associated with the development of modern market capitalism, beginning in Europe and spreading through processes of dispossession and privatizing enclosure, often under directly colonial terms. This restructuring of relations to the land has objectified, in a generalized way, modern ideas of ownership as well as land improvement or "development," through the human reworking of a natural world conceived as passive material for human labor (Polanyi, 2001).

These features clearly demonstrate that "nature" is currently unavoidably embedded in a complex political and economic reality that involves the definition of values through different processes of decision-making to establish who does what and for what purposes. Equally importantly, such a complex reality involves ethics and morals, and is thus ultimately about the ways society-human-nature relations are framed and organized (Ankersmit, 1996; Weber, 1946). Although governments have the legal power to control and regulate society–nature interactions, in practice governance and politics are exercised by public and private actors at multiple scales, from the state, the market, and civil society (Gebara et al., 2014). This is the main reason why it is important to put the question of "who defines nature, for what purposes and in whose interests?" at the center of the cosmopolitics debate. With the melting of the icecaps, the acidification of the oceans, deforestation, widespread pollution of the earth and waters, the accelerated extinction of species, and generally the demonstrably cumulative impact that modernity, its capitalist system, and its vision of development is having, we are now in an ethical position with no precedent for various decision-makers.

It is impossible, however, to physically disentangle the social and the natural. In reality, all there is—to borrow political ecologist Erik Swyngedouw's (1999: 443) apt neologism—is "socionature." Or in Latimer's and Miele's (2013) words "naturecultures." What does this mean? It is certainly not a denial of the material reality of those things we routinely call natural—be they trees, rivers, animals, or anything else. Rather, it is an insistence that the physical opportunities and constraints nature presents societies with can only be defined relative to specific sets of economic, cultural, and technical relations and capacities (Castree and Braun, 2001). In other words, the same "chunk" of nature—say the Amazon forest—will have different physical attributes and implications for societies depending on how those societies perceive and use it. In this sense, such attributes of nature and the ways different societies view them are contingent upon social practices; they are not fixed. As Slater (2002) metaphorically puts it, the Amazon is a shape-shifter, as that of a kaleidoscope, constantly turning, causing images to vary when seen by different persons and from diverse angles. For her the different stories from the different inhabitants of the Amazon, or what she calls the "insiders" and "outsiders" (of African, Japanese, Arab, Sephardic Jewish, indigenous, Portuguese, and nordestino origins) "are clues to our own thinking, and to others' conception of a rainforest that is both place and icon, natural creation and ongoing human invention" (p. 203).

Critically, what I want to address here is how the relation between nature and culture (and with this the human and the nonhuman) is enacted and performed as division and dichotomy. Such dichotomy proves inadequate for addressing the problems that face not just science but also more general understandings about how humans live in relation to the rest of the world they co-inhabit (Venn, 2010). The idea that reforming humans' relations to nature must involve learning to think of humanity not as disembedded from it, but as relationally constituted within it, has been once voiced in various iterations and for some time in Western contexts. By using unexpected analogies, Humboldt's essay *Naturgemalde*, for example, unpeeled a previously invisible web of life. Connection was the basis of Humboldt's thinking, and his fundamental view of nature as a whole (Humboldt and Bonpland, 2009). Back in the nineteenth century, Friedrich Schelling's (1988) philosophy of nature—the *Naturphilosophie*—was a call for "the necessity to grasp nature in her unity." He rejected the idea of an irreconcilable chasm between the internal and the external—between the subjective world of the self and the objective world of nature. Instead, Schelling emphasized the vital force that connected nature and human beings, insisting that there was an organic bond between the self and nature. "I myself am identical to nature," he claimed.

Following Schelling's ideas, Humboldt's nature encompassed humanity rather than the other way around; the natural world was the source of order, the whole. The following sub-section looks to Humboldt's ideas in an attempt to better incorporate the thoughts of the so-called "rediscoverer of the Americas" (Wulf, 2015) into the cosmopolitics debate. I choose to focus on Humboldt's work instead of other approaches to being "embedded" or "placed" in the world (e.g., Heidegger, Deleuze) because Humboldt's views, based in his unique empirical experiences in the tropics, are better linked to the human and other-than-human interactions within the cosmos and can, therefore, provide useful insights into the practice of cosmopolitics.

Coupling Humboldt's Ideas with the Emerging Cosmopolitics

The word cosmos was first used by the Greeks, Pythagoras, and later Plato and Aristotle, to refer to the order of the universe. Early evidence suggests that Pythagoras presented a cosmos structured according to moral principles and significant numerical relationships (Burkert, 1972). The term only became part of modern language two millennia later, however, with Humboldt's five-volume treatise, *Kosmos*. Humboldt's ideas influenced modern and somewhat holistic perceptions of the universe as one entity, interacting with the cosmos in a "harmoniously ordered whole" (Humboldt, 1997: 24). His science of the cosmos "recalls the mind of the inhabitant of the Earth" (Humboldt, 1997: 68), meaning that its horizon is much wider than any nation or region; it embraces "the assemblage of all things with which space is filled, from the remotest nebulae to the climatic distribution of those delicate tissues of vegetable matter which spread a variegated covering over the surface of our rocks" (Humboldt, 1997: 71). Humboldt is seen as the founder of the natural sciences, creator of a grand theory through which he sought to link all physical elements of the world, including every kind of being, in a web of interdependence. Humboldt's work thus provides fruitful ideas for the cosmopolitics debate.

Cosmopolitics emerges as a way to transform the war that has so far ruled silently through a singular biopolitics of improvement into what Isabelle Stengers calls a politics where "cosmos refers to the unknown constituted by these multiple, divergent worlds and to the articulation of which they would eventually be capable" (Stengers, 2010: 995). Cosmopolitics puts the cosmos at the center of politics, meaning that every being included in the horizon of the cosmos is a political being, or a being that politics should consider (de la Cadena, 2010). The main question is: how do we do it? It is certainly a challenge to the way governance has been implemented in recent years; Humboldt's "inhabitants of the Earth"—or what de la Cadena (2015) calls "earth-beings"—have mostly been absent from the hegemonic political sphere. The recent appearance of earth-beings in indigenous protests (e.g., anti-mining movements in Peru and Ecuador, celebratory events in Bolivia), however, may reveal a moment of rupture in modern politics, and an emergent "indigeneity" governance (de la Cadena, 2008). It marks an insurgence of indigenous movements, a growing acknowledgement of their ideas with the capacity to significantly disrupt prevalent political formations, and reshuffle hegemonic antagonisms, first and foremost by rendering illegitimate (and, thus, denaturalizing) the exclusion of indigenous practices from nation-state institutions (de la Cadena, 2010).

For Latour (2010) the "things" that indigenous movements are currently "making public" in politics are not simply nonhumans; they are sentient entities whose material existence—and that of the worlds to which they belong—is threatened by the neoliberal wedding of capital and state. Indigenous rights have failed in the face of strong pressure from the effects of neoliberal capitalism, and Amazonian countries are still structurally dependent on extractive industries and transnational capital. What emerges is the domination of powerful sectors in an economy structured within our capitalist society and political systems (Gebara, 2018). In cosmopolitics, politics should not only be composed of power relations and silenced antagonisms based on the prevalent economic system—it should, as Rancière has already argued, be "made up of relationships between worlds" (2003: 42). By articulating the form of a democratic political gesture, Rancière (2006) refuses the reduction of the political to the economic (or to its concomitant biopolitical sphere), while recognizing that the space of the genuinely political can only exist as an edge that must be repeatedly performed anew, by construing political action as the invention of new spaces of political contestation in an ongoing making and unmaking of politics.

Cosmopolitics could then be seen as one of Rancière's new spaces of political contestation that might then incorporate efforts by scholars and activists to imagine, produce, and govern more liveable worlds in which other-than-human beings are also respected by different societies and cultures. Cosmopolitics would also confront the failures of present political and scientific systems to adequately represent and radically alleviate widespread suffering among diverse beings, of which humans make up only a fraction (Kirksey and Helmreich, 2010). Such cosmopolitics, therefore, requires acknowledgement of the plural. As Schmitt phrases it (1996), the political world is a pluriverse, not a universe. However, the pluriverse, a world where many worlds fit (Escobar, 2018), the multiple worlds that Schmitt deemed crucial to enable the possibility of the political, has disappeared. Instead, a single world has made its appearance, inhabited by many peoples and cultures more or less distanced from a single "Nature" (Descola, 1996; Haraway, 1991; Latour, 1993; Viveiros de Castro, 2004). In treating the cosmos as an ecology of practices, meaning an acknowledgement that no practice can be defined as "like any other" but always as diverse, just as no living species in ecology is like any other, Stengers (2010: 10–11) conceives cosmopolitics as the process of affirming Humboldtian approaches to the environment and peoples, by tracing the practical interactions of multiple organisms, as needed to comprehend nature as a living diverse whole. Similarly, when Latour (2004a) argues that democracy needs to "be extended to nonhumans" (p. 223) by drawing from "ecology and social sciences" (p. 226), he is also echoing Humboldt's intense preoccupation with multispecies' relationships.

A truly mature response to the challenges of forest and natural resources governance would likely result in an embrace of Humboldt's ideas around cosmos and views of nature. A close reading of Humboldt's major writings reveals the extent to which he developed a socially conscious ecology, an almost postmodern awareness that nature and culture are inextricably linked, yet he also felt a profound respect for nature's diversity. If, as Tzvetan Todorov has argued, cross-cultural understanding depends on "an affirmation of the other's exteriority" (1982: 250), perhaps healthy cosmopolitics depends on a similar construction of humans' relationship to the natural world. The next section will elaborate on the antagonistic constructions of human-nature relationships that become evident when comparing contemporary structures in forest and natural resources governance with indigenous cosmologies.

THE ANTAGONISM BETWEEN CONTEMPORARY CONSERVATIONISM AND INDIGENOUS COSMOLOGIES

My intention in this section is to lay out a comparison of the dominant Western rationality behind contemporary conservationism with indigenous cosmologies so that we are able to think about practical differences in how diverse cultures understand and relate to the natural environment.

Contemporary Structures for Forest and Natural Resources Governance

Contemporary structures for forest and natural resources conservation aspire to incorporate all aspects of valued external nature within rational and linear practices compatible with a capitalist economy. Under the narrative of "green growth" development, such structures approach nature as new commodity units, or a "set of assets" (Helm, 2014: 6) that can be valued in economic terms (e.g., payments for ecosystems services, REDD+, natural capital). This is what scholars have called the commodification of nature—the abstraction of nature's inherent vitality by capital into units of value (Jackson et al., 2017; Jackson and Palmer, 2015; Singh, 2015; Sullivan, 2009, 2013). This privileges market frameworks, whereby natural entities are conceptualized so that they can be counted, and thereby valued, in capitalist market economic terms (Sullivan, 2017). This framing of nature as part of different markets tends to occur in a discursive field in which the exercise of power is "rationalized" (where behavior is explained in logical terms) and embodied in routine action and normative orientations, so that certain forms of exercise of power in preference to others can be considered rational (Gebara and Agrawal, 2017).

The central premise of such way of framing nature is based on the ecological modernization movement, where environmental interests have been increasingly incorporated into social relations, institutions, and everyday practices (Mol, 2010). Ecological modernization theory presumes capitalist dynamics and relations of production, and does not see challenging them as necessary for overcoming ecological problems (Mol, 2002). In this context, environmental changes are said to occur through (i) political modernization, involving transformations in the respective roles of "the state" and "the market," including the "increasing importance and involvement of economic and market dynamics, institutions and agents in environmental reforms" (Mol, 2010); (ii) transformations in technology and technological development; and (iii) broader social transformations within "greening" consumption practices and in global civil society (Mol, 2002, 2010). It is the successful integration of such Western ecological rationality into these arenas and processes that allows environmental modernization theorists to argue that the capitalist world-system and economic growth are ultimately potentially compatible with ecological sustainability.

The ways conservationism and development are framed by ideas fundamentally based in capitalist thinking seem to reflect an assumption, or perhaps a desire, that we all inhabit a world that is rationalized, experienced, and accessed in the same way (Escobar, 2018; Sullivan, 2017). This perspective displays little appreciation of either the historical conquests shaping capitalism's particular truth regime, within which this type of thinking is set (Chakrabarty, 2000), or of the significant alienations of peoples from natures that have established the privileges and enclosures under which captured "natural capital assets" have been depleted (Federici, 2004). It minimizes and depoliticizes the problems of a world structured by neoliberalism, and destabilized by the equality-preventing dominance of rentier capitalism (Piketty, 2014; Storm, 2013; Vitali et al., 2011).

Over the last few hundred years, diverse cultural landscapes have become globally connected, through their contact with the particular expansionary economic system associated with merchant capitalism, and the ensuing capital accumulations permitting the multiple radical transformations caused by industrialization (Sullivan, 2017). The Amazon is no exception, as it is persistently perceived by policy-makers as an infinite frontier of development for capital accumulation. With policies historically determined by colonialism and the substantial presence of local elites (Pochmann, 2009), forest governance in the Amazon is still oriented around a development model that looks to global commodity markets and well-qualified entrepreneurs with the necessary capital for large-scale investments. This orientation continues despite a growing consensus about its negative ecological consequences, social impacts, and economic risks (Bebbington, 2013; Delgado, 2017; Fearnside, 2017; Gebara, 2018; Hogenboom, 2015; May et al., 2016; Munarriz, 2008; Pokorny et al., 2013).

As argued by Fairhead et al. (2012), these deep-rooted historical processes provide a vital set of conditions of possibility—meaning the set of factors and relations that turn some framings into an accepted narrative—which those seeking to profit from the new commoditization of nature have been able to inhabit. The authors argue that "such contemporary dynamics of appropriation of nature involve new (re)configurations of actors and relationships; the reinvention of regulatory processes and the construction of novel justificatory discourses" (Fairhead et al., 2012: 249). As already argued by Gibson-Graham (2006), we are always being (re)produced and are (re)producing ourselves, our space, and our economies. Here I am interested in a process-driven change of framings, which does not fix subjectivity or practices but offers room for multiple and even radical forms of thought and actions to emerge and so provide for the possibility for transformation at perhaps, larger scales than the ones addressed here. Proposing an alternative approach to the contemporary dynamics in how nature is understood; the next sub-section draws on indigenous cosmologies to discuss the antagonism between these and current structures for forest and natural resources governance.

Indigenous Cosmologies

The topic of indigenous cosmologies can be said on many fronts to be "in the air" (de la Cadena and Blaser, 2018; de la Cadena, 2015, 2010, 2008; Guenther, 2006; Sarivaara et al., 2013; Viveiros de Castro, 2008), as it is all around us in ways that provide a general background that informs my work here. In fact, through different material and immaterial practices of caring for the forest and its living beings, in the sense of having a commitment to and respect for the lives of such beings, indigenous people have not only enriched their landscapes but also cultivated (sometimes with strong resistance) what it is called indigenous cosmologies. Cosmological ideas emerge out of the necessity of "being" through time, of connecting the existential realms of nature and human beings (Viveiros de Castro, 2008). "Being" itself is the experience of encountering nature and others (seemingly of all ontic and non-ontic types) (Haraway, 2008) and becoming the other. The question of being is posed directly by Maturana and Verden-Zoller (2008: 126):

> Our ancestors in non-patriarchal cultures lived in a systemic dynamic interconnectedness with a cosmos that they were aware of and able to integrate. (. . .) In that way of living they were not concerned with controlling the different aspects

> of their existence. They just lived them; and they did so through the conservation of practices that both conserved and realized their harmonious participation in the cosmic dynamic of their daily living in the human community to which they belonged.

This is an essential part of Stengers's (2010a, b, 2011) and Latour's (2004a, b) studies of cosmopolitics, though they emphasize the institutional forms of self-representation that they know best: science and politics. While both authors' historical and empirical accounts focus centrally on institutions of knowledge production—the sciences—they rather insistently narrate the knowledge practices of these institutions in political terms. Latour's cosmopolitics, for example, is an effort to rethink our systems for representing the so-called "humans" and "nonhumans" with the objective of producing peaceful worlds that do not lend ontological primacy to the human/nonhuman distinction or, presumably, any other "great divide" (Latour, 1993: 97–100). But how possible is it really to come up with a cosmopolitics that truly avoids dualistic propositions in the relationship of humans with the natural world? The next sub-section gives some practical examples of Amazonian cosmologies to give insights in response to such a question.

Amazonian Cosmologies

Viewing (and living) the world as relational and animate is central to indigenous cosmologies and ways of being (Cajete, 2000; Escobar, 2016; Kimmerer, 2013). For different indigenous cultures, viewing and living the world as relational means having a deep awareness of their interconnectedness with the world, which often nurtures attitudes of humility, gratitude, and wonder at the contingency of life. Indigenous cosmologies are based in indigenous traditional knowledge. This knowledge is local and context-specific, transmitted orally or through imitation and demonstration, adaptive to changing environments, collectivized through shared social memory, and situated within numerous interlinked facets of individual and other-than-human beings' lives (Mistry, 2009). Such knowledge is increasingly important, given the growing global challenges of ecosystem degradation and climate change (Ford et al., 2016).

Different cultures in the Amazon provide us with examples that could inspire and enable a cosmopolitics rooted in a respectful and intuitive relationship with the natural world. As Escobar (2018) points out, Amazonian ethnology shows how elders throughout the region used to hold—some still do, though more precariously—a complex shamanic knowledge of the entire Amazon basin, without ever having traveled far from their own areas, and despite their different locations and languages, from Colombia and Peru to

Brazil. This knowledge was grounded in a tight relation between the level of thought (*pensamiento*) and that of practice. Practices—concerning community spaces, water, and plant worlds, cultivation and food, fishing and hunting, healing, and so forth—simply enacted what already existed in thought. There used to be agreement among the various groups on how to live and manage the territory. All of this points to the existence of a lived knowledge of the cosmos out of which entire worlds were (and to some extent still are) constructed. Much of this came under attack with colonialism, evangelization, and development, and even more so today with extractivism in indigenous territories (Escobar, 2018). The following are some examples of such lived knowledge or, in other words, indigenous cosmologies.

For one, among peoples of the Jê linguistic family, who live around the eastern part of the Brazilian Amazon, the cosmos is conceived as inhabited by different humanities that have always existed—the subterranean, the terrestrial, the underwater, and the celestial (Fausto, 1992). The origins of time are indifference and disorder, coexistence, and the interpenetration of those domains. Astros, like the Sun and the Moon, are primordial twins who live adventures in the Earth and here they leave their legacy, before leaving for their eternal abode. For the Jê linguistic family, social structures—the naming of individuals, war, shamanism—have their origins and essences described and exposed in myths (Leach, 1983), which make explicit references to subsistence activities and social practices in general.

The Kayapó, a tribe that belongs to the Jê linguistic family, perceive their village as the most socialized space of the Kayapó universe. The surrounding forest is considered an antisocial space, where men can turn into animals or spirits, fall ill without reason or even kill their relatives; it is inhabited by half-animal, half-human beings. The further from the village, the more antisocial the forest becomes, and the more dangers are associated with it. As there is always the danger that the "social" will be appropriated by the domain of nature, escaping from human control, the Kayapó seek a ritualistic relationship with the natural, transforming it into the social through songs of healing and ceremonies that establish a constant exchange between man and the world of nature (Turner, 1991). This means that any human intervention in nature is accompanied by rituals. For example, the installation of new growing areas is preceded by a dance that bears many structural similarities to the war ritual. Installing new growing areas is a symbolic war indeed, not against a human enemy, but against the natural world. Upon returning from the hunt, Kayapó men sing for the spirits of their prey, which they themselves killed, so that they remain in the forest. Thus, Kayapó rituals consist of a very particular language expressing and actualizing fundamental values of their culture, also reflected in their social organization and cosmological views. Each ritual translates a part of this cosmological vision and establishes a connection between man and nature, in which it is above all the man-animal relationship that is reinforced (Turner, 1966).

Tukano-speaking peoples living in the upper Rio Negro region, in the northwest of the Brazilian Amazon, believe that mythical ancestors created the world at the beginning of time (Lopes da Silva and Grupioni, 1995). The first ancestors of each of the various peoples in the region emerged from the bowels of a large ancestor serpent, who made the course of the river, thus determining each group's territories, their specific attributions, and hierarchical relationships between them. In many of their cosmologies, the relationships between humans and other beings are conceived through the idea of predation, in a metaphor that symbolically and logically approaches hunting, war, sex, and commensality. Even now on the upper Rio Negro, the shaman seems to be in charge of ensuring that streams and volumes of vital energy shared by humans and animals remain at adequate levels. Excessive slaughter of animals would trigger epidemics and evil effects among humans, provoked by the protective spirits of animals. A vital balance in memories, and conviviality with the idea of death, are daily experiences in the appreciation and conduct of life (ISA, 1998). Similarly, the Yawanawá people, living on the borders of Brazil, Bolivia, and Peru, view themselves and nature as part of "an extended ecological family that shares ancestry and origins" (personal communication, Nani Kateyuve, April 2018). They see human awareness of the "total interconnectedness and integration of all life" as central to the flourishing of life in their ecologies. The pre-cosmological world described by many of their myths is one entirely saturated with personality or personhood. A Yawanawá story begins, "At that time there was nothing, but there were already people" (Carid Naveira, 1999). The emergence of species and the stabilization of the trophic processes described in different indigenous myths did not extinguish this universal personhood origin; it merely put it in a dangerous state of nonappearance, that is, of latency or potentiality. Every being a human confronts during the process of producing their own life may suddenly push past its "other side" (common language in indigenous cosmologies, meaning the re-appearance of the universal personhood), reverting to a humanoid background condition. This causes the human condition to cease being "special" and to become, instead, the default mode or generic condition of any species (Viveiros de Castro, 2007).

The Wari, also known as the Pakaa Nova, living in Rondônia in the Brazilian Amazon, define humanity by the possession of a spirit or soul. Animals endowed with spirit are considered as "people," "human." They have a human body that the shaman can see; they live in houses, drink maize beer, and eat their food roasted and boiled. All "human" animals have culture, the same culture of the Wari (Vilaça, 1998). Along the same lines, the Kaxinawá, also living in the Brazilian Amazon, in Acre, believe the person is formed by flesh (or body) and *yuxin* (soul). Animals have a bodily side and a *yuxin* side, as do plants. Not by chance, it was the squirrel that taught man the art of planting (the squirrel is characterized by storing food for a long time, which is necessary for planting). Likewise, the nail monkey taught the human to copulate (ISA, 1998).

As a whole, these different cosmologies are indigenous theories of the world—of the order of the world, of movement in the world, in space and time—which reflect foundational conceptions of life in which humanity is only one of many characters in lived experiences (Lallemand, 1978). These notions define the place that humans occupy in the universe, expressing the permanent interdependence and constant reciprocity in the exchanges of energies and vital forces, of knowledges, abilities, and capacities that give characters the source of their renewal, perpetuation, and creativity. In everyday life, these conceptions guide, give meaning, and allow indigenous people to interpret events and ponder decisions. They are expressed through the symbolic language of the dramaturgy of rituals. Music and body ornaments, among other types of expression, allow contact with other cosmic dimensions, with other moments in the world, and with the process of life (and death) (Lopes da Silva, 1992).

Central to the different indigenous cosmologies described here is the definition of humanity and its place in the cosmic order, which is seen as diverse moments of a continuous reciprocal process of producing life and the world as opposed to a domain that is inhabited and controlled by beings of another nature. In Lopes da Silva's (1984) view, in different Amazonian cosmologies there is order, classification, logical opposition, hierarchy, and inclusive and exclusive categories in the cosmos as conceived. But there is also movement and a constant play with time; either to suppress it, allowing living human beings a possible reunion with the past, ancestors, and origins; or to make it the axis of existence itself, when time is destined to be completed and fully constituted after death, in the eternal overcoming of the limitations of the human condition (Lopes da Silva, 1984). This playing with time is important to show its relativity and the different aspects of life's temporality.

Amazonian cosmologies then provide practical insights to the cosmopolitics debate in the sense that such cosmologies include the animist assumption that "other beings besides the humans, notably animal species, are provided with intentionality and conscience, and in this sense are persons, i.e., in terms of social relations" (Viveiros de Castro, 2008: 13). A shaman, a spiritual indigenous guide, may then be seen as what Viveiros de Castro calls a "cosmopolitic diplomat," meaning someone "capable of seeing other species as they see themselves." He explains that "like humans, the shaman of each species play the role of cosmopolitic diplomats, operating in an arena where different 'socionatural' interests confront each other" (Viveiros de Castro, 2008:

17). Thus, the shaman builds versions of the world from inside, from the community environment and is a producer of knowledge, notably cosmological explanations about phenomena of all kinds, including the phenomenon of environmental change, which is essential to the construction of the narratives present in current indigenous movements (Overing, 1990). An interethnic leader, who may or not may be a shaman, also performs this construction through interethnic political discourse when indigenous groups have the opportunity to present their views in political arenas. This, in return, produces an opportune interethnic political exchange and actualizes indigenous cosmologies through an intertwining of traditional cosmological elements and current historical and political contexts (Nascimento Araujo, 2015), which are relevant to the practical implementation of cosmopolitics. Based on the indigenous cosmologies presented here and the concept of "earth beings," the next section will discuss how the rediscovery of the "earth beings" is critical for a practical political reframing and response to contemporary conservation narratives and practices.

REDISCOVERING THE EARTH BEINGS

In the last section, I made reference to a collection of indigenous cosmologies that help us to de-center as well as rethink the human and our own natures as emergent, rather than fixed and immutable. Instead of the dominant conception of "the subject" as a "standalone cognitive actor acting upon the world," this perspective is significant in helping us to think in terms of fluid agencies emerging from active engagement with the world (Ruddick, 2017). This logic allows the creation of a phenomenological path toward perceiving human and other-than-human beings as earthed and embedded with, not in, their environment. Such emergent subjectivities may help in rediscovering human and other-than-human beings as "earthbound people" (Latour, 2013), "earth beings" (de la Cadena, 2015), "terrans" (de la Cadena and Blaser, 2018), or "the inhabitants of the Earth," to use Humboldt's words. The aim behind such concepts, in this sense, is to provide a new set of relationships between all beings on the Earth that are mainly based in animism.

Anthropologist Tim Ingold (2006: 9) attributes this stance of animism and relationality not to indigenous cosmologies, but to "a way of being that is alive and open to a world in continuous birth." Contrary to the Western conception of relationality as the imputation of life to inert objects, Ingold (2006: 10) describes relationality as the animacy of the lifeworld, meaning "the dynamic, transformative potential of the entire field of relations within which beings of all kinds, more or less person-like or thing-like, continually and reciprocally bring one another into existence." In nature-society studies, an engagement with these ideas leads to a deeper appreciation of the interconnectedness of all life and the need to attend not only to the political, ecological, cultural, and economic dimensions, but also to the affective and emotional aspects at scales ranging from the psyche to the global.

A growing body of work is engaging with such ideas of *relationality* to further the critical and deeply political narratives of our times, namely, rethinking our ways of being human and belonging (Gibson-Graham, 2011), and providing ways of making legible and valuing alternative socio-natures that allow the recognition of earth beings and challenge "conventional nature-society thinking and practice" (Jackson and Palmer, 2015). These include the call for an ethic that nurtures the concept of "earth beings," or, in other words, the "being-in-common" of "all being(s), human and non-human, animate and inanimate, processual and fluid as well as categorical and definite in conception," and models of doing research as a "process of co-transformation that re/constitutes the world" (Gibson-Graham and Roelvink, 2010: 320). These ideas have also been mobilized in manifestos for conservation based on ideas of "pluriverse" and "multispecies abundance"—that is, more diverse, interdependent and autonomous forms of life and ways of living together (Collard et al., 2015; Escobar, 2018; Tsing, 2015; see also Weber, 2013).

The prospect that a political system constructed to represent human interests and constituencies might be able to know and care for the ontologically disparate worlds and experiences of other-than-human beings, however, may amount to the present epoch's predominant ideological delusion. As already argued by Karl Marx (1973) decades ago, our political systems cannot act ecologically, because they were not built to do so. Instead, they are built on neoliberal rationality that tends to commodify all aspects of our worldly relations and activity and the general development of marketized, mass societies. Marxian approaches to the planetary ecological crisis and the socio-ecological transformation necessary for its resolution have evolved rapidly in recent decades, creating the basis for a much more powerful, collective struggle for a Great Transition, in which the Holy Trinity of god, humans and neoliberalism that generated "consumerism, individualism, and domination of nature" are replaced with "a new triad: quality of life, human solidarity, and ecological sensibility" (Raskin, 2006).

Cosmopolitics, then, might also mean an activist reconstitution of earth beings in a plural world. The plural, what we hold together and—presumably—what holds us together, requires reconstruction in rigorously nonhumanist and anti-correlationist terms. Cosmopolitics thus entails producing co-habitable worlds and corollary forms of governance with other earth beings that do not share the cognitive, bodily, and metaphysical forms of human beings. Political systems are still artifacts of an eighteenth- and nineteenth-century romanticism that presumed the absolute externality, immutability and permanence of nature (Watson, 2013). Instead, more pluralistic and heterogeneous forms of governance would allow for the rediscovery of earth beings, leading to a path toward perceiving human and other-than-human beings as earthed and embedded with, not in, their environment. Political subjects then need to be reframed so we could rework politics toward an understanding of the "things of the world" in terms of the "forces, bonds and interactions" in which they "speak" to us (Serres, 1995: 39).

CONCLUSIONS

In an epoch when nature is less natural than at any time in human history—an age when even the human body is becoming subject to social reengineering, and artificial intelligence may dominate human spheres of action—connecting recent developments in Western political and environmental sciences with indigenous cosmologies may further the opportunity to perceive nature in ways that are useful to the cosmopolitics movement. Given the long-term experience indigenous peoples have with cosmopolitical practice, the possibility for intensive interethnic communication is stronger today than it may ever have been before. Indeed, Latour and Stengers' cosmopolitics concept has evolved as a new term to describe the global indigenous movement, because it is based on ideas of relationality. The major point here, in other words, is that without a rethinking of the intersecting capitalist logics in which contemporary conservationism is based, we are likely to only continue reproducing the generative conditions of the current ecological crisis. We will continue to conceive and experience our relationship with nature, what society and nature are and mean, and so on, in ways that set us up to be limited by the hegemonic narrative in which the current ecological crisis is grounded.

What makes these ways of being and perceiving nature limited, is the fact that they rarely address the fundamental socioeconomic processes transforming today's natural diversities. Instead, they represent a type of rationality and knowledge that leads to governance structures geared to ameliorating environmental problems, without ever addressing the deeper causes responsible for those problems in the first place. A primary contribution of the cosmopolitics movement should then be to demythologize this idealistic mystification of the human-nature relationship—as adopted by economic exploitation—and to begin the construction of a new and pluralistic political system. If we keep perceiving nature and humans as having only a commodity and exchange value, the acquiescence of science and politics to that same perspective can only lead to a deepening dehumanization within society, and to a further exploitation of nature (Miller, 1978).

We may be tempted to believe that the solution will appear from the marriage of capitalism with technology, while the mantra of progress and development through the domination of nature continues to be chanted. Or we may start learning from indigenous cosmologies, acknowledging the plural, respecting other-than-human beings. A first step toward this end is to stop folding all of the natural world into the Western discursive and calculative value-frame of apparently substitutable capital(s) that seems to encourage a reduction of ways of perceiving, thinking about, and relating with cultural and natural diversities.

BIBLIOGRAPHY

Ankersmit, F. 1996. *Aesthetic Politics.* Stanford University Press: Stanford.

Bebbington, D. H. 2013. Extraction, Inequality and Indigenous Peoples: Insights from Bolivia. *Environmental Science & Policy*, 33, 438–46.

Bellacasa, M. 2011. Matters of Care in Technoscience: Assembling Neglected Things. *Social Studies of Science*, 41(1), 85–106.

Brown, R. 1993. Modern Science: Institutionalization of Knowledge and Rationalization of Power. *The Sociological Quarterly*, 34, 153–68.

Burkert, W. 1972. *Lore and Science in Ancient Pythagoreanism.* Harvard University Press: Cambridge.

Cajete, G. 2000. *Native Science: Natural Laws of Interdependence.* Clear Light Publishers: Santa Fe.

Carid Naveira, M. 1999. Yawanawa: Da Guerra à Festa. MSc dissertation. UFSC: Florianópolis.

Castree, N. and Braun, B. 2001. *Social Nature: Theory, Practice, and Politics.* Blackwell Publishers: Malden, MA.

Chakrabarty, D. 2000. *Provincializing Europe: Postcolonial Thought and Historical Difference.* Princeton University Press: Princeton.

Clot, S., Andriamahefazafy, F., Grolleau, G., Ibanez, L., and Méral, P. 2015. Compensation and Rewards for Environmental Services (CRES) and Efficient Design of Contracts in Developing Countries. Behavioral Insights from a Natural Field Experiment. *Ecological Economics*, 113, 85–96.

Collard, R. -C., Dempsey, J., and Sundberg, J. 2015. A Manifesto for Abundant Futures. *Annals of the Association of American Geographers*, 105(2), 322–30.

Damasio, A. 2003. Mental Self: The Person Within. *Nature*, 423, 227.

de la Cadena, M. 2008. Alternative Indigeneities: Conceptual Proposals. *Latin American and Caribbean Ethnic Studies*, 3(3), 341–49.

de la Cadena, M. 2010. Indigenous Cosmopolitics in the Andes: Conceptual Reflections beyond "Politics." *Cultural Anthropology*, 25(2), 334–70.

de la Cadena, M. 2015. *Earth Beings: Ecologies of Practice Across Andean Worlds.* Duke University Press: Durham.

de la Cadena, M. and Blaser, M. 2018. *A World of Many Worlds.* Duke University Press: Durham.

Delgado, A. C. 2017. The TIPNIS Conflict in Bolivia. *Contexto Internacional*, 39(2), 373–92.

Descola, P. 1996. *In the Society of Nature: A Native Ecology in Amazonia*. Cambridge University Press: Cambridge.

Escobar, A. 2016. Thinking-feeling with the Earth: Territorial Struggles and the Ontological Dimension of the Epistemologies of the South. *Revista de Antropología Iberoamericana*, 11(1), 11–32.

Escobar, A. 2018. *Designs for the Pluriverse: Radical Interdependence, Autonomy, and the Making of Worlds*. Duke University Press: Durham.

Ewing, J. 2017. Hollow Ecology: Ecological Modernization Theory and the Death of Nature. *Journal of World Systems Research*, 23(1), 126–55.

Fairhead, J., Leach, M., and Scoones, I. 2012. Green Grabbing: A New Appropriation of Nature? *Journal of Peasant Studies*, 39(2), 237–61.

Fausto, C. 1992. Fragmentos de História e Cultura Tupinambá: Da Etnologia Como Instrumento Crítico de Conhecimento Etno-Histórico. In: *História dos Índios no Brasil*, Carneiro da Cunha, M. (ed.). Cia das Letras/Edusp/SMC: São Paulo, pp. 381–96.

Fearnside, P. 2017. Deforestation in Brazilian Amazonia. In: *Oxford Bibliographies in Environmental Science*, Wohl, E. (ed.). Oxford University Press: New York. doi: 10.1093/acrefore/9780199389414.013.102.

Federici, S. 2004. *Caliban and the Witch: Women, the Body and Primitive Accumulation in Medieval Europe*. Autonomedia: New York.

Fischer-Kowalski, M., Krausmann, F., and Pallua, I. 2014. A Sociometabolic Reading of the Anthropocene: Modes of Subsistence, Population Size and Human Impact on Earth. *The Anthropocene Review*, 1(1), 8–33.

Ford, J. D., Cameron, L., Rubis, J., Maillet, M., Nakashima, D., Willox, A. C., and Pearce, T. 2016. Including Indigenous Knowledge and Experience in IPCC Assessment Reports. *Nature Climate Change*, 6(4), 349–53.

Gebara, M. F., Fatorelli, L., May, P., and Zhang, S. 2014. REDD+ Policy Networks in Brazil: Constraints and Opportunities for Successful Policy Making. *Ecology and Society*, 19(3), 53.

Gibson-Graham, J. K. 2006. *A Postcapitalist Politics*. University of Minnesota Press: Minneapolis.

Gibson-Graham, J. K. 2011. A Feminist Project of Belonging for the Anthropocene. *Gender, Place and Culture*, 181, 1–21.

Gibson-Graham, J. K. and Roelvink, G. 2010. An Economic Ethics for the Anthropocene. *Antipode*, 41(1), 320–46.

Guenther, M. 2006. The Concept of Indigeneity. *Social Anthropology*, 14(1), 17–32.

Harari, Y. N. 2014. *Sapiens: A Brief History of Humankind*. Harper & Row: New York.

Haraway, D. 1991. *Simians, Cyborgs, and Women: The Reinvention of Nature*. Routledge: New York.

Haraway, D. 2008. *When Species Meet*. University of Minnesota Press: Minneapolis.

Helm, D. 2014. Taking Natural Capital Seriously. *Oxford Review of Economic Policy*, 30(1), 109–25.

Heyne, P., Boettke, P. J., and Prychitko, D. L. 2014. *The Economic Way of Thinking*. 13th ed. Pearson.

Hogan, L. 1996. *Dwellings: A Spiritual History of the Living World.* Simon & Schuster: New York.

Hogenboom, B. 2015. Latin America's Transformative New Extraction and Local Conflicts. *ERLACS*, 99, 143–51.

Humboldt, A. 1997. *Cosmos.* Johns Hopkins University Press: Baltimore.

Humboldt, A. and Bonpland, A. 2009. *Essay on the Geography of Plants*. University of Chicago Press: London.

Ingold, T. 2006. Rethinking the Animate, Re-Animating Thought. *Ethnos*, 71(1), 9–20.

INPE (Instituto Nacional de Pesquisas Espaciais). 2018. *Monitoramento da Cobertura Florestal da Amazônia por Satélites: Sistemas PRODES, DETER, DEGRAD—2004–2018*. INPE: São José dos Campos.

IPCC (Intergovernmental Panel on Climate Change). 2014. *Climate Change 2014: Mitigation of Climate Change*. IPCC: Geneva.

ISA (Instituto Socio-Ambiental). 1998. *Livro de História: Professores Indígenas do Parque Indígena do Xingu.* Vol. 1. Instituto Socioambiental/MEC: São Paulo/ Brasília.

Jackson, S. and Palmer, L. 2015. Reconceptualizing Ecosystem Services: Possibilities for Cultivating and Valuing the Ethics and Practices of Care. *Progress in Human Geography*, 39(2), 122–45.

Jackson, S., Palmer, L., McDonald, F., and Bumpus, A. 2017. Cultures of Carbon and the Logic of Care: The Possibilities for Carbon Enrichment and its Cultural Signature. *Annals of the American Association of Geographers*, 107(4), 867–82.

Kalberg, S. 1980. Max Weber's Types of Rationality: Cornerstones for the Analysis of Rationalization Processes in History. *American Journal of Sociology*, 85(5), 1145–79.

Kimmerer, R. 2013. *Braiding Sweetgrass: Indigenous Wisdom, Scientific Knowledge and the Teachings of Plants*. Milkweed Editions: Minneapolis.

Kirksey, S. E. and Helmreich, S. 2010. The Emergence of Multispecies Ethnography. *Cultural Anthropology*, 25(4), 545–76.

Lallemand, S. 1978. Cosmogonia. In: *A Construção do Mundo*, Augé, M. (ed.). Edições 70: Lisbon, pp. 25–42.

Latimer, J. and Miele, M. 2013. Naturecultures? Science, Affect and the Non-human. *Theory, Culture & Society*, 30(7/8), 5–31.

Latour, B. 1993. *We Have Never Been Modern.* Trans. C. Porter. Harvard University Press: Cambridge.

Latour, B. 2004a. *The Politics of Nature: How to Bring the Sciences into Democracy*. Trans. C. Porter. Harvard University Press: Cambridge.

Latour, B. 2004b. Why Has Critique Run Out of Steam? From Matters of Fact to Matters of Concern. *Critical Inquiry*, 30(2), 225–48.

Latour, B. 2005. *Reassembling the Social: An Introduction to Actor-Network-Theory*. Oxford University Press: Oxford.

Latour, B. 2010. From Realpolitik to Dingpolitic or How to Make Things Public. In: *Making Things Public: Atmospheres of Democracy*, Latour, B. and Weibel, P. (eds.). MIT Press: Cambridge, pp. 14–43.

Latour, B. 2013. Facing Gaia: A New Enquiry into Natural Religion. Gifford Lectures, University of Edinburgh. Available at: https://www.youtube.com/watch?v=gsZCS5Zicx4.

Leach, E. 1983. O Gênensis Enquanto um Mito. In: *Antropologia*, Matta, R. (ed.). Ática: São Paulo, pp. 57–69.

Lopes da Silva, A. 1984. Mitos e Cosmologias Indígenas no Brasil: Breve Introdução. In: *Índios no Brasil*, Grupioni, L. D. B. (ed.). Secretaria Municipal de Cultura: São Paulo, pp. 75–82.

Lopes da Silva, A. 1992. A Expressão Mítica da Vivência Histórica: Tempo e Espaço na Construção da Identidade Xavante. In: *Anuário Antropológico/82*. Tempo Brasileiro/Universidade Federal do Ceará: Rio de Janeiro/Fortaleza, pp. 200–14.

Lopes Da Silva, A. and Grupioni, L. 1995. *Mito, Razão, História e Sociedade: A Temática Indígena na Escola*. MEC/MARI/Unesco: Brasília.

Lovejoy, T. E. and Nobre, C. 2018. Amazon Tipping Point. *Science Advances*, 4(2), 2340.

Malhi, Y. 2017. The Concept of the Anthropocene. *Annual Review of Environment and Resources*, 42, 25.1–25.28.

Manschot, H. and Suransky, C. 2014. From a Human-centered to a Life-centered Humanism: Three Dimensions of an Ecological Turn. In: *Everyday Humanism*, McGowan, D. and Pinn, A. (eds.). Equinox Publishing: Sheffield, pp. 125–37.

Marcuse, H. 1982. Some Social Implications of Modern Technology. In: *The Essential Frankfurt School Reader*, Arato, A. and Gebhardt, E. (eds.). Continuum: New York, pp. 138–62.

Marx, K. 1973. *Grundrisse*. Trans. M. Nicolaus. Penguin: Harmondsworth.

May, P., Millikan, B., and Gebara, M. F. 2016. *The Context of REDD + in Brazil: Drivers, Agents, and Institutions*. 3rd ed. Center for International Forestry Research (CIFOR): Bogor.

Miller, A. 1978. *A Planet to Choose: Value Studies in Political Ecology*. Pilgrim Press: New York.

Mistry, J. 2009. Indigenous Knowledges. In: *International Encyclopedia of Human Geography*, Kitchin, R. and Thrift, N. (eds.). Elsevier: Oxford, pp. 371–36.

Mol, A. 2002. Ecological Modernization and the Global Economy. *Global Environmental Politics*, 2(2), 92–115.

Mol, A. 2010. Ecological Modernization as a Social Theory of Environmental Reform. In: *The International Handbook of Environmental Sociology*, Redclift, M. R. and Woodgate, G. (eds.). Edward Elgar Publishing: Cheltenham, pp. 63–76.

Moore, J. 2017. The Capitalocene, Part I: On the Nature and Origins of Our Ecological Crisis. *The Journal of Peasant Studies*, 44(3), 594–630.

Munarriz, G. 2008. Rhetoric and Reality: The World Bank Development Policies, Mining Corporations, and Indigenous Communities in Latin America. *International Community Law Review*, 10(4), 431–43.

Nascimento, A. F. 2015. Interethnic Discourse and Cosmology in Indigenous Association Practices in the Amazon: Hutukara Yanomami Association. *Electronic Journal of Studies in the Tropic*, 14(2), 53–62.

Nobre, C., Sellers, P. J., and Shukla, J. 1991. Amazonian Deforestation and Regional Climate Change. *Journal of Climate*, 4(10), 957–88.

Noss, R., Nash, R., Paquet, P., and Soulé, M. 2013. Humanity's Domination of Nature is Part of the Problem: A Response to Kareiva and Marvier. *BioScience*, 63(4), 241–42.

Overing, J. 1990. The Shaman as a Maker of Worlds: Nelson Goodman in the Amazon. *Man, New Series*, 25(4), 602–19.

Peterson, N. and Isenhour, C. 2014. Moving Beyond the "Rational Actor" in Environmental Governance and Conservation. *Conservation and Society*, 12(3), 229–32.

Piketty, T. 2014. *Capital in the Twenty-first Century*. Harvard University Press: Cambridge.

Pochmann, M. 2009. *Qual Desenvolvimento? Oportunidades e Dificuldades do Brasil Contemporâneo*. Boitempo: São Paulo.

Polanyi, K. 2001. *The Great Transformation: The Political and Economic Origins of Our Time*. Beacon Press: Boston.

Porter, R. 1980. The Terraqueous Globe. In: *The Ferment of Knowledge: Studies in the Historiography of Eighteenth-century Science*, Rousseau, G. and Porter, R. (eds.). Cambridge University Press: Cambridge, pp. 285–324.

Pretty, J. 2013. The Consumption of a Finite Planet: Well-being, Convergence, Divergence and the Nascent Green Economy. *Environmental and Resource Economics*, 55(4), 475–99.

Rancière, J. 2003. The Thinking of Dissensus: Politics and Aesthetics. In: *Reading Rancière: Critical Dissensus*, Bowman, P. and Stamp, R. (eds.). Continuum: New York, pp. 1–17.

Rancière, J. 2006. *Hatred of Democracy*. Trans. S. Corcoran. Verso: London and New York.

Raskin, P.D. 2006. *The Great Transition Today: A Report from the Future*. Tellus Institute: Boston.

Read, J. 2015. Relations of Production. *Historical Materialism*, 233, 201–14.

Rifkin, J. 2009. *The Empathetic Civilization: The Race to Global Consciousness in a World in Crisis*. Penguin: New York.

Robertson, M. 2012. Measurement and Alienation: Making a World of Ecosystem Services. *Transactions of the Institute of British Geographers*, 37, 386–401.

Ruddick, S. 2017. Rethinking the Subject, Reimagining Worlds. *Dialogues in Human Geography*, 7(2), 119–39.

Sarivaara, E., Maatta, K., and Uusiautti, S. 2013. Who is Indigenous? Definitions of Indigeneity. *European Scientific Journal*, 1, 369–78.

Schelling, F. W. J. 1988. *Ideas for a Philosophy of Nature*. Trans. E. Harris and P. Heath. Cambridge University Press: Cambridge.

Serres, M. 1995. *The Natural Contract*. Trans. E. MacArthur and W. Paul-son. Michigan University Press: Ann Arbor.

Singh, N. 2015. Payments for Ecosystem Services and the Gift Paradigm: Sharing the Burden and Joy of Environmental Care. *Ecological Economics*, 117, 53–61.

Slater, Candace. 2002. *Entangled Edens: Visions of the Amazon*. University of California Press: Berkeley, pp. 332.

Smith, N. 1984. *Uneven Development: Nature, Capital and the Production of Space*. Blackwell: Oxford.

Stengers, I. 2010a. *Cosmopolitics I*. Trans. R. Bononno. University of Minnesota Press: Minneapolis.

Stengers, I. 2010b. The Cosmopolitical Proposal. In: *Making Things Public: Atmospheres of Democracy*, Latour, B. and Weibel, P. (eds.). MIT Press: Cambridge, pp. 994–1003.

Stengers, I. 2011. *Cosmopolitics II*. Trans. R. Bononno. University of Minnesota Press: Minneapolis.

Storm, S. 2013. Why the West Grew Rich and the Rest Did Not, Or How the Present Shapes Our Views of the Past. *Development and Change*, 44(5), 1181–206.

Sullivan, S. 2009. Green Capitalism, and the Cultural Poverty of Constructing Nature as Service-Provider. *Radical Anthropology*, 3, 18–27.

Sullivan, S. 2013. Nature on the Move III: (Re)countenancing an Animate Nature. *New Proposals: Journal of Marxism and Interdisciplinary Inquiry*, 6(1–2), 50–71.

Sullivan, S. 2017. On "Natural Capital," "Fairy Tales" and Ideology. *Development and Change*, 48(2), 397–423.

Swyngedouw, E. 1999. Modernity and Hybridity: Nature, Regeneracionismo, and the Production of the Spanish Waterscape, 1890–1930. *Annals of the Association of American Geographers*, 89(3), 443–65.

Todorov, T. 1982. *The Conquest of America: The Question of the Other*. Trans. R. Howard. Harper & Row: New York.

Tsing, A. 2015. *The Mushroom at the End of the World: On the Possibility of Life in Capitalist Ruins*. Princeton University Press: Princeton.

Turner, T. 1966. Social Structure and Political Organization Among the Northern Cayapó. PhD thesis. Harvard University: Cambridge.

Turner, T. 1991. *The Mebengokre Kayapó: History, Social Consciousness and Social Change from Autonomous Communities to Inter-Ethnic System*. Chicago University Press: Chicago.

Venn, C. 2010. Individuation, Relationality, Affect: Rethinking the Human in Relation to the Living. *Body & Society*, 16(1), 129–61.

Vilaça, A. 1998. Fazendo Corpos: Reflexões sobre Morte e Canibalismo entre os Wari' à Luz do Perspectivismo. *Revista de Antropologia*, 41(1), 9–67.

Vitali, S., Glattfelder, J. B., and Battiston, S. 2011. The Network of Global Corporate Control. *Plos ONE*, 6(10), e25995.

Viveiros de Castro, E. 2004. Perspectival Anthropology and the Method of Controlled Equivocation in Tipití. *Journal of the Society for the Anthropology of Lowland South America*, 2(1), 3–22.

Viveiros de Castro, E. 2007. *O Medo dos Outros*. Trans. B. Perrone-Moisés. Available at: https://jyvukugi.files.wordpress.com/2012/02/o-medo-dos-outros-eduardo-viveiros-de-castro-1af2.pdf.

Viveiros de Castro, E. 2008. Xamanismo Transversal: Lévi-Strauss e a Cosmopolítica Amazônica. In: *Lévi-Strauss: Leituras Brasileiras*, Queiroz, R. C. and Nobre, R. F. (eds.). UFMG: Belo Horizonte, pp. 79–124.

Waal, F. 2010. *The Age of Empathy: Nature's Lessons for a Kinder Society.* Harmony Books: New York.

Walls, L. D. Introducing Humboldt Cosmos. *Minding Nature*, 2(2), 3–15.

Watson, M. C. 2013. Derrida, Stengers, Latour, and Subalternist Cosmopolitics. *Theory, Culture & Society*, 31(1), 75–98.

Weber, A. 2013. *Enlivenment: Towards a Fundamental Shift in the Concepts of Nature, Culture and Politics.* Heinrich-Böll-Stiftung: Berlin.

Weber, M. 1946. *Politics as a Vocation: Essays in Sociology.* Oxford University Press: New York.

Wulf, A. 2015. *The Invention of Nature.* John Murray Publishers: London.

Chapter 11

Territorial Conflicts on Brazilian Amazonian Frontiers

A Research and Public Policy Framework

Paul E. Little

This chapter presents a research methodology for the social scientific study of contemporary frontiers in Brazilian Amazonia. A brief comparative review of the broader frontier literature will provide historical and theoretical context for understanding contemporary Brazilian Amazonian frontiers. An anthropological definition of territory will be used to contextualize ethnographic engagement with frontier conflicts, and the major frontier social actors will be identified and situated within their respective fields of power. This will be followed by an analysis of the complex and contradictory relationships that each territorial actor maintains with different government agencies, and a delineation of the principal distributive, legal, and normative issues involving natural resource governance on contemporary frontiers. The conclusions will explore the possibility of achieving a definitive closure of frontiers in Brazilian Amazonia and the installation of a post-frontier dynamic.

COMPARATIVE FRONTIER ANALYSIS

Expanding frontiers have been a key mechanism in historical change processes for millennia. The comparative study of frontier dynamics offers a privileged site for understanding socio-environmental change within broader political and economic contexts (Lattimore, 1962).[1] Comparative frontier history must, on the one hand, take into account the fact that frontiers "contain so many cultural and environmental variables—including moments in time—that they are always unique" (Weber and Rausch, 1994: xxii), while, on the other, identify common themes such as the treatment of native peoples, laws regarding settlement, demographic characteristics, the role of women, the presence of nationalism, and the exploitation of environmental resources (Nugent, 1994). Gerhard (1959) explores many of these issues in his comparative analysis of the frontiers of the U.S. West, the trading expansions of French Canada in the present province of Quebec, the Anglo settlement of Upper Canada (Ontario), German colonization in Eastern Europe, the Australian advance inland, Boertrekking into the interior of South Africa. and Russian transmigrations eastward toward the Ural Mountains and beyond into Asia.

The concept of an expanding frontier, as used here, refers to a region or zone where two or more distinct cultures, societies, ethnic groups, or modes of production come into contact, and which is characterized by conflicts over control of natural resources and human territories, in what Guy and Sheridan (1998: 15) aptly call "contested ground," or "zones of intense negotiation as well as violence—zones where social identities [are] far more fluid that they [are] in more established areas where states exercise greater political, economic, and social control." A general overview of the social scientific literature on frontiers reveals various approaches to understanding this phenomenon from the fields of geopolitics, economic history, and anthropology.

Geopolitical approaches to frontiers focus on the efforts by empires and nation-states to expand their domains through the conquest and incorporation of marginal lands into their political and economic cores. Lattimore (1951) has written extensively about millennial frontier dynamics in China and the Middle East, where contact zones between "the steppe and the sown" have exhibited both conflict and cooperation through the back and forth movement between steppe confederations and the Chinese State, while Wheeler (1955) chronicles centuries of incursions by the Roman Empire into frontier regions in Northern Europe, Africa, and Asia.

Frederick Jackson Turner's thesis (1893) of the role of the western frontier in U.S. nation building dominated geopolitical thinking on frontiers for the first half of the twentieth century. Webb (1952) expanded this thesis to include all the demographic expansions of European peoples from the sixteenth century onwards in what he calls the "Great Frontier." Many analyses modeled on the Turner thesis, however, reveal a strong Eurocentric bias and suffer from acute ethnocentrism in their neglect and/or denigration of the indigenous peoples whose territories were being invaded, as exemplified in Webb's characterization of a moving frontier "where a civilized people are advancing into a wilderness, an unsettled area, or one sparsely populated by primitive people" (Webb 1952: 3).

By the end of the twentieth century, a new wave of postmodern deconstruction of celebratory frontier narratives came to the fore. Watts (1992: 116–17) views frontiers in their symbolic, ideological, and material dimensions and characterizes them as being part of the "cutting edge of state-sponsored forms of accumulation" that "create their own territorial form of law and (dis)order" and, in this way, is able to draw parallels between the nineteenth century U.S. western frontier and twentieth-century Amazonian frontiers. A "New Western History" of the United States emerged that did not focus on the frontier exploits of Anglo-American males, but rather on impacts on women, Native Americans, Hispanics, and other minorities in contexts of class conflict and social inequality (Limerick et al., 1991; Cronon et al., 1992). Limerick (1987: 15), one of the leaders of this revisionist vision, eschews the concept of the frontier in favor of that of "conquest," which she considers to be "the literal, territorial form of economic growth." McNeill (1983) continues to speak of the Great Frontier, but now the imposition of peonage and slavery on frontiers and the destruction of non-European cultures and societies are highlighted.

Another significant body of literature uses geopolitical analysis to address postcolonial nation-building frontiers in the Global South, both historical and contemporary. In the nineteenth century, the nation-building expansion of the newly independent Latin American countries produced cattle, mining, agricultural, and rubber frontiers (Hennessy, 1978; Weber and Rausch, 1994). Mañach (1975) argues that these frontiers formed an essential part of the creation of national societies in Latin America, as exemplified by Argentina, Brazil, Chile, and Mexico. Geiger (2015) gives an excellent review of contemporary postcolonial frontiers with particular attention to cases from South and Southeast Asia. Meanwhile, Indonesia's diverse archipelago has been a source of highly innovative ethnographies of frontier dynamics in a context of intense globalization (Peluso, 1992; Tsing, 2005; Li, 2014). The contemporary Amazonian frontiers to be discussed here form part of this cohort of postcolonial, nation-building frontiers.

The primary focus of economic history approaches to frontiers is on the essential role of natural resource use and capital accumulation in the process of fostering economic growth. Barbier (2011: 7) asserts that "since the Agricultural Transition over 12,000 years ago, exploiting new sources, or 'frontiers,' of natural resources has often proved to be a pivotal human response to natural resource scarcity." Findlay and Lundahl (1999) provide a medium-term perspective on "resource-led growth" by focusing on its "Golden Age" (1870–1914) and find that these frontiers were generally favorable to sustained growth via industrialization. Barbier (2011: 20) offers historical evidence to support his "frontier expansion hypothesis," which postulates that contemporary frontiers in the Global South generate few economic rents and that these "rents have not been reinvested in more productive and dynamic sectors." At the same time, he argues that the so-called BRIC countries (Brazil, Russia, India, and China), due to the sheer scale of their populations, economies, and resource endowments, are "beginning to reap economy-wide benefits from exploiting their vast sources of land and natural resources" (Barbier 2011: 611), while facing a looming "Age of Ecological Scarcity" (pp. 663–729).

A far more critical economic perspective is found in neo-Marxian dependency theory, in which frontiers are placed within the framework of capital accumulation on a world scale (Amin, 1976) and the ecologically unequal exchange between cores and peripheries of the world-system (Martinez-Alier, 2002). World-system theory analyzes frontiers by drawing connections between long-term global processes and micro interactions on specific frontiers (Hall, 2009). There is considerable disagreement, however, regarding when the world-system originally coalesced; proposals include 5,000 (Frank and Gills, 1993), 800 (Abu-Lughod, 1989), and 500 (Wallerstein, 1983) years ago.

An interventionist economic perspective on frontiers posits the need for specific government intervention to promote national economic growth. Topics of the interventionist literature include avoiding the "staple trap" by strengthening the national manufacturing sector (Sachs and Warner, 1995) and moving beyond the "resource curse" whereby national economies dependent on oil, gas, and mineral extraction have poorer development outcomes than countries with fewer natural resources (Shaffer and Ziyadov, 2012). Weiss (1998: 13) rebuts the "myth of the powerless state" and argues that in the global contest between competing capitalisms "state capacity, far from becoming irrelevant, has acquired new significance in a changing world economy."

Anthropological approaches to frontiers have incorporated the perspectives of indigenous peoples and other social groups victimized by frontier expansion. The pioneering study by Mexican anthropologist León-Portilla (1959) presents an "Aztec account of the Conquest of Mexico," thereby reversing centuries of triumphalist narratives of European invasions of indigenous peoples' lands. A volume of collected essays (Albert and Ramos, 2002) narrates the variegated ways indigenous peoples in northern Amazonia have sought to "pacify the White Man" by reconstructing the histories of contact with Europeans from the perspective of their cosmologies. Ethnohistorians have recently investigated indigenous "historicities," that is, "the cultural schemata and subjective attitudes that the make the past meaningful," including indigenous cartographies, rituals, genealogies, and collective memories (Whitehead, 2003: xi).

The chronicling of indigenous peoples' resistance and reinterpretation is another key marker of historical-anthropological approaches to frontiers. Karasch (1998: 129), describing the Brazilian frontier of Goiás from 1750 to 1890, indicates that the survival of indigenous peoples—Kayapó, Xavante, Xerente, and Canoeiro—"demanded multiple forms of relationships with the invaders, most of whom were non-white" and indicates that the choice by the Canoeiro of implacable resistance led to their near extinction as a group. Describing what happened two centuries after this, Turner (1991: 304) provides a first-hand ethnographic reading of the dramatic "change in social and cultural self-consciousness" that occurred in the 1970s and 1980s among the Kayapó, who, through the adoption and adaptation of visual media, gained control over the "technology of representation" which granted them the "power of conferring value and meaning on themselves in the eyes of the outside world, and reflexively, in new ways, in their own eyes as well" (Turner 1991: 306).

BRAZILIAN AMAZONIAN FRONTIERS

Amazonian frontier history covers a period of several thousand years, beginning with the territorial migrations of Arawak, Ashanika, Carib, Gê, and Tupi-Guarani indigenous language families in the centuries before the arrival of the Europeans; these migrations are now the subject of intense archeological revision (Heckenberger, 2005; Lopes, 2017). With the arrival of Spanish explorer Yáñez Pinzón at the mouth of the Amazon River in 1500 and his subsequent capture of thirty-six Potiguar Indians (Hemming, 1978), a new era of frontier expansions began. Over the five centuries since, numerous frontiers have come and gone, led by global market demand for agave, vanilla, wild cocoa, resins, quinine, ivory nut, and sarsaparilla, only to be eclipsed by the grandest global boom of the late nineteenth century, the rubber boom, which provoked major economic, demographic, and environmental transformations throughout the entire region (Dean, 1987). In a study of the Maracá River basin, a tributary that joins the Amazon near its mouth, eight distinct waves of frontier development were identified over a five-century period (Little, 2000).

Geopolitically, Becker's (1982) analysis of contemporary Amazonian frontiers as a national "resource frontier" ushered in a renewed interest in Amazonian frontiers by delineating the powerful economic vectors that transform Amazonian economies. More recently, Browder and Godfrey (1997) describe the social and economic heterogeneity of "disarticulated urbanization" on the "populist frontier" in Rondônia state and the "corporativist frontier" in southern Pará state and argue for greater theoretical pluralism in urban frontier theory.

Economically, Foweraker's (1981) groundbreaking analysis of the pioneer frontier in Brazil between 1930 and 1980 provides an excellent introduction to the economic, social, and political forces that promote frontier expansion among migrants to the Amazon that moves the frontier through the three stages of noncapitalist, precapitalist, and capitalist development. Lisansky (1990: 16) humanizes this analysis through a community case study of the "survival strategies" of settlers on the Mato Grosso frontier, which include "informal economic activities, dyadic and group cooperation, and alliances through co-parenthood" as well as "immigration, intensification and diversification of household labor, various types of reciprocity, and emigration."

In anthropology, Cardoso de Oliveira (1964. 1976), in a series of pioneering studies of Brazilian Amazonia, develops the concepts of the "contact situation" and the resulting "interethnic friction," which reveals the hierarchical ethnic dynamics that frontiers produce. Velho (1972) builds upon this analysis by conceptualizing Amazonian frontier expansion during the Brazilian military dictatorship as "authoritarian capitalism." Ramos (1998) details the varied frontier expansions that Brazilian indigenous peoples faced in the second half of the twentieth century, culminating in the 1990s with the emergence of a "knowledge frontier," characterized by attempts to exploit indigenous knowledge to fuel national and international development, and a "genetic frontier" involving activities such as "collecting blood from an Indian person in the most recondite corner of Amazonia for the benefit of those who suffer from genetic diseases somewhere in Europe or North America" (pp. 218–21).

Ethnographies of Amazonian peoples have long provided a rich and variegated understanding of cultural diversity and the complexities of frontier life, particularly with regard to the hundreds of indigenous societies that inhabit the region (Jackson, 1975). Fisher (2000), for example, chronicles the ways that the Xikrin Kayapó peoples employed native social systems to gain access to manufactured goods through strategic cooperation and trade with frontier logging and mining industries. Nonindigenous Amazonian peoples have recently received sustained ethnographic treatment, as attested by a flowering of studies of Amazonian peasant (*caboclo*) societies (Slater, 1994; Harris, 2000; Adams et al., 2009). Other nonindigenous groups that have gained ethnographic study include Cleary's (1990) pioneering study of itinerant gold miners and Hoelle's (2015) ethnography of "rainforest cowboys," small-scale ranchers that have installed a cattle culture in the Brazilian state of Acre. Campbell (2015), in a study of "land futures," details the ways that land speculators, small-scale agricultural settlers, and large-scale ranchers conceive of property and "conjure" it through papers, appeals to government officials, and the manipulation of landscapes and memories of occupation. Raffles (2002) offers an innovative, long-term "natural history" of Amazonia, which narrates the many ways that human activity—by native peoples, explorers, naturalists, canal builders, agroextractivists, miners, loggers, and traders—has transformed Amazonian landscapes over the past centuries.

A RESEARCH ENGAGEMENT WITH FRONTIER TERRITORIAL CONFLICTS

A political ecology approach, which has emerged within the social sciences as an important means of understanding the interplay between political and ecological issues within conflictive situations, will be honed here through the introduction of a research methodology that analyzes the dynamics of territorial conflicts and contributes to the public debate that aims to resolve them (Schmink and Wood, 1987; Little, 2010). The key analytical tasks are the identification of the forces driving socioeconomic and environmental change, the delineation of their main points of interaction and power asymmetries, and the depiction of the overriding normative issues that are points of contention in public debate and policy formulation. The goal is to produce reliable, strategic knowledge of complex situations that synthesizes multiple interests and points of view from inside frontier dynamics (Li, 2003; Little, 2006).

Frontier conflict, as used here, is understood not only in its political usage of legal standoffs and violent confrontations, but also in an anthropological sense that includes its ritual, cosmological, and identity underpinnings. This multi-actor approach places all social actors involved in the conflict within the same public space of debate and requires the researcher to interact with each social actor in order to be able to represent them ethnographically and present their respective legal standing and political claims. In the case of Brazilian Amazonia, the use of this methodology performs the critical function of giving increased visibility to historically marginalized groups, such as indigenous peoples, *quilombos*, riverside communities, rubber tappers, agricultural smallholders and itinerant gold miners, who are often underrepresented in frontier analysis.

From a normative perspective, the social scientific study of frontier conflicts seeks to move beyond the description of their causes to include an understanding of the processes of conflict resolution. In this way, the issues of public policy formulation and frontier governance become part of the object of research. This normativity, however, is not designed to depict binary clashes between "good" social actors who are defending the rainforest and "bad" ones who are destroying it. Rather, it can address normative issues through a rights-based foundation in which all social actors are treated as citizens who have civil, ethnic, and environmental rights.

The focus on territorial conflict as a specific subset of frontier conflicts is a further strategic choice of this framework. The concept of territory is most commonly used in the social sciences to refer to nation-states, which have sovereign control over their respective national territories. This understanding of territory is represented by the political world map, on which each country is clearly delimited and shown in just one color (Malkki, 1992), but it is not very useful for comprehending the conflicts on Amazonian frontiers, which occur within the territory of a single state. The use of an anthropological concept of territory serves to decenter the territorial question away from the state and toward social groups.

Territory, as defined here, refers to geographic spaces that are imbued with cultural meanings and ritual significance by a social group that delimits, uses, and defends them. Each territory is founded in a specific cosmography, or "the collective, historically contingent identities, ideologies, and environmental knowledge systems developed by a social group to establish and maintain human territory" (Little, 2001: 5). Territory is conceptually distinct from "land" since it is established by the collective effort of a social group that identifies with it and protects it. For this reason, there can be a "no man's land," but there can never be a "no man's territory." The concept of territory unites biophysical attributes with political entities within a single analytical framework, thereby bridging the natural/cultural divide, which is prevalent in much research on the Amazon.

Frontier territories have specific geographic coordinates and, as such, they are the source of natural resources and material sustenance needed by members of the social group. These territories are grounded in territorial claims made within the political sphere where power depends, in large part, on whether each group's social identities are recognized (either formally or informally) as legitimate. Amazonian frontier territories are also historical entities. Pacheco de Oliveira (2018: 7) describes how territorialization takes place through "complex processes, occurring at various scales and propelled by actors driven by distinct and even contradictory interests and ideologies." Taking Brazil as a historical example, he identifies six different "forms of territorialization" that indigenous peoples have used since the arrival of European invaders through a "multiplicity of indigenous, regional histories with very distinct temporalities and spatial circumstances" (p. 14).

A historical framework allows the researcher to detect the ways that more recent frontier territories, together with their respective political claims, are superimposed upon existing ones, creating situations of territorial conflict within distinct spheres of power. The spatial nature of conflicting territorial claims has its most visible political expression in the legal conflicts between differing land tenure regimes in Brazil. Territorial conflicts are also expressed in the economic sphere through clashes between modes of production, in the environmental sphere through biophysical impacts, and in the cultural sphere through struggles over social identities.

Territorial perspective provides a new way of delimiting social actors, one which highlights land tenure issues and natural resource use. Among the criteria to be considered in the delimitation of territorial groups are a sense of belonging to a specific geographic space; a collective memory of the occupation of that space; group territorial behavior; social organization for territorial management; and a shared vision of development.

MAPPING TERRITORIAL GROUPS ON MICRO-REGIONAL FRONTIERS

The delimitation of specific frontier regions for this research framework is made at a micro-regional level. Micro-regional frontiers are usually classified according to the leading economic drivers that produce different types of frontier: extractive frontiers (Bebbington, 2012); settlement frontiers (Geiger, 2008); agroindustrial frontiers (Ioris, 2017); and energy frontiers (Carvalho, 2006). It needs to be noted that each of these types of frontiers are also sites of forest destruction and timber extraction and, as such, can be subsumed under the broader category of deforestation frontier. Meanwhile, the emergence of a socio-environmental movement, with its focus on new sustainable development activities, has created what can be called ethno-environmental frontiers (IEB, 2012).

Figure 11.1 identifies ten active micro-regional frontiers in Brazilian Amazonia: the agroindustrial (soybean) frontiers of 1) northern Mato Grosso

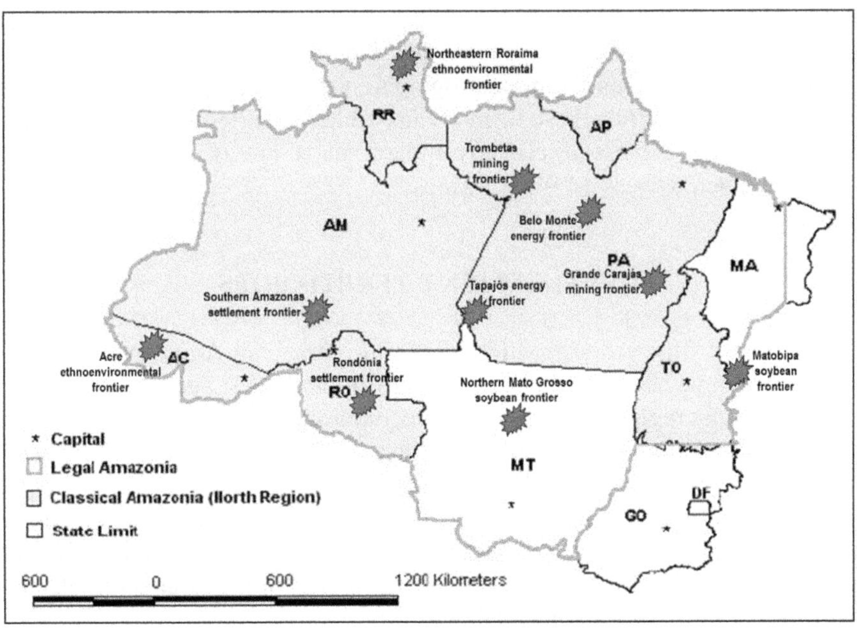

Figure 11.1 Micro-regional Frontiers in Brazilian Amazonia (2020). *Source*: Food and Agriculture Organization of the United Nations (FAO).

and 2) the Matopiba[2] region; the extractive (mining) frontiers of 3) Grande Carajás and 4) the Trombetas River basin; the (hydroelectric) energy frontiers of 5) the Belo Monte dam and 6) the Tapajós River basin dams; the agricultural settlement frontiers of 7) Rondônia and 8) southern Amazonas; and the ethno-environmental frontiers of 9) Acre and 10) northeastern Roraima.

Once the scalar delimitation of the micro-regional frontier to be studied is made, the next research task is to conduct a mapping of its principal territorial groups, where a political ecology approach is once again incisive. Each micro-regional frontier harbors a unique mix of the territorial groups that will need to be delineated as the research unfolds. Schmink and Wood (1992: 344), for example, analyze the "contest for resources between social groups capable of mobilizing varying forms of power" on the southern Pará Amazonian frontier as a constitutive aspect of historical change, whereby "groups of social actors formulate a common agenda based upon self-identity in the course of defending and furthering their particular interests" (Schmink and Wood 1992: 349). This actor-based, sociocultural mechanism for identifying frontier territorial groups does not imply the existence of a tightly knit, homogeneous social group, but rather posits that members have a common set of claims to and uses of specific territories.

The participatory mapping techniques that Almeida employs to spatialize the territorial claims of marginalized traditional peoples throughout Brazilian Amazonia provides an essential guide to this effort. His "New Social Cartography of Amazonia" project,[3] has identified hundreds of traditional communities and given them a cartographic voice by mapping their diverse territorial claims. When these formerly invisible territories are published on a map, they often contest official government maps, provoking what Almeida (1994) characterizes as a "War of Maps."

DEVELOPMENT TERRITORIES

Over the past fifty years, frontier creation in Brazilian Amazonia has been led by economic actors implementing a developmentalist agenda designed to "integrate" this region into the national economy through the construction of large infrastructure projects. In the 1970s and 1980s, under Brazil's military dictatorship, nationalist developmentalism promoted the expansion of the settlement frontier, notably along the Trans-Amazonian Highway; the large-scale extraction of the region's mineral resources, such as the Grande Carajás mining complex and railroad; and the construction of the large hydroelectric dams of Tucuruí, Samuel, and Balbina.

By the beginning of the twenty-first century, a neodevelopmentalist variant of this agenda had emerged, fueled by a fifteen-year-long commodity boom

(2000–2014) and supported by strong state financing from the National Social and Economic Development Bank (BNDES) and by Chinese development and export/import banks (Little, 2014). This new variant included the strategic aim of expanding the international reach of Brazilian construction companies through the building of large-scale infrastructure projects throughout Latin America and Africa, a system that prospered throughout the boom, only to descend into crisis with a decline in commodity prices, the fallout from several international corruption scandals involving bribes from Brazilian companies and the impeachment and removal from office of President Dilma Rousseff.

These efforts have established an industrial mode of production in the rainforest: in the use of the soil, with massive planting of soybeans; in the use of the subsoil, with the extraction of minerals, petroleum, and natural gas; in the use of the rivers, with the construction of large-scale hydroelectric dams; in the use of the fish stocks, with large fleets of modern fishing boats; and in the use of the forests, with the introduction of eucalyptus plantations for the production of cellulose. In general, industrial territorialities tied to the developmentalist agenda tend to be expansive (often at a highly accelerated rate), provoke large socio-environmental impacts, and fluctuate according to the ups and downs of the global commodity market. Each industrial mode of production has specific territorial attributes and claims and produces agribusiness, industrial mining, and enclave territories.

Large-scale, mechanized soybean production experienced vertiginous expansion during the first two decades of the twenty-first century, engendering the Mato Grosso and Matopiba soybean frontiers, creating massive agribusiness territories that dominate the regional landscape and have resulted in a dramatic increase in deforestation rates. These territories are founded upon privately owned farms whose economic interests are defended by the rural landholding elite that has formed a powerful ruralist lobby in Congress and controls several executive ministries.

Industrial mining projects for bauxite, iron ore, and gold, among other minerals, have generated industrial mining territories, which exhibit characteristics that are quite different from agribusiness ones. They depend upon licenses obtained from the Brazilian government, which owns all subsurface minerals. Industrial mining territories are widely dispersed throughout the Brazilian Amazon region as companies locate wherever mineral deposits are found and have a temporal component, since once the mineral deposits are depleted, the company abandons the area, often leaving a legacy of toxic soils and water and a devastated landscape. Since access to minerals can only be gained through above-ground sites, mining territories usually overlap with other types of territory, thus creating legal conundrums regarding vertical versus horizontal territorialities.

The construction of the large-scale hydroelectric dams, such as the Santo Antônio and Jirau dams in Rondônia, the Belo Monte dam in Pará, and the Teles Pires and São Manoel dams on the Pará/Mato Grosso border, established enclave territories that house thousands of workers and day-laborers and are controlled by consortia that are building the dams. After the construction of the dams is complete and they begin operation, many workers do not leave the frontier region, but rather settle in nearby cities, which swell in size in a matter of months and dramatically change the demographic dynamic of the frontier.

State-financed infrastructure construction not only facilitates the installation of large-scale economic activities such as soybean production and industrial mining, but also promotes the arrival of a diverse set of new social actors onto the frontiers who establish a host of small-scale territories. Land speculators are usually the first to enter a new frontier region, where they clear plots of land and sell the timber, and then sell the plot (illegally) to incoming settlers and cattle ranchers. The settlement of forested lands by agricultural colonists often occurs on unclaimed public lands (*terras devolutas*) held by the federal government, since they appear to be "available" for taking. The end result is the presence in Brazilian Amazonian frontiers of thousands of squatters (*posseiros*) who do not have legal titles for their land, but do have a more limited set of squatters' rights as outlined in Brazilian law. The population of itinerant placer gold miners (*garimpeiros*) can swell into the thousands within weeks of news of recently discovered gold deposits, only to disappear months later when the deposits become depleted, spawning the most fleeting of all territories in the Amazon.

SOCIO-ENVIRONMENTAL TERRITORIES

Contestations of hegemonic development territories have come from indigenous peoples, traditional communities, and environmentalists, though each of these groups are involved in a distinct set of territorial conflicts and, as such, will be analyzed separately below.

Indigenous Peoples

A historical perspective on territorial dynamics inverts the temporal frame of most frontier narratives, which tend to highlight the newest economic actors to arrive on a frontier and to characterize them as more "modern" than social actors tied to older "anachronistic" modes of production and territorial claims (cf. Webb, 1952). Indigenous peoples' territories become the starting point for analysis rather than an afterthought, and offer a baseline upon which all subsequent territorialities are superimposed.

Since indigenous peoples were widely dispersed throughout the Amazon region in 1500 (Beckerman, 1991), the arrival of European invaders and the subsequent creation of hundreds of micro-regional frontiers over the ensuing five centuries means that every Amazonian frontier is founded on occupied indigenous lands. Indigenous peoples, for their part, have resisted these invasions in numerous ways and continue to do so today, though this historical process is filled with examples of ethnocide (destruction of indigenous cultures) and genocide (extermination of indigenous societies) (Ribeiro, 1970).

Today, indigenous peoples are increasingly pursuing their own alternative forms of development, under what has been called an "ethnodevelopment" paradigm (Stavenhagen, 1985), whereby their collective needs and interests as an ethnic group serve as the guiding principle. Ethnodevelopment is founded upon the political position of self-determination, in which decisions about the group's future are made and implemented by each indigenous society in accordance with the precepts of food security, territorial defense, and sustainable livelihoods.

With the promulgation of the new Federal Constitution in 1988, indigenous peoples in Brazil were guaranteed their "original rights to the lands they traditionally occupy" while making the federal government responsible for "demarcating, protecting and conserving" indigenous peoples' territories (Article 231). This was a major stimulus for Amazonian indigenous peoples to pressure the federal government to demarcate and register their respective indigenous lands. By 2018, 27 percent of Brazilian Amazonia, or 115 million hectares, was parceled into 419 indigenous lands that are currently at various stages of the formalization process.[4]

This strengthening of the territorial rights of indigenous peoples over the past three decades has placed the issue of respect for different property-rights regimes at the forefront of any discussion of contemporary frontier conflicts. A 2009 Supreme Court ruling (STF, Pet 3388), which ratified the federal registry of the 1,747,464-hectare Raposa do Sol indigenous land in Roraima state, established a "Time Limitation" (*Marco Temporal*) for the formal registration of "new" indigenous lands, thereby severely limiting the ongoing processes of identifying and demarcating indigenous lands (Carneiro da Cunha and Barbosa, 2018).

Another area of contention regarding property-rights regimes on Brazilian frontiers today involves attempts by large landowners and the powerful ruralist lobby in the National Congress to restrict the process of identifying indigenous lands and to limit the official recognition of the territorial rights of traditional peoples, using the argument that these lands should be available for private, industrial-style production. This position is strongest on the expanding soybean frontiers of Mato Grosso and the Matopiba region, but is also promoted by the mining lobby, which is eager to gain access to indigenous lands for the extraction of minerals.

Resurgent Traditional Peoples

Brazilian Amazonian frontiers have a long history of processes of ethnocide of indigenous peoples and subsequent ethnic restructuring through processes of ethnogenesis (Hill, 1996). The resurgence of indigenous peoples' identities in rural Amazonian communities has become a powerful new ethnopolitical force in Brazilian Amazonia. After decades of suppressing their indigenous identities in the face of racism and discrimination, rural communities in the Tapajós River basin are now asserting their identities as an indigenous people and seeking to have their indigenous lands recognized and demarcated (Ioris, 2014). Other cases, such as the emergence of Kokama and Baré indigenous identities, reveal how frontier dynamics can reshuffle social identities and, in the process, foster the resurgence of those that were formerly submerged.

Today, Brazilian Amazonia is also home to hundreds of agroextractivist[5] groups, who are not recognized as indigenous peoples but who maintain many "traditional" modes of adaptation, such as collective use of communally held lands, long-term occupation and sustainable use of their territories, and a cultural/ritual attachment to their homeland enshrined in collective memory. Traditional peoples offer a stark contrast to industrial processes of development since their modes of adaptation are based in the small-scale, extensive use of forest and aquatic resources, such as the agroforestry practices of *quilombos*, the agroextractivist practices of rubber-tappers and Brazil-nut collectors, and the artisanal fishing techniques of *ribeirinhos*. These modes of production protect and sustainably use the Amazon's rich biological diversity and generate great sociodiversity, that is, the presence of numerous modes of adaptation that are tailored to the ecological conditions of specific Amazonian ecosystems.

One important example of the emergence of a nonindigenous, traditional people as an Amazonian political actor is that of the rubber-tappers. Under the inspired leadership of Chico Mendes, a rural labor leader from the state of Acre, the rubber-tappers fended off loggers who were clear-cutting their forests and forged an alliance with the international environmental movement (Mendes, 1989).[6] A new federal territorial entity—Extractive Reserve—was subsequently created to guarantee their collective rights. Other agroextractivist groups, such as Brazil-nut collectors and babassu coconut gatherers and processors, organized and requested collective titles for their extractive reserves. Amazonian riverine communities have recently established a group identity as *ribeirinhos,* which has the positive connotation of using collective, environmentally sound practices, as opposed to fishermen (*pescadores*), a term that tends to be associated with the individualistic predation of natural resources (Esterci, 2002). These diverse social groups have constructed a political identity as a "traditional people" and received partial federal recognition through the National Policy for the Sustainable Development of Traditional Peoples and Communities (Nogueira, 2017).

The political assertion of the rights of *quilombo* residents to their traditionally occupied territories can be considered to be part of the ethnogenesis process. *Quilombos* are communities that are descended from runaway slave communities dating back to Brazil's colonial period, or rural Afro-descendent communities that remained on lands abandoned by former slave owners, and have recently increased their political strength. Their territorial rights were recognized by the 1988 Federal Constitution[7] and titles to their collectively held lands are now being granted by the National Institute for Colonization and Agrarian Reform (INCRA). There are currently hundreds of *quilombos* in Brazilian Amazonia, but only a small percentage of them have received formal legal titles to their collective territories.

The struggle of these numerous resurgent social groups to gain official recognition for their ethnic identities is hindered by the popular understanding of the term "traditional" as an unchangeable, pre-modern condition, whereby it would be anomalous for an indigenous people to "emerge" in the twenty-first century. From a social scientific perspective, however, the construction and assertion of social identities is an ongoing process that is constantly modernizing to meet changing political realities and is therefore not fixed in any specific historical epoch, meaning that all social identities can be considered "coeval" (Fabian, 1983).

Environmentalists

The involvement of the international environmental movement in Amazonian affairs starting in the 1980s added a new social actor to the frontier mix, one which contested, at least partially, the developmentalist agenda in its call for an end to indiscriminate deforestation of the Amazon rainforest. Environmentalists promote an alternative paradigm of "sustainable development" in which the use of renewable resources provides the potential for long-term, nonpredatory economic growth. This concept, popularized at the 1992 Earth Summit in Rio de Janeiro, is used in the hope of reorienting attitudes toward development by considering the needs of future generations when making decisions about natural resource use.

And though deforestation rates in Brazilian Amazonia dropped significantly over a ten-year period from their highs in 2003–2004, they have since experienced a dramatic increase and the total deforested area is now approaching 20 percent of the original forest cover. Some scientists warn of reaching a tipping point that could lead to ecosystem collapse and the transformation of the rainforest into savanna, in a phenomenon known as "Amazon dieback" (Vergara and Scholtz, 2010).

The preservationist current of this movement pushes for the creation of new protected areas and the extension of existing ones in order to preserve the rich biodiversity of the Amazon rainforest. Since 1980, the total area of protected land in Brazilian Amazonia more than quadrupled and now 22 percent of the region is under some type of formal protection (RAISG, 2012). The conservation of biodiversity, however, has sparked a growing interest in bioprospecting for genetic material to be used by international biotechnology firms, thus producing a genetic frontier in Amazonia, one that targets indigenous environmental knowledge for its ability to find "shortcuts" to useful material (Hayden, 2006).

By the mid-2000s, climate change activists of the international environmental movement had successfully lobbied for the issues of deforestation and land-use change in tropical forests to be discussed in the greenhouse gas emission negotiations of the United Nations Framework Convention on Climate Change (UNFCCC) and the Amazon rainforest occupied, once again, a prominent place in international affairs, this time due its immense amount of stored carbon. Major efforts were made to establish working models for REDD+[8] payments and for an international carbon market, creating the embryo of what could become a carbon frontier with social groups vying to receive international payments for protecting stored carbon (i.e., trees) from deforestation. A bevy of "carbon cowboys" have already appeared on Amazonian frontiers, attempting to swindle indigenous peoples with REDD+ market schemes. The Brazilian government, for its part, established a US$ 1 billion Amazon Fund with Norwegian (and subsequently German) funds, which are only disbursed after reductions in Amazonian deforestation rates are confirmed.

Environmentalists have advocated for the application of social and environmental safeguards by multilateral and national development banks to prevent or mitigate negative socio-environmental impacts, though these are mainly designed to minimize the impact of the industrial mode of production rather that alter the way that production occurs (Santos, 2014). Environmentalists have also proposed the use of Strategic Environmental Assessments, which consider the impacts of multiple projects in different economic sectors at a regional level, identify potential cumulative and synergistic impacts, employ a long-term framework of evaluation and delineate low-carbon alternatives to existing development projects.

CLASH OF DEVELOPMENT PARADIGMS

The term "development paradigm" refers to the overarching framework of interrelated actions that guides the economic activities of an ethnic group or a nation-state. While most frontier dynamics are led by a dominant social actor, usually allied with a particular developmentalist activity, the ways that less powerful social actors contest that dominance and project their own claims will vary from frontier to frontier. The varied group-level interactions within the frontier dynamics, in which multiple territorial actors wield widely divergent levels of political and economic power, creates a field of public debate noted for shifting alliances that are responding to rapidly changing social and economic realities. Starting in the 1980s, Brazilian Amazonia has seen the articulation of diverse social groups into two opposing, makeshift political "alliances": a development alliance and a socio-environmental alliance. These *ad hoc* groupings do not represent formally constituted political entities, but instead refer to a tendency among social actors toward political alignment when frontier conflicts become aggravated and polarized.

One constant between the developmentalist surge of the 1970s and 1980s and the neodevelopmentalist surge of the 2000s and 2010s is the treatment of the Amazon region as a resource frontier in which economic benefits are exported out of the region in the name of either national development or export-led growth. The development alliance includes a variety of powerful economic actors—large-scale soybean farmers; Brazilian construction companies; multinational mining companies; development banks—that promote the precepts of economic growth, industrialization, and meeting global demand for commodities.

Political coordination between environmentalists, indigenous peoples, and traditional communities became a potent force by the end of the 1980s and led to the consolidation of a vibrant socio-environmental movement of Amazonian peoples who employed novel forms of protest over the destruction of their forests and rivers (Santilli, 2005). These counterhegemonic groups are now proposing a socio-environmental paradigm centered upon the notion of "Amazoncentric" development where the protection of the Amazon biome and its peoples are given top priority. The case of the rubber-tappers reveals how their emergence as a political force was strengthened by the alliance with the international environmental movement. The historic 1989 Altamira meeting, where over 600 Kayapó warriors, supported by Sting and other international environmentalists, came together to stop the construction of the Kararaô dam on the Xingu River in Pará state was another major victory of this movement. These and other cases form the basis of an indigenous-environmental alliance in favor of protecting the Amazon rainforest (Conklin and Graham, 1995), which has experienced many highs and lows in political cohesion since then.

The rapid industrialization of the Amazon is producing social and environmental impacts of unprecedented magnitude and geographic scale. The construction of dozens of large dams throughout the pan-Amazon region, for example, is causing major disruptions in river flows and fish migrations, with many unknown consequences. Other long-term biophysical impacts are the loss of biodiversity, genetic erosion, increased greenhouse gas emissions and more frequent extreme climate events such as floods and droughts.

The unequal distribution of the benefits from these development activities and their negative impacts raise the issue of environmental justice. The groups most affected by these negative impacts are local Amazonian peoples whose livelihoods depend upon natural resources that are being depleted, degraded, or contaminated at alarming rates. Risk assessments and cost-benefit analyses carried out prior to the initiation of large-scale projects, for example, rarely take into account the needs of local groups whose very survival may be threatened by the project. Negative impacts are also felt by the thousands of migrants who are lured to the frontier by the promises of the development agenda but often end up with precarious, insalubrious livelihoods in newly established Amazonian shantytowns. Infrastructure projects—roads, waterways, hydroelectric dams—offer an important entryway, and potential jobs, for urban migrants, agricultural settlers, construction workers, and traders. These diverse social groups become part of the long-term national development goal of creating "living borders" in Amazonia populated by people with national (as opposed to ethnic) identities. While these groups have a clear

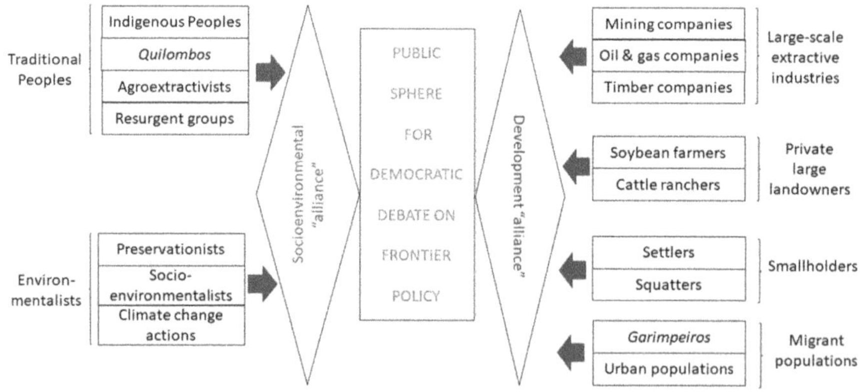

Figure 11.2 Template of Key Territorial Actors on Brazilian Amazonian Frontiers. *Source*: Image created by author.

stake in the developmentalist agenda, they nonetheless represent its poorest, most marginal segment.

It remains an open question whether an expansionist-developmentalist paradigm can co-exist with a socio-environmental paradigm, or whether the former, given its enormous economic and political power, will eventually overwhelm the latter. Figure 11.2 presents a template for the analysis of key territorial actors in contemporary Brazilian Amazonian frontiers. Though this template is a highly schematic representation of the complex social dynamics outlined above, it serves the function of ensuring that all the major social actors present on the frontier are considered in research and public policy formulation on contemporary Amazonian frontiers.

MAKING PUBLIC POLICY WITHIN A SCHIZOPHRENIC STATE

A characteristic attributed to many contemporary frontiers is the weak presence of the state, which is used to explain the administrative chaos found on frontiers. That notion is countered here with the argument that a close analysis of all of the major social actors reveals that the state is present on the frontier in many ways and forms, and that it has differentiated, unequal impacts on these social actors. Far from being a monolithic entity, the presence of the state—at federal, state, and municipal levels—is exercised through multiple agencies pursuing contradictory ends.

The contradictory nature of the state does not prevent certain state sectors from exercising hegemony in frontier contexts. The developmentalist agenda, given the large amounts of capital investment and strong government support behind it, is almost always the leading force for frontier change throughout Brazilian Amazonia. The powerful Ministry of Agriculture and Ranching (MAPA) has stimulated the rapid expansion of soybean production onto previously forested lands. The Ministry of Mines and Energy (MME) promotes industrial mining projects throughout the region and, since the state receives royalty payments from these projects, has a vested interest in the expansion of mining activities. The state-run BNDES bank has been a major funder of large-scale infrastructure projects during the first two decades of the twenty-first century.

However, the state also supports nonhegemonic social actors such as indigenous peoples, *quilombos*, socio-environmentalists and small family farmers. Though the government agencies that work with these groups generally have much less political clout than the MAPA, MME, and BNDES, they are still able to provide legally sanctioned support to those Brazilian citizens covered by their remit. Another site of governmental action that tends to work against the hegemonic forces of the state in Brazil is the Public Prosecutor's Office (*Ministério Público*), which is tasked with defending the "diffuse" interests of Brazilian citizenry, such as the maintenance of a clean, healthy environment and the protection of the nation's ethnic minorities. Those state agencies and departments that have official punitive power—the Armed Forces; the Federal Police; the Environmental Enforcement Agency (IBAMA)—do not act in a monolithic manner and must mediate between opposing social and economic agendas.

Each of the territorial actors outlined in figure 11.2 has a specific tie to the state apparatus that is tasked with assisting them and defending their interests in a phenomenon that Fontaine (2007), in his analysis of Ecuadorian Amazonian frontiers, calls the "schizophrenic State." Figure 11.3 builds on the previous template by identifying the corresponding government entities that are responsible for protecting the interests of each social group and highlights (in the bold-lined rectangle in the middle of the figure) the contested intra-state sphere of frontier policy development and enforcement.

Since frontiers are part of a continuous process of society-making in which new social actors emerge and existing social actors must respond to rapidly changing social and economic circumstances, they are public spaces of interaction and negotiation. The contradictory, schizophrenic state thus becomes the principal site of public policy formulation. A guiding concept of a public policy approach is that of "frontier governance," or the way that the use and management of natural resources are debated within an expanded public space that involves all territorial actors in the decision-making process. This provides a gateway to the eventual resolution of territorial conflicts through ongoing negotiations of competing territorial actors, mediated by competing

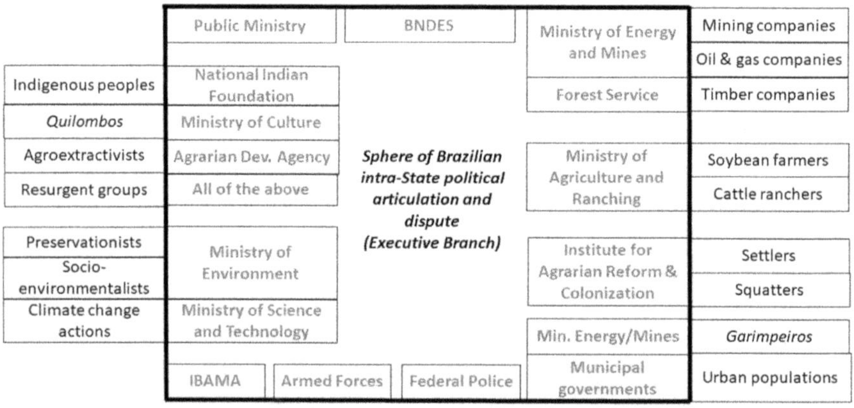

Figure 11.3 Intra-state Sphere of Frontier Public Policies (With Their Accompanying Social and Economic Actors). *Source*: Image created by author.

state agencies, and the subsequent implementation of new public policies for Amazonian frontier regions.

FRONTIER GOVERNANCE: GETTING TO THE POST-FRONTIER?

Establishing effective techniques for landscape management, grounded in both ecological science and social agreements, can promote the goal of developing a long-term, integrated vision for Amazonian landscapes. This vision would incorporate policies for the adequate protection and use of biological diversity and water cycles, for the establishment of a structure for monitoring natural resource use that would ensure proper enforcement of existing laws and policies, and for addressing the technical and social issues related to climate change, while still allowing for intensive economic production activities on ecologically appropriate lands. Fine-grained studies of "human-environment interaction" (Brondizio and Moran, 2012) utilize methodologies that correlate data obtained at multiple scales—from satellite images to on-the-ground conversations with farmers—which can provide guidance in how to achieve long-term sustainability of land use practices. This frontier governance structure would be responsible for guaranteeing a more just distribution of both the economic benefits and the negative environmental impacts of frontier activities.

If efforts to improve forest governance are able to reduce the levels of territorial conflict and attenuate the clash of development paradigms, some authors consider this to represent a "closing" or "taming" of a frontier and postulate the existence of a "post-frontier" where the rule of law respects the rights of all and, in turn, is respected by all without recourse to violence. Santos-Granero and Barclay (2000) assert that after 150 years of transformations since the Peruvian government's first efforts to integrate the Amazonian region of Loreto into national life, the Loretan frontier has finally been "tamed," that is, "no longer governed by the logic of frontier expansion" (p. 308) due to the "extension of civil rights, the emergence of a regionalist ideology, and the consolidation of a regional identity" (p. 299). This argument dovetails with the analysis of Cronon, Miles, and Gitlin (1992: 16) who, in reference to the U.S. western frontier, affirm that "the transition from frontier to region was a process of creating new, stable geographic identities in the midst of landscapes that people chose to perceive as homes."

Hvalkof (2008), using cases from the Peruvian Amazon, argues that "the status quo can be changed and a more equitable social order established" (p. 276) but warns that this can only be achieved if there is a "general legal and regulatory framework in place, a hierarchical order of enforcement and sufficient resources" (p. 277). Haarstad (2012) admits that economic dependence on natural resource extraction can restrict policy options for governments and civil society organizations but recognizes that "new rights can provide discursive rallying points and organizational resources that sharpen claims and mobilize campaigns" thereby creating a "new terrain for interest negotiation and conflict" (p. 244). Larsen (2015) posits the existence of a "post-frontier paradox" wherein "sustainability institutions coexist side by side with degradation and marginalization of communities" and "technological fixes to social and environmental problems at the resource frontier displace rather than resolve the nature of sustainability problems" (pp. 149–51).

A strong argument against the notion of any definitive closing or taming of a frontier is their cyclical dimension; Geiger (2008: 163) characterizes frontiers as "durable and resurgent phenomena," which can close and reopen in response to the expansion and contraction of states and to cycles of exploitation of valuable resources. The frontier will only wither away with the achievement of a "*negotiated autonomy* of groups which have, by force of an unprecedented break with the policies of the past, been recognized as a true part of the nation's patrimony" (Geiger 2008: 181). Market-based approaches to frontier governance assert that incentives to companies to modify their behavior are a necessary step for reducing expansionist pressure on the Amazon and can forge a common path that allows for economic growth to be achieved while still achieving conservation goals (Nepstad et al., 2014). The experiments with deforestation-free beef and a moratorium on soybean expansion, however, have shown mixed results and reveal just how difficult it is to reign in market forces on the frontier.

It is not only the cyclical nature of commodity markets and their associated impact on frontier dynamics that produces what Little (2001) calls the phenomenon of "perennial frontiers," but also the fact that new types of frontiers, tied to new technological developments, can emerge unexpectedly. As mentioned above, Amazonia is facing a new set of demands, such that biotechnological advances are generating genetic frontiers and climate change agreements may eventually engender carbon frontiers. In addition, the clash between horizontal frontiers—involving disputes over control of the land surface—and vertical frontiers—involving access to minerals and hydrocarbons housed in the subsoil—can become the source of new frontier conflicts.

The efforts to establish a renewed resource governance structure for Brazilian Amazonian frontiers must be addressed within the new far-right political and economic climate in Brazil. With the election of former army captain Jair Bolsonaro to the Brazilian Presidency in 2018, amid an electoral campaign marked by violence and recrimination, a new political alignment is forming that favors the construction of roads, railways, and waterways and proposes a massive expansion of Amazonian lands dedicated to soybean production, this time led by private capital investments, though probably with continued support from Chinese development and export/import banks. This new political alignment is marked by strong anti-indigenous and anti-environmentalist discourses, which have put the socio-environmental alliance on alert to prevent any rollback of the environmental and territorial rights and policies that have been consolidated at national and state levels of government.

The unprecedented wave of fires that raged across the Amazon in 2019 provoked an international outcry but produced few on-the-ground changes to existing development policy. Other clashes regarding frontier policy include disputes over environmental licensing of development projects, revoking the prohibition on mining in indigenous lands, unfettered access to genetic material and its associated traditional knowledge, and maintaining the deforestation goals in Brazil's commitment to the 2015 Paris Climate Change Agreement. The way that these points of contention are understood and addressed will shape the next wave of Brazilian Amazonian frontier expansion.

Engaged research on territorial conflicts will not provide definitive answers to these complex issues, but they can participate in the effort to attenuate frontier conflicts by articulating the diverse claims of territorial actors, by analyzing the effectiveness of existing public policies and by presenting alternative models of development that could lead to more effective frontier governance. They can also provide crucial inputs to the comparative discussion of frontier dynamics with which this chapter began. Ethnographies of different Amazon frontier conflicts can be compared across countries and between types of frontiers so as to gain a deeper, more nuanced understanding of frontier dynamics. These comparative insights can then be employed in the broader debates about the role of frontiers in contemporary processes of globalization.

NOTES

1. Each of the topics discussed in this chapter has a vast literature that supports it. The references provided here are what I consider to be foundational or paradigmatic texts in their respective fields of expertise.

2. The Matopiba region encompasses portions of the states of Maranhão, Tocantins, Piauí, and Bahia and straddles the border between the Amazon rainforest biome and the Cerrado savanna biome.

3. More information available at: http://novacartografiasocial.com.br.

4. More information available at: https://pib.socioambiental.org.

5. Agroextractivism is an adaptive method of Brazilian rural communities based upon collective-use rights to nontimber forest products combined with family agricultural and horticultural plots.

6. Chico Mendes was assassinated at this home in 1988 by rural landowners and subsequently became a martyr for the socio-environmental movement.

7. Article 68 of the Transitory Constitutional Dispositions Act recognized the definitive tenure rights of "remnant communities of quilombos" and delegated to the federal government the responsibility for emitting their respective land titles.

8. REDD+ stands for Reduced Emissions from Deforestation, Forest Degradation and Land-use Change.

BIBLIOGRAPHY

Abu-Lughod, J. 1989. *Before European Hegemony: The World System AD 1250–1350*. Oxford University Press: New York.

Adams, C., Murrieta, R., Neves, W., and Harris, M. (eds.). 2009. *Amazon Peasant Societies in a Changing Environment: Political Ecology, Invisibility and Modernity in the Rainforest*. Springer.

Albert, B. and Ramos, A. (eds.). 2002. *Pacificando o Branco: Cosmologias do Contato no Norte-Amazônico*. UNESP: São Paulo.

Almeida, A. W. B. 1994. *Carajás: A Guerra dos Mapas*. Falangola: Belém.

Amin, S. 1976. *Unequal Development*. Monthly Review Press: London.

Barbier, E. B. 2011. *Scarcity and Frontiers: How Economies Have Developed through Natural Resource Exploitation*. Cambridge University Press: Cambridge.

Bebbington, A. (ed.). 2012. *Social Conflict, Economic Development and Extractive Industry: Evidence from South America*. Routledge: London.

Becker, B. K. 1982. *Geopolítica da Amazônia: A Nova Fronteira de Recursos*. Zahar: Rio de Janeiro.

Beckerman, S. 1991. A Amazônia Estava Repleta de Gente em 1492? In: *Origens, Adaptações e Diversidade Biológica do Homem Nativo da Amazônia*, Neves, W. (ed.). NPEG/CNPq/SCT/PR: Belém, pp. 143–59.

Brondizio, E. S. and Moran, E. F. (eds.). 2012. *Human–Environment Interactions: Current and Future Directions*. Springer Scientific Publishers: Dordrecht.

Browder, J. D. and Godfrey, B. J. 1997. *Rainforest Cities: Urbanization, Development, and Globalization of the Brazilian Amazon*. Columbia University Press: New York.

Campbell, J. 2015. *Conjuring Property: Speculation and Environmental Futures in the Brazilian Amazon*. University of Washington Press: Seattle.

Cardoso de Oliviera, R. 1964. *O Índio e o Mundo dos Brancos*. Difusão Européia do Livro: São Paulo.

Cardoso de Oliviera, R. 1976. *Identidade, Etnia e Estrutura Social*. Livraria Pioneira Editora: São Paulo.

Carneiro da Cunha, M. and Barbosa, S. (eds.). 2018. *Direitos dos Povos Indígenas em Disputa*. UNESP: São Paulo.

Carvalho, G. O. 2006. Environmental Resistance and the Politics of Energy Development in the Brazilian Amazon. *Journal of Environment and Development*, 15(3), 245–68.

Cleary, D. 1990. *Anatomy of the Amazon Gold Rush*. University of Iowa Press: Iowa City.

Conklin, B. A. and Graham, L. R. 1995. The Shifting Middle Ground: Amazonian Indians and Ecopolitics. *American Anthropologist*, 97(4), 695–710.

Cronon, W., Miles, G., and Gitlin, J. 1992. Becoming West: Toward a New Meaning for Western History. In: *Under an Open Sky: Rethinking America's Western Past*, Cronon, W., Miles, G. and Gitlin, J. (eds.). W.W. Norton: New York, pp. 3–27.

Dean, W. 1987. *Brazil and the Struggle for Rubber: A Study in Environmental History*. Cambridge University Press: Cambridge.

Esterci, N. 2002. Conflitos Ambientais e Processos Classificatórios na Amazônia Brasileira. *Boletim Rede Amazônia*, 1(1), 51–62.

Fabian, J. 1983. *Time and the Other: How Anthropology Makes its Object*. Columbia University Press: New York.

Findlay, R. and Lundahl, M. 1999. Resource-Led Growth—A Long-term Perspective: The Relevance of the 1870–1914 Experience for Today's Developing Economies. *Working Papers No. 162*. UNU – World Institute for Development Economics Research: Helsinki.

Fisher, W. H. 2000. *Rainforest Exchanges: Industry and Community on an Amazonian Frontier*. Smithsonian Institution Press: Washington, D.C.

Fontaine, G. 2007. *El Precio del Petróleo: Conflictos Socio-Ambientales y Gobernabilidad en la Región Amazónica*. Flacso Ecuador/IFEA/Abya-Yala: Quito.

Foweraker, J. 1981. *The Struggle for Land: A Political Economy of the Pioneer Frontiers in Brazil from 1930 to the Present Day*. Cambridge University Press: Cambridge.

Frank, A. G. and Gills, B. K. 1993. The 5,000-year World System: An Interdisciplinary Introduction. In: *The World System: Five Hundred Years of Five Thousand?* Frank, A. G. and Gillis, B. K. (eds.). Routledge: London and New York, pp. 3–55.

Geiger, D. 2008. Turner in the Tropics: The Frontier Concept Revisited. In: *Frontier Encounters: Indigenous Communities and Settlers in Asia and Latin America*, Geiger, D. (ed.). IWGIA: Copenhagen, pp. 75–215.

Geiger, D. (ed.). 2015. *Colonization and Conflict: Contemporary Settlement Frontiers in South and Southeast Asia*. LIT Verlag: Berlin.

Gerhard, D. 1959. The Frontier in Comparative View. *Comparative Studies in Society and History*, 1(3), 205–29.

Guy, D. J. and Sheridan, T. E. (eds.). 1998. *Contested Ground: Comparative Frontiers on the Northern and Southern Edges of the Spanish Empire*. University of Arizona Press: Tucson.

Haarstad, H. 2012. Conclusions. In: *New Political Spaces in Latin American Natural Resource Governance*, Haarstad, H. (ed.). Palgrave Macmillan: New York, pp. 239–48.

Hall, T. D. 2009. Puzzles in the Comparative Study of Frontiers: Problems, Some Solutions, and Methodological Implications. *American Sociological Association*, 15(1), 25–47.

Harris, M. 2000. *Life on the Amazon*. Oxford University Press: Oxford.

Hayden, C. 2006. Bioprospecting: The "Promise" and Threat of the Market. *NACLA Report on the Americas*, 39(5), 26–31.

Heckenberger, M. J. 2005. *The Ecology of Power: Culture, Place, and Personhood in the Southern Amazon, A.D. 100–2000*. Routledge: London.

Hemming, J. 1978. *Red Gold: The Conquest of the Brazilian Indians*. Harvard University Press: Cambridge, MA.

Hennessy, A. 1978. *The Frontier in Latin American History*. Edward Arnold: London.

Hill, J. D. (ed.). 1996. *History, Power, and Identity: Ethnogenesis in the Americas, 1492–1992*. University of Iowa Press: Iowa City.

Hoelle, J. 2015. *Rainforest Cowboys: The Rise of Ranching and Cattle Culture in Western Amazonia*. University of Texas Press: Austin.

Hvalkof, S. 2008. Colonization and Conflict on the Amazon Frontier: Dimensions of Interethnic Relations in the Peruvian Montaña. In: *Frontier Encounters: Indigenous Communities and Settlers in Asia and Latin America*, Geiger, D. (ed.). IWGIA: Copenhagen, pp. 291–86.

IEB. 2012. *Corredores Etnoambientais na Amazônia Ocidental: Principais Resultados do Projeto Garah Itxa—2009–2012*. Instituto Internacional de Educação do Brasil: Brasília.

Ioris, A. A. R. 2017. *Agribusiness and the Neoliberal Food System in Brazil: Frontiers and Fissures of Agro-Neoliberalism*. Routledge: London.

Ioris, E. 2014. *Uma Floresta de Disputas: Conflitos sobre Espaços, Recursos e Identidades Sociais na Amazônia*. UFSC: Florianópolis.

Jackson, J. 1975. Recent Ethnography of Indigenous Northern Lowland South America. *Annual Review of Anthropology*, 4, 307–40.

Karasch, M. 1998. Interethnic Conflict and Resistance on the Brazilian Frontier of Goiás, 1750–1890. In: *Contested Ground: Comparative Frontiers on the Northern and Southern Edges of the Spanish Empire*, Guy, D. and Sheridan, T. (eds.). University of Arizona Press: Tucson, pp. 115–34.

Larsen, P. B. 2015. *Post-Frontier Resource Governance: Indigenous Rights, Extraction and Conservation in the Peruvian Amazon*. Palgrave Macmillan: New York.

Lattimore, O. 1951 [1940]. *Inner Asian frontiers of China*. 2nd ed. American Geographic Society: New York.

Lattimore, O. 1962. *Studies in Frontier History: Collected Papers, 1928–1958*. Oxford University Press: London.

León-Portilla, M. 1959. *La Visión de los Vencidos: Relaciones Indígenas de la Conquista*. Universidad Nacional Autónoma de México: Mexico City.

Li, T. M. 2003. Situating Resource Struggles: Concepts for Empirical Analysis. *Economic and Political Weekly*, 29, 5120–8.

Li, T. M. 2014. *Land's End: Capitalist Relations on an Indigenous Frontier*. Duke University Press: Durham.

Limerick, P. N. 1987. *The Legacy of Conquest: The Unbroken Past of the American West*. W.W. Norton: New York.

Limerick, P. N., Milner II, C. A. and Rankin, C. A. (eds.). 1991. *Trails: Toward a New Western History*. University of Kansas Press: Lawrence.

Lisansky, J. 1990. *Migrants to Amazonia: Spontaneous Colonization in the Brazilian Frontier*. Westview Press: Boulder.

Little, P. E. 2000. O Rio Maracá e o Delta do Rio Amazonas: Entre Isolamento e a Globalização. *Revista Ethos*, 1(1), 63–81.

Little, P. E. 2001. *Amazonia: Territorial Disputes on Perennial Frontiers*. Johns Hopkins University Press: Baltimore.

Little, P. E. 2006. Ecologia Política como Etnografia: Um Guia Teórico e Metodológico. *Horizontes Antropológicos*, 25, 85–103.

Little, P. E. 2010. A Prática Brasileira da Ecologia Política: Aportes da Antropologia. In: *Horizontes das Ciências Sociais: Antropologia*, Duarte, L. F. (ed.). ANPOCS/ Editora Barcarolla: São Paulo, pp. 341–65.

Little, P. E. 2014. *Mega-Development Projects in Amazonia: A Geopolitical and Socioenvironmental Primer*. Red Jurídica Amazónica/ARA/DAR: Lima.

Lopes, R. J. 2017. *1499: O Brasil Antes de Cabral*. HarperCollins: Rio de Janeiro.

Malkki, L. 1992. National Geographic: The Rooting of Peoples and the Territorialization of National Identity among Scholars and Refugees. *Cultural Anthropology*, 7(1), 24–44.

Mañach, J. 1975. *Frontiers in the Americas: A Global Perspective*. Teachers College Press: New York.

Martinez-Alier, J. 2002. *The Environmentalism of the Poor: A Study of Ecological Conflicts and Valuation*. Edward Elgar: Chelternham and Northhampton.

McNeill, W. 1983. *The Great Frontier: Freedom and Hierarchy in Modern Times*. University of Princeton Press: Princeton.

Mendes, C. 1989. *Fight for the Forest: Chico Mendes in his Own Words*. Latin American Bureau: London.

Nepstad, D., et al. 2014. Slowing Amazon Deforestation through Public Policy and Interventions in Beef and Soy Supply Chains. *Science*, 344(6188), 1118–23.

Nogueira, M. 2017. *Gerais a Dentro a Fora: Identidade e Territorialidade entre Geraizeiros do Norte de Minas Gerais*. IEB Mil Folhas: Brasília.

Nugent, W. 1994. New World Frontiers: Comparisons and Agendas. In: *Where Cultures Meet: Frontiers in Latin American History*, Weber, D. J. and Rausch, J. M. (eds.). Scholarly Resources: Wilmington, pp. 72–85.

Pacheco de Oliveira, J. 2018. Fighting for Lands and Reframing the Culture. *Vibrant: Virtual Brazilian Anthropology*, 15(2), 1–21.

Peluso, N. L. 1992. *Rich Forests, Poor People: Resource Control and Resistance in Java*. University of California Press: Berkley.

Raffles, H. 2002. *In Amazonia: A Natural History*. Princeton University Press: Princeton.

RAISG. 2012. *Amazonía Bajo Presión*. Red Amazónica de Información Socioambiental Georreferenciada; Instituto Socioambiental: São Paulo.

Ramos, A. 1998. Development Does Not Rhyme with Indian, or Does it? In: *Indigenism: Ethnic Politics in Brazil*, Ramos, A. (ed.). University of Wisconsin Press: Madison, pp. 195–221.

Ribeiro, D. 1970. *Os Índios e a Civilização*. Editora Civilização Brasileira: Rio de Janeiro.

Sachs, J.D. and Warner, A. 1995. *Economic Convergence and Economic Policies*. Harvard Institute for International Development: Cambridge, MA.

Santilli, J. 2005. *Socioambientalismo e Novos Direitos: Proteção Jurídica à Diversidade Biológica e Cultural*. Fundação Peirópolis: São Paulo.

Santos, A. D. 2014. Os Megaprojetos, o BNDES e o Tema das Salvaguardas Socioambientais. In: *Os Novos Desafios da Política Ambiental Brasileira*, Little, P. (ed.). IEB Mil Folhas: Brasília, pp. 344–66.

Santos-Granero, F. and Barclay, F. 2000. *Tamed Frontiers: Economy, Society, and Civil Rights in Upper Amazonia*. Westview Press: Boulder.

Schmink, M. and Wood, C. H. 1987. The 'Political Ecology' of Amazonia. In: *Lands at Risk in the Third World: Local Level Perspectives*, Little, P. D. and Horowitz, M. M. (eds.). Westview: Boulder, 38–57.

Schmink, M. and Wood, C. H. 1992. *Contested Frontiers in Amazonia*. Colombia University Press: New York.

Shaffer, B. and Ziyadov, T. (eds.). 2012. *Beyond the Resource Curse*. University of Pennsylvania Press: Philadelphia.

Slater, C. 1994. *Dance of the Dolphin: Transformation and Disenchantment in the Amazonian Imagination*. University of Chicago Press: Chicago.

Stavenhagen, R. 1985. Etnodesenvolvimento: Uma Dimensão Ignorada no Pensamento Desenvolvimentista. *Anuário Antropológico*, 84, 13–56.

Tsing, A. L. 2005. *Friction: An Ethnography of Global Connection*. Princeton University Press: Princeton.

Turner, F. J. 1893. The Significance of the Frontier in American History. *Annual Report of the American Historical Association*, 1893, 197–227.

Turner, T. 1991. Representing, Resisting, Rethinking: Historical Transformations of Kayapo Culture and Anthropological Consciousness. In: *Colonial Situations: Essays on the Contextualization of Ethnographic Knowledge*, Stocking, G. (ed.). University of Wisconsin Press: Madison, pp. 285–313.

Velho, O. 1972. *Frentes de Expansão e Estrutura Agrária*. Zahar: Rio de Janeiro.

Vergara, W. and Scholtz, S.M. 2010. *Assessment of the Risk of Amazon Dieback*. World Bank Studies: Washington, D.C.

Wallerstein, I. 1983. *Historical Capitalism*. Verso: London.

Watts, M. J. 1992. Space for Everything (A Commentary). *Cultural Anthropology*, 7(1), 115–29.

Webb, W. P. 1952. *The Great Frontier*. The Riverside Press: Cambridge.

Weber, D. J. and Rausch, J. M. (eds.). 1994. *Where Cultures Meet: Frontiers in Latin American History*. Scholarly Resources: Wilmington, DE.

Weiss, L. 1998. *The Myth of the Powerless State: Governing the Economy in a Global Era*. Polity Press: Cambridge.

Wheeler, M. 1955. *Rome beyond the Imperial Frontiers*. Penguin Books: Harmondsworth.

Whitehead, N. L. 2003. Introduction. In: *Histories and Historicities in Amazonia*, Whitehead, N. (ed.). University of Nebraska Press: Lincoln, pp. vii–xx.

Chapter 12

Slow Violence and Slow Seeing in *Beyond Fordlândia*

Marcos Colón

Contemporary environmental questions present fundamental challenges related to representation. Principally, how can researchers and ecocritics elaborate appropriate narratives, images, and symbols, which depict ongoing environmental violence and the catastrophic effects of slow violence? Slow violence is "a violence that occurs gradually and out of sight, a violence of delayed destruction that is dispersed across time and space, an attritional violence that is typically not viewed as violence at all" (2011: 2). My continuous conversations with Rob Nixon about slow violence and the Amazon inspired me to direct the documentary *Beyond Fordlândia* (2017, 75 min). Slow violence is central to my social reflection and intervention into the environmental discussion of the Amazon through the use of filmic language.[1]

Slow violence persists through time and remains invisible to the eyes of the common observer. It is closely linked to capitalist processes and predatory strategies of neoliberal economic development dependent on the exhaustive use of natural resources. This violence produces and reproduces procedures of renewed primitive accumulation on indigenous peoples and lands, because it adheres to models of progressive linear development that commodify natural resources and the environment as products of the market (Nixon, 2011; Atiles-Osoria, 2013).

The fact that many people living through this tragedy remain ignorant of its continuous nature and do not know how to report their suffering, destruction, and death of their environments through a sociocultural and political lens, affects my thinking as a researcher and documentarist. Surprisingly, there are cases in which corrupt legal systems of slow governmental violence permanently cover up this destruction and indefinitely impede its public disclosure (Bernucci, 2017: 17). Therefore, representing and narrating the distinct features of slow violence, stands as the greatest challenge undertaken by ecocritics and more concretely by myself in *Beyond Fordlândia*. Nixon (2011) reminds us that the crux of the matter passes through strategies of representation:

> How can we convert into image and narrative the disasters that are slow moving and long in the making, disasters that are anonymous and that star nobody, disasters that are attritional and of indifferent interest to the sensation-driven technologies of our image-world? How can we turn the long emergencies of slow violence into stories dramatic enough to rouse public sentiment and warrant political intervention, these emergencies whose repercussions have given rise to some of the most critical challenges of our time? (p. 3)

Throughout my personal and professional relationship with Nixon, he has encouraged me to intervene in the environmental crisis by creating representations that connect historical and sociocultural phenomena located in the relationship between Amazonian realities and the environment. To achieve this, I rethought the informational bases for my perceptions of ecological crises threatening life, and the violence practiced against the environment. This meant bringing a critical and dissenting disciplinary content to bear upon dominant discourses and traditional canons by integrating immediate data and information from sensationalist news agencies into a theoretical corpus dealing with long-term global-scale processes. As one can see in previous applications of a postcolonial ecocritical approach,[2] such an integration of knowledge has led us to reinterpret immediate structural violence that produces environmental and human crises. A simultaneously political and theoretical understanding broadens the possibilities of transdisciplinary scientific, epistemic, and artistic interlocution,[3] by facilitating interactions between literature, sociology, and science.[4]

Nixon calls us to action through pathways that expand epistemic subjects' participation in a polysemic dialogue capable of affecting political policies. Confronting slow violence should be considered part of a continuous far-reaching strategy, on all fronts and in all dimensions of the environmental and social effects that arise from ecological crisis. It calls for the construction of an innovative and dissident theoretical framework, related to the concept of "ecology of thought" (Morton, 2010). This multidisciplinary lens finds itself capable of dealing with theoretical and practical tasks and of "correlating processes and structures of imperialism, neoliberalism, violence, environmental degradation and environmental struggles," that arise from the "abandonment of bioregionalism, susceptible to being applied under new purposes of organizing material life and the emergence of a new transnational ethic of space" (Nixon, 2011: 26). These simultaneously political and theoretical reflections were intricately connected to *Beyond Fordlândia*'s genesis because they moved me to rethink the foundations of the ecological crisis and violence inflicted upon Amazonian biomes and ecosystems.

My documentary, *Beyond Fordlândia,* presents an intergenerational environmental report on the Amazonian experience of Henry Ford, decades after its failure.[5] The businessman ambitiously proposed to take civilization to the heart of the Amazonian Forest by planting of 800 thousand hectares of rubber trees and constructing an entire city dedicated to producing rubber on the banks of the River Tapajós, one of the main tributaries of the Amazon. Beginning in 1927, the film narrates and illustrates how the ill-fated project of producing rubber transitioned into the successful cultivation of soy by agribusiness. This industry's involvement and intervention in the exploitation of land, the forest, and of man, has produced irreversible environmental and social harm. The film illustrates the nexus between the exploration of the past (Fordlândia and Belterra), and present (Santarém, Itaituba, Miritituba, and Lago Maicá), made viable by specific material and symbolic manifestations of slow violence, which threaten rural, urban, and indigenous populations with a longstanding cultural and historical connection to the region. Or, in the words of Hecht (2019: 376), "Beyond Fordlândia takes us into terrains that you might have thought you knew but foreshadow those more devastating than you might have imagined."

CONCEPTS AND CATEGORIES: THE PERCEPTION OF SEEING

I can abstract from Nixon that a concept's capacity for political mobilization lies in the possibility of transforming it into an operative category that guides one's actions. This means that concepts have a dual function in their interpretation of reality. They guide understanding and offer an intelligent platform of intervention, with a language, images, strategies, and other interactive resources. This mixture of theory and practice allows us to affirm that the concept of slow violence is at the heart of *Beyond Fordlândia*'s interpretation of the Amazon and the horizon of my axiological positions and decisions. Ecocritical thought and cinema took me to the Amazon that I yearned to know and uncover while filming *Beyond Fordlândia.* This movement of life and ideas spanned memories and meanings that induced the reclaiming of information and perceptions transmitted in my family environment during childhood. It enabled me, above all, to value sensitive contact and adjust my anxious outlook to the rhythm of culture and of the adaptability of the biotic and abiotic Amazonian worlds, without losing sight of the strategic position that I occupy as a political subject and agent of knowledge witnessing slow violence in the Amazon. At distinct moments impressions of my inaugural, some would even say seminal, encounter with the "region" were registered through an ethnographic orientation. In *Beyond Fordlândia*, I represent and document my insertion into a land of imagination, Amazonia.

Although for Nixon slow violence remains a theoretical concept used to analyze environmental writing,[6] it served as an analytical category that connects ecocritical thought with the Amazon while I filmed the documentary *Beyond Fordlândia*. As a result, from my perspective the film serves as an allegorical-explicative object of various similar situations around the world. It casts a slow seeing eye of suspicion over the contemporary methods that the conquistadors of the present leave exposed in their relationship with the past. *Beyond Fordlândia* aims to transform the indifferent spectator into a critical observer of processes and methods devastating the region, which are not immediately apparent, because they are covered by a slow violence that generates long-term impacts. Whether it is from reality to fiction, from the experimental to the filmic, *Beyond Fordlândia* favors illustrating the dimensions of slow violence in Amazonia carried out by social elites, who reorganize the region's space and wish to construct a nationally homogenous image of Brazil. This slow violence erodes spaces of authenticity, nature, indigenous peoples and cultures.

While describing slow violence, Nixon (2011: 8) makes reference to abandonment through forgetting: "We need a deeper understanding of the slow violence of delayed effects that structures so many of our most consequential forgettings." Abandonment through forgetting is a specific way of feeling distance and articulating differences and inequalities. My filmic perception depicts the social and human conditions of the Amazon's alienated and poor populations excluded, isolated, and distant from the country's urban centers of economic and political activity.

My filmic representation grants human and nonhuman agents the condition of subjects capable of social action, an *actance*.[7] Ignorance of remote spaces and populations is a way of not looking, of turning one's back, of denying the issue's importance, and obscuring past and present slow structural violence. *Beyond Fordlândia* seeks to rectify this social and environmental harm by establishing connections between the past and present and enabling the viewer to reflect on what remains from the past and question what is remembered, forgotten, denied, or archived in the present of the Amazon rainforest. As Nixon reminds his readership through Faulkner's words "the past is never dead. It's not even past" (p. 7). Unlike mainstream environmental narratives, *Beyond Fordlândia* proceeds in a radically distinct manner. It confers importance on traditionally marginalized Brazilians, such as Indians, rubber tappers, caboclos, and North-easterners, whose narratives integrate humans and nonhumans. The documentary bestows on them specific and collective identities and repositions and represents them on another narrative plane. This filmic realm offers them additional means of visibility and speakalibity.[8]

Beyond Fordlândia establishes a relationship between two logical planes: that of ecocriticism and that of a filmic production that renders intelligible urgent ecological concerns. My documentary depicts the concept of slow violence by reflecting on the alienation and exclusion of the most genuine elements of local and indigenous populations and biotic and abiotic agents that are outside the production of national culture and the dominant canons. The documentary is the environmentalism of the poor, "those people lacking resources who are the principal casualties of slow violence" (Nixon, 2011: 4). The film's narrative traces the invisible "territory" of an ecological and human violence, which has been continuous since colonization.

Nixon's ecocritical thought enables a broadening of interpretations on the Amazonian region to a project of appreciation of constituents untouched by Brazil's cultural ethos.[9] This multifaceted theory enables one to exercise a filmic imagination capable of documenting reality and leaving clues as to another way of man relating to his environment. At the same time, these contributions enable innovative reflections on the nature-culture relationship, which *Beyond Fordlândia* achieves with respect to the Amazonian region. Throughout the film, I highlight a continuous displacement between epistemic contributions, which take on the function of operative categories, be it in theoretical production or in the intervention into cultural and environmental policies addressed in the documentary.

Beyond Fordlândia invigorates an epistemic framework for the comprehension of Amazonian space through an innovative esthetic of artistic imagination, produced largely with insights arising from research of Brazilian intellectuals on the logic of scientific (Euclides da Cunha) and artistic creation (Mário de Andrade). Added to this framework is an intersubjective social action approach that draws connections between the material and symbolic life of Brazil's various cultures. The film's esthetic and ideological project establishes bridges and incisions between the mythic, artistic, philosophical, scientific, and political imagination. This is necessarily the case because a documentary does not transparently express reality, nor does it capture the objective truth. It is no longer Cinéma Vérité.[10] Instead, the producer discovers and creates his own truth, be it with the filmed material, or with what does not appear on screen. In the bias of "interpretation," I remember the discourse analysis of Foucault (1988: 91), where he often encounters an "unsaid," in that which is said, in an attempt to recover the image inside and outside the screen. In the end, everything is the fruit of choices that are consolidated in the act editing materials. Thus, a documentary invariably adopts fictional technique akin to genre film. *Beyond Fordlândia* tells the "truth" from the archives of Ford and the "truth" of the chosen testimonies (Costa, 2017: 29).

Placing these two perspectives in opposing dialogue, expresses my cinematic perception of the truth, which I call "*Cinéma d'aperception*" (Cinema of apperception). This term is based on Walter Benjamin's essay "The Work of Art in the Age of Mechanical Reproduction," where he states that perception in film "has brought about a similar deepening of apperception" (Benjamin, 1973: 235). Later on, he explains that this apperception tends to favor the mutual interpenetration of art and science.[11]

In light of issues related to interpretive bias, I intend to clarify my position on the production and direction of *Beyond Fordlândia*. I decided to see Amazonian[12] Nature&Culture[13] slowly by adjusting my notion of time and space in accordance with the customs of traditional, riparian, quilombola and indigenous populations stigmatized on various regional levels. These individuals, collectives and biomes are the most recurrent epistemic subjects in the Amazon; they are experienced interpreters of the slow violence devouring humans and non-humans in the region. Riparians, extractivists, rural landholders, cattlemen, pantaneiros, quilombolas and the indigenous carry with them the condition of human beings stripped of social conditions necessary for the maintenance of life. They are reduced by agribusiness to the labor performed on their land, waters, forests, *cerrados*, fields mountains, and the peripheral fringes of the cities, villages, and settlements, while their environments continue to decline and are deprived of support.

IMAGINARY AND LITERAL HIGHWAYS

I was attracted to the Amazon rainforest by the passion and despair of my grandfather.[14] He shared with me his environmental concerns and the importance of the Amazon region to the world's ecosystem. He imbued me with the same level of enthusiasm and wonder that he attained from the Amazonian writings of Euclides da Cunha. This author[15] opened the imaginative highways and byways of the Amazon, which I now cross physically and textually in *Beyond Fordlândia*. Throughout the documentary, I penetrate the Brazilian Amazon through both metaphorical and literal highways. These thoroughfares not only changed my sense of time and space, but also changed how one lives by affecting the pace and rhythm of life of those dispersed along its course.

The Amazon's roads, in conjunction with other infrastructural manifestations, produce an array of inorganic meeting points that ignore the lives of humans, forests, and the wider ecosystems. These physical networks built by humans in a situation of political and economic power serve as an instrument of possession, conquest, and demarcation of place and passage. In the Amazon, the *Bandeirantes* trails[16] marked out the location for roadways in movements of colonial reiteration, in national actions of territorialization of the space of the nation-state and the productive forces sustaining it. Since the rubber years, the forces of Brazilian nationalization have called for the economic development of the region through the exploitation of its natural resources and potential.[17] However, since the nineteenth century the Brazilian state has acted upon the Amazon as a dominator and beckoned international capital to finance fluvial and maritime navigation in the name of economic, scientific, and diplomatic expeditions. These provisional fronts of exploration and prospecting came to define policies of settlement and displacement of the region's indigenous populations. Rigid and flexible networks of infrastructure connect intentions and desires for intervention into spaces and cultures. In the Amazon there is a constant construction of economic and populational dynamics that are imposed upon those closest to the rhythm of nature. As Laurance et al. (2001: 438) stated, in "Brazilian Amazonia [. . .] 95% of all deforestation and burnings occur within 50 km of highways or roads." The roads alter physical and biotic worlds through the mobilization of "traffic" through the forests, river biomes, and ecosystems of the Amazon. This traffic that occupies supposedly "empty" space continually demonstrates the power of government agents and economic forces over the region. Without a comprehensive focalization on culture, the results of these "developmentalist" mega projects generate long-term impacts, even when they immediately resort to technological and scientific interventions, such ventures are soon revealed as anachronic and with no sustainable policy of social inclusion in the region.

At the same time, the Amazon roads are also instruments that open up a different sense of perception that brings us to pathways otherwise ignored. Access to primitive societies, which adapted over the course of 500 years of coexistence in the humid tropic,[18] make the Amazon an archive of nature and culture allied to its condition as a living subject that expresses the impulses of living beings inside its enormous biome. The humans and nonhumans housed in the Amazon are completely interdependent on nationalist socio-political systems that separate nature and culture.

As an apprentice tourist surrounded by the dense forest of this vast green territory, my path to knowing the Amazon was sensitive, yet rational.[19] I entered a world known only in appearance. Slow violence has dimensions of meaning in the Amazon that remain hidden at first glance. The forest does not give itself easily to its pioneers, it offers only superficial, incomplete understanding of its complexities. Knowing and recognizing the Amazon forms an important step that requires adaptation of an urban rhythm to the rhythm of this green giant. "Scientific equipment is not enough," states Moraes (2011). According to the author from Manaus, "to comprehend, assimilate and express the complexity of the gigantic Amazonian chaos of its nature, the narrator needs to be gifted with a true talent, assisted by all the strengths of spirit and will, besides simultaneously possessing the faculty to perceive, at first glance, the specific, sensitive circumstances that explain influences past and present" (p. 19). These intentions prepared me to look, learn, and open my imagination to new possibilities while traveling the Amazon's imaginary and literary pathways.

The Trans-Amazonian Highway crossed today is not the same as the metaphor for the national glory peddled at the time of its construction. The impacts arising from deforestation and the fragmentation of biomes in Transamazonia repeat the same "privileged spectacle of capitalist civilization in the jungle" (Hardman, 1988: 25), experienced in the construction of the Madeira-Mamoré railroad. The process of slow violence is evident in both devastating interventions that aggregate nothing for biotic and abiotic agents. The Amazonian writings of Euclides da Cunha indicated the problematic relationship between nature and culture seen from the perspective of great ventures like that of Madeira-Mamoré. The intervention of Ford in the region underscored the complicated relationship between the center and the periphery with lenient mediation by the state. The connection between the colonial past, rubber-tapping, and agribusiness demonstrates the material networks of capitalism sustaining slow violence today. This continuation of the past in the present had immediate implications for my outlook and how I registered this perspective in the filmic narrative.

The grandeur and diversity of the Brazilian Amazon[20] enables one to configure it as a nation in itself—a nation full of consonances and dissonances, stitched within its fabric that offer lessons on how humans and nonhumans can understand each other and live harmoniously within a common geographical territory with differing cultures. The Amazon, more than any definition I may outline, is a place of insinuations in which a linear sense of perception is strained by blurring the record of time. It is a "suspect paradise" for its capacity to hide and disguise its dangers, states Leopoldo Bernucci. The images presented in *Beyond Fordlândia* demonstrate the inhospitable Amazonian space, its various rhythms, rites, and cadences of infinite diversity and dynamics, as well as the destructive human and ecological trail of slow violence that butchers the region.

By reassembling the past, present, and future hypotheses, I expose these rhythms and connect apparently disjointed themes that acquire meaning when interpreted through the visible networks of nature and society and invisible networks of the slow violence caused by interventionism in the region. As Berger (1977: 11) points out, "The past is never there waiting to be discovered, to be recognized for exactly what it is. History always constitutes the relation between a present and its past." He concludes that "fear of the present leads to mystification of the past. The past is not for living in; it is a well of conclusions from which we draw in order to act. Cultural mystification of the past entails a double loss. And the past offers us fewer conclusions to complete in action." The relating of these connections through an audio-visual medium is not only chronological or linear, but also constructed through my intellectual disposition, which confers logical and historical dimensions upon the region. The entwining of such dimensions highlights concerns, which transcend and project beyond the Amazon's fragile borders. Heeding the words of Nunes (1998: 24), my aim in this chapter is to illustrate "a reality that (perhaps) one cannot touch but touches us: every photographic image opens up the expansive network of our imagination, drawing us in with its suggestive force and moving us." In summary, entwining mosaics of resources, rhythms, and images suggested by the Amazon, provides us with a unique cartography constituted by a multitude of social, economic, political, historical, cultural, and environmental landscapes.

TRAVEL IMPRESSIONS

I arrived in Belém (Pará, Brazil) in 2016, during a winter evening in June, so I was unable to contemplate the beauty imagined in my previous readings. This moment of bewildered contemplation occurred the following day upon leaving on a flight from Belém to Santarém. I arrived at 10:00 am and my destination from there would be the city of Fordlândia. The plane had some places free by the windows, which allowed me to get a better view. As soon as the plane took off, I began taking photos as quickly as possible. I took an uncountable number of images of the majestic silhouette formed in the middle of that "Green Hell," as Alberto Rangel would have described Amazonia in 1908. There at the window, bewildered by beauty I could see the forest vanishing into an infinite sea of scattered clouds that slowly dispersed until the plane disappeared into a haze. The vision of the forest and its record in my literary imagination marked out a place of memory, already sensorially embodied within me, and opened a special file with records of collective knowledge to be narrated in *Beyond Fordlândia*. Reviewing this moment, I retain the image and carry it with me until today. Upon returning home, I continued to deliberate over the feeling that captivated me on that day in the June of 2016. Reflecting on my journey, I recall a passage by Benjamin (2009: 27–28):

> The power of a country road is different when one is walking along it from when one is flying over it by airplane. The airplane passenger sees only how the road pushes through the landscape, how it unfolds according to the same laws as the terrain surrounding it. Only he who walks the road on foot learns of the power

> it commands, and of how, from the very scenery that for the flier is only the unfurled plain, it calls forth distances, belvederes, clearings, prospect at each of its turns like a commander deploying soldiers at a front.

The experience portrayed by Benjamin encapsulates the essence what has captured the imagination of travelers, explorers, scientists, and adventurers for centuries. Anthropologists, in a broad sense, struggle to record and understand the other side of humanity and natural life through the resources and systems of meaning at their disposable. As a result, their memories of the journey drawn from fleeting glimpses, and their creation of taxonomies of unknown and misinterpreted phenomena may be more fantastic than rational. However, when seen from the ground these myths are readily dispelled by the realities imposed by the environment. This distorted perception is precisely what Euclides da Cunha (1909: 45) draws one's attention to, "Rather than admiration and enthusiasm, what usually comes over someone beholding the Amazon at the point where the Tajapuru's vibrant confusion flows full into the great river is a sense of disillusionment." The urgent concern over Brazilian thought regarding the Amazonian territory and on how its populations are seen, are conveyed in a tradition of scientific narratives that bridges the gap between the literature and the disciplinary discourses. Both literary and the scientific narratives present perspectives on the Amazon, therefore one cannot study the region without examining other thinkers and naturalists who registered distinct perceptions in the form of scientific discourse on the region. There is no way to discard the past or the present. The pre-scientific notions found in Cunha's work(s), inspire a way of seeing experiences (including mine), or at least of critically reading the region, albeit in a fragmented manner. Cunha inaugurated a new way of envisioning human and nonhuman relationships in Amazonia later taken up by Leopoldo Bernucci, Francisco Foot Hardman, and others. Demonstrating his wonder for the region, Cunha (1909: 25) explains:

> The sheer volume of water is unmatched and therefore capable of inducing that wonderment of which Wallace speaks. But since, from early on in life, each of us had drawn an ideal Amazonia in our minds [. . .] we experience a common psychological reaction when we come *face to face with the true Amazon*: we see it as somehow lacking with respect to the subjective image we have long held of it. (Emphasis added)

The sense of "disillusionment" upon coming "face to face with the real Amazon" to which Euclides would allude, gained its own meaning during my trip to Fordlândia on the BR-163—an extension of the better known Trans-Amazonian Highway. The "Quixotic Trans-Amazon project" presently remains incomplete with partially paved stretches that symbolize the destruction wrought upon the region due to ecological managerialism (Skidmore, 2014: 202). Although the construction of the highway, by the military government in the 1960s, intended to integrate the North of Brazil with the rest of the country, the project failed. From the 1970s onward, in the case of Brazilian Amazonia, the relationship between the new and the old became conflictual rather than a simply complimentary coexistence. Although during the first half of the twentieth century, policies aimed to preserve the Amazon from international greed, from the 1960s onward, the objective became one of productive exploitation and integrating it into a national context, in the name of eliminating its "primitive" character (Paes Loureiro, 2019: 1).

For the next eight hours the bumpy ride on BR-163 gave me front row seats to the spectacle of deforestation produced by corporate soybean farming. I was taken to an encounter with a web of Amazonian human ecology: settlers, natives, riparian dwellers, fishermen, truck drivers, and local farmers; invisible agents that cross my eyes and need to be seen slowly, without the rush imposed by the dynamic of the roads and highways. During the three months in which I traveled across the Amazon, documenting, observing, and recording the voices of actors affected by and involved in the production of slow violence, I witnessed first-hand an environmental catastrophe not seen as violence. Nixon conceptually describes what the Amazon region has experienced and continues to experience, a slow and gradual process of deforestation in favor of soy production. The soybean crop, planted for the most part illegally, is one of the engines, that drives the alarming rate of deforestation[21] in Brazil, as in the MATOPIBA.[22] The Amazon rainforest pays the price for Brazil being the largest soy producer in the world. Nixon urges us to consider this process of slow violence not only imaginatively and theoretically, but also from a political perspective, through a combative discourse, where the central premise is above all representational. In *Beyond Fordlândia*, his challenge led me to expose the scars of predatory colonization found in present policies affecting regions and cultures and inspired me to make visible the environmental and cultural disasters brought by slow violence. The film represents this slow violence by revealing how Ford's project prepared the ground for the predation of soy in the present. This retrenches the impacts of colonization on Amazonian territories, peoples, and cultures. *Beyond Fordlândia* questions and provokes thought on Amazonia today by presenting connections between the past and the present and enabling reinterpretations that reveal the invisible slow violence in the region. Social thought, be it in the literature or social sciences, establishes connections between the past and present to reinterpret their interrelationship; this precisely corresponds to the contribution made by *Beyond Fordlândia*.

Beyond Fordlândia, considers Nixon's (2011) characterization of "development" and "progress" (p. 144) and their multiple designations, in connection with the concept of slow violence. Some of these designations include "progress narratives" (pp. 86, 190), "glorious progress" (pp. 92, 100), "Partners in Progress" (p. 112), "self-declared standard-bearers of progress" (p. 120), "act of progress" (p. 136), "triumphal national progress" (p. 161), "progress discourse" (p. 186), and "colonial progress narratives" (p. 198). "Progress narratives" are at the center of questions related to the long-term environmental damage that fall upon the "the environmentalism of the poor" (p. 4). He explains that "Inside these progress narratives, the cults of speed, novelty, and spectacle can seem to generate their own innate morality" (p. 207). *Beyond Fordlândia* evaluates the Nature&Culture relationship that operates in these "progress narratives" currently affecting the Amazon. My documentary, underlines the importance of slowly seeing how this process of slow violence inflicted on nature, peoples, and cultures alters the way of life, the organization, the strategies of survival, and authenticity of Amazonian spaces and biomes. A slow, detailed look at the advent of the implantation of soy in the region finds repetitions of the same discursive practices that sustain "progress narratives," used by Ford during the 1920s.

If the neoliberal era has intensified the assault on resources, it has also intensified resistance (or as Michael Foucault emphasizes "Where there is power, there is resistance"), whether through isolated struggles specific to a given place or through activism that has reached across national boundaries in an effort to build trans-local alliances. The questions that resonate are: does one only resist upon seeing a problem? If one cannot see a problem, does he or she not resist? Bosi (2002: 118) states, "Resistance is an originally ethical concept, and not esthetic. Its deepest meaning appeals to the will power that resists another power, external to the subject. To resist is to oppose one's own force and forces beyond ourselves. The close cognate is in/sist, the familiar antonym is de/sist." For Nixon (2011), the pivotal demand continues to be

> How to convert into dramatic form urgent issues that unfold too slowly to qualify as breaking news—issues like climate change and species extinction that threaten in ecologies of the aftermath slow motion. (p. 210)

Beyond Fordlândia, in its narrative plot, positions itself against the neoliberal narrative in motion in Brazil, with massive campaigns on the major television channels that succeed in making the aggressive environmental actions of agribusiness in the North of Brazil invisible. The slogan "*Agro é tech, agro é pop, agro é tudo*"[23] [Agro is tech, Agro is pop, Agro is everything] is catchy and obscures the real implications of the soybean industry in the very heart of the Amazon. At the same time, it has raised questions and provoked discussion on the invisible effects of this expansion for individuals, communities, societies, and the local environment. However, one should be asking why the ecological implications of this profitable agroindustry, which are as numerous as the uses of the seed itself, remain invisible in the public sphere.

Beyond Fordlândia seeks to exhibit this gradual slow violence affecting Amazonia, which is invisible in the public sphere and not contemplated in public policies, whether regional or national. Among the many examples dealt with in the narrative of the film, Dercy Gondinho stands out for her powerful and condemnatory narrative. Gondinho, a resident of the lower Amazonas region, reports on the impact of the arrival of agribusiness and its destructive economic effect on the region. During her testimony, which exemplifies what Nixon defines as "environmentalism of the poor," she explains how slow violence reduces traditional peoples and biomes to poverty and devastation:

> Boa Esperança was a center for tapioca flour production. In (2003) they came, the first buyers of the areas, and the people started selling their land. We'd never seen R$ 30, R$ 50 thousand. We were made to feel like we were rich to get that value, and many sold. When they arrived there at the head of the mountains, as I always say, there was no money left. But the land was already in the hands of the soybean planters. So, with this the production of tapioca flour diminished greatly, as did the earnings of the people who came in search of work here in the village. Because as a center, as a small business, it employed a lot of people, many people here that were not manioc flour producers were given jobs, I mean, it had a flour house, it had the farms, so it gave work to me and my family, for us to earn a living, because we did the jobs of pulling up, shaving, washing, mixing, and even toasting the flour. So, it was our bread and butter, that was how we got paid here in Boa Esperança. (Dercy Gondinho, *Beyond Fordlândia*, 2017)

SEEING AMAZONIA SLOWLY

Rob Nixon critically examines the words of Aldo Leopold when he states: "we can be ethical only toward what we can see" (p. 16). He further enquires, "But what happens when we are unsighted, when what extends before us remains invisible? How, indeed, are we to act ethically toward human and biotic communities that lie beyond our sensory ken?" Being more incisive, the author provokes "How do we both make slow violence visible yet also challenge the privileging of the visible?" Applying the words of Leopold to the Amazonian context, and considering the geographical displacement of the Amazonian forest from the great centers of national political decisions in Rio de Janeiro and São Paulo, one can witness the devastation wrought by the mentality of "of being ethical only toward what we can see." Bennett (2010), coming from a perspective of political theory, analyzes questions of environmental sustainability, human rights, and political questions of economic justice; updating the discussion, she explains what she calls "the ethical turn":

> The ethical turn encouraged political theorists to pay more attention to films, religious practices, news media rituals, neuroscientific experiments, and other noncanonical means of ethical will formation. In the process, "ethics" could no longer refer primarily to a set of doctrines; it had to be considered as a complex set of relays between moral contents, aesthetic-affective styles, and public moods (p. xii).

Rob Nixon, going deeper into the discussion and questioning Leopold on how to be ethical when one does not see, clarifies

> We are accustomed to conceiving violence as immediate and explosive, erupting into instant, concentrated visibility. But we need to revisit our assumptions and consider the relative invisibility of slow violence. I mean a violence that is neither spectacular nor instantaneous but instead incremental, whose calamitous repercussions are postponed for years or decades or centuries. I want, then, to complicate conventional perceptions of violence as a highly visible act that is newsworthy because it is focused around an event, bounded by time, and aimed at a specific body or bodies. Emphasizing the temporal dispersion of slow violence can change the way we perceive and respond to a variety of social crises, like domestic abuse or post-traumatic stress, but it is particularly pertinent to the strategic challenges of environmental calamities.[24]

I consider that the "invisibility" perceived by Nixon is born from a particular manner of observation. Perhaps the issue is not whether the object—the person, the phenomenon, or the process—is immediately visible or not, but instead how they are seen, which shapes and determines actions that lead to results. If properly seen and exposed, they may produce different sensory-driven reactions that may feed propagandistic processes harmful to life (agro is tech) or serve to interrupt those same neoliberal processes through an education that makes a call to action.

More so than recognizing ongoing slow violence, I propose a change in the mode of perception, which I call "slow seeing." I use slow seeing (be it the slow seeing of film, the slow seeing of the observer, or the slow seeing of the subject) as an operational category guiding actions, or a *modus operadi*, whereby the images and narratives of destruction can be reframed to make the information not only actionable but ecologically transformational. More than building a particular Amazonian mode of perception[25]—or cannibalizing the visual beauty of the forest and its flora and fauna—slow seeing seeks to present concerns that elucidate the present and that glimpse another outlooks toward nature-culture relationships of the region. As Berger (1977: 10) synthesizes "although every image embodies a way of seeing, our perception or appreciation of an image/sight depends also upon our own way of seeing." With this in mind, slow seeing proposes a "more relational point of view" (to use a term of Viveiros de Castro), with the aim of rethinking the "Anthropocene" label.

This label developed by physicist Paul Crutzen and ecologist Eugene Stoermer appeared in the 2000s. The authors argue that the Holocene is already a thing of the past and that both mankind and the Earth have entered a new geological era triggered by boundless human actions. Humanity has become a force of nature; now, it can overwhelm the Earth, transform it. It creates new climates over which it has no control. It changes the pH of the oceans, unknowingly altering life forms and ecosystems. Humans melt the ice flows without due care. They create unprecedented atomic isotopes and fossilized plastics. In summary, mankind produces problems that it cannot solve. The term "Anthropocene" ended up being designated to foresee what is happening through an official narrative.

However, this label based on a nonnormative outlook does not prescribe action, it merely informs the current situation. My perspective, on the other hand, takes into consideration the pedagogical dimension of seeing the Amazon slowly. By slow seeing, one can observe the Amazon with fresh critical eyes. Seeing nature slowly allows one to identify solutions embedded in the process of change. If one looks slowly at the environment, he or she can identify spaces and places with opportunities for critical and theoretical intervention. To intervene, then, means to change the course of events and shape the future. Seeing slowly draws one away from the violent speeds of the roads and highways and engages him or her, instead, in the possibility of a new method of Amazonian perception and eco-political practice.

Seeing slowly is pegged by the pace of nature and the esthetics of its visual landscape. It enables us to find other approaches and reinterpret the relationship of man versus nature. It enables us to reconfigure Amazonia and the unfortunate interventions in the region that are consolidated and reiterated over its territories, peoples, and biomes. Seeing slowly may even indicate other modes of coexistence between humans and nonhumans, between nature and culture, between culture and power. Seeing slowly a diversity of situations and relationships reveals other means of intervening ethically, affectionately, aesthetically, morally, and politically in other models and relational and public mechanisms. Rethinking the past and present history of the Amazon forces us to reflect on the roads that led us down the path of destruction that imperils daily experiences in the region. Not only that, it also provides one with better tools capable of engaging and changing the stories currently being written. Slow seeing is a means of resistance to the ongoing slow violence.

In *Beyond Fordlândia*, the analytical category of slow seeing permits one to observe the Amazon from a different point of view. By looking at peripheral rather than central areas and taking on the vision of those hidden and invisible in the "environment of the poor" one comes to see a set of actions from these individuals' perspectives, and adopts a different methodology of intellectual and sentimental comprehension toward the central problems faced by the Amazonian peoples. Methodology that enables the inversion of this unorthodox vision shows a broader dimension of Amazonia and enables me to understand, the environmental problems of the poor and sensitive issues affecting the region. Slow seeing facilitates a distinct understanding of the Amazonia, where the Nature&Culture relationship is not antagonistic from a distinct perspective. This methodological practice exposes the nation-state's negligence in the region and agribusiness' most predatory actions. This plane of observation enables us to rethink Amazonian topics typically seen through the prism of high-speed capitalism. From this perspective, it is not the slow violence, it is not the dominant powers, but rather the traditional populations and the positioning of their region itself before Brazil that should guide the decisions and the futures of Amazonian societies.

The question that should be of interest is how to represent slow violence where there are no obvious well-defined guilty parties or none that can be readily seen because the origin of evil seems to be light years away in time. There is little success in representing the cadential chaos caused by this slow violence. Chaos, in Greek and in Spanish, is defined simply as disorder. However, modern science offers a broader view. The concept of chaos can be observed in many practical aspects of life, including finance and climate. Understanding of chaos involves critically thinking on dynamic local and national relationships between biotic and abiotic agents in the Amazonian region of Brazil. This is the magic accomplished by mythic, artistic, and scientific narratives. With this thread guiding *Beyond Fordlândia*'s interpretation, I have constructed a simultaneous narrative of memory and archive. I have produced and recovered historical and logical connections with ethical and esthetic choices in the documentary, a rational but sensitive hybrid of reality and imagination. Seeing the Amazon slowly I have discovered the vicious implications of continual slow violence.

Here I once again return to the questions of Nixon—how can one make the long slow occurrences of violence, whose impact has led to some of the present's most critical and urgent challenges, dramatic enough to arouse public opinion and justify political intervention? A possible solution is to rethink the role of historical memory alongside normalized and naturalized dominant discourses on nature and culture. Another alternative is to narrow the gaps between nature and society, through what Latour calls "hybrids."[26] There is an urgent need to reopen the files, rethink the orthodoxies alluded to by Pierre Bourdieu in order to change the current praxis of environmental discourse in Brazil and more specifically in Amazonia.

Seeing slowly is my method of analysis and narration. It can be defined on two levels of change. One is internal and related to my own creative process; that changed me through my work. Another change refers to the knowledge that one produces on people, things and the surrounding environment. Seeing slowly is a way of inverting the priorities of intervention found in external views of the Amazon. It is to see the care that its fragile ecosystems require. It is to establish deep, continual, nonvisible connections between slow violence and the environmental impact suffered by the Amazon. In the words of Berger (1977: 13), "We only see what we look at. To look is an act of choice. As a result of this act, what we see is brought within our reach—though not necessarily within arm's reach. To touch something is to situate oneself in relation to it."

My meeting with Mr. Avelino Campos impacted my perception, changing my filmic eye, and exemplifies my understanding of seeing slowly. Gito, a friend who fights against the invasion of the soy industry, took me to the Maguari Community,[27] where he lives, located forty minutes (by car) from Belterra, to meet Avelino Campos, a resident who arrived in the region when he was eight years old. Ninety years after his arrival, he shared with us his impressions on how his place and the surroundings have changed. He talked easily and had vivid memories of facts and events. I tried my best to keep track of all the information he passed on, which was a lot. During part of the time we spent together, Mr. Avelino talked of how he and his father started to supply food to Henry Ford's workers in Belterra. Besides the numerous accounts of his experiences, what I took from that visit was his reports of the transformations that had taken place, which he sensed as being nothing more than slow violence he observed in the open air. He proudly showed us his private Igarapé (a small stream of water that arises in and crosses the forest and finally flows into a river), which supported different species of fish. Then he told us, "This Igarapé is not for business but for when we have the need; we come here to get something to eat." He walked with us through part of his property and showed us the different kinds of plants that he conserved. He said that some species were rare and hard to find nowadays—"This one is called Andiroba, these over there are called Papunha, and this one is called Sapitinga, and this one is Pau-Rosa," and the list goes on. He told us, pointing to the Pau-Rosa, that it is the best perfume fixative ever found. Coco Channel uses it extensively. He told us that its market price is around $25.00 a pound. Nowadays there is a lack of Pau-Rosa in the region due to the high price and demand. I paused for a minute, considering what he had just told me, as I continued strolling through his property. I was mesmerized by his passion, love, and care for preservation of the environment. However, he also shared with me his hardships:

> Since the soy industry arrived here, he said, everything has changed. The wildlife has disappeared, and we cannot hunt anymore . . . . Some people that buy on credit have debts because of this, they think that this (the advance of the soy industry in the region) is getting results, but for me this is a loss. Because no one here uses anything produced by them there. Of the soybean, nothing. There is a soybean farmer who has four tractors, each tractor is operated by two people, a helper and an employee, and that's it; and the others just stand there watching. A huge piece of land out of sight. At a certain moment they turn on the tractor again with two funnels throwing poison, so that the soybean ripens quickly, and the insects are driven off. Today, we do not find any more Juriti in the jungle—not one of those pigeons. They disappeared because they died poisoned. (Avelino Campos, *Beyond Fordlândia*, 2017)

The account of Mr. Campos has great implications for him and for those nearby, those who share the same space but do not perceive the forest's ongoing destruction. What Mr. Campos observes is the slow violence of environmental change—the gradual loss of biodiversity, the subtle transformation of food systems, and the tectonic rise of agro-industrialization. The voice of this epistemic subject is contained in *Beyond Fordlândia*. The film complements individual perceptions and intersubjective understandings of the way in which the violent interventions of yesterday feed those of today. When one casts his or her eyes over the natural world, to look at "nature," it is true that nature is dynamic and diverse. It changes continuously and randomly (more than I dare to say), so the only way one can attempt to understand it is to observe and record; step back, observe and record again, then again, later, and once more, later. Are the changes that occur between these observations the only facts one can comment on? The experience of Avelino Campos captures the act of seeing slowly, tracing the "highways" and "roads" of the environmental changes suffered by the region in which he dwells. Therefore, one must ask his or herself for the sake of the Amazon and its future: Is sharing intimate stories about the changes all that can be done? Or, can one use these stories to intervene and change the future?

In *Beyond Fordlândia*, I executed this conceptual and political proposal against slow violence by producing a methodology for action. This documentary allows one to relearn how to see the Amazon and slowly promotes a new encounter between nature and culture with new epistemic and ethical bases. *Beyond Fordlândia* communicates a point of view (via ecocritical thought), narrates a human story/adventure in the Amazon (of the American Henry

Ford vs. Amazonian populations) and makes connections between the rubber and the prospection of soy of yesterday, with the extensive and intensive agribusiness of soy of today. The changes emphasized in the film through dialogues with local people (Indians, farmers, union leaders, soy producers, the church, and others) present an understanding as to what occurred and the way "my truth" captures the collective imagination.

The film emphasizes the social processes of the past and the present, the colonial interventions of yesterday and today, the forms of totalitarian economic appropriation, the domination of territories and peoples, the impacts of destruction on the natural resources, and on the forms of traditional usage, and the physical and symbolic manipulation manifested during these processes in the present on structures constructed in the past and unremoved during "regional development." Confirming my filmic intuition, Pacheco Oliveira (2016: 16) explains that, "contrary to the idealizing self-representations, supported by belief, on a civilizing mission of the white man, the current researcher needs to date back to the past in another way, seeking to comprehend the appearance of structures generating wealth, inequality and territorial expansion of that we identify as the real Brazil."

In showing the production of physical and cultural violence and its symbolic processes, *Beyond Fordlândia* seeks to pull apart the narratives that affect innocent people who do not immediately perceive slow violence's destructive, toxic, and perhaps irreversible, alteration of social life, cultures, biomes, ecosystems, and forms of survival. These processes create a dynamic of occupation and sociability dependent on a political logic of economic that produces irreversible consequences on Amazonian cultures, biomes, and ecosystems. These territories, peoples, and natural conditions are subverted, exhausted, and dismissed, thus leaving spatial dispersion, physical destruction, and sociocultural desolation over the affected reality.

My documentary depicts a cartography of violence in the Amazon—its geographical expansion, the social segments it affects (farmers, riparians, municipalities, Indians, etc.), its damage to collective interests and to public policies, and its ideologies of illusion that result in local's disillusionment. The film also highlights the connections of the development of capitalism in its centers of production within the spaces of its colonial and neocolonial conquests. This cartography includes the subversion of political systems and governments to obtain special legal protections that abusively increase profit. The predatory subtraction of sources of wealth and appropriation of the power by manipulative multinational corporations significantly affect marginalized groups eluded by the ideal of progress. *Beyond Fordlândia* casts a light on past cycles of capitalist exploitation, as seen in the timber, rubber, energy, and automobile industry of yesterday and the agribusiness of today. The film makes such connections by recovering local voices desolated by such processes and forgotten by history.

Categorizing records of history and collective memories is a political commitment to inter-subjectively and dialectically activate movements of remembering and forgetting, which are both of great significance in their impact on the Amazon. Records of Indian genocide and erasing of the cultural memory are of equivalent importance to great environmental catastrophes. Even if they have yet to be exhaustively examined using ecocritical thought, these astonishing events are more remembered than the continual occurrence of slow violence. It is over this phantasmagorical memory of the past that *Beyond Fordlândia* documents my slow seeing, which is expressed in the voice, the images, and extreme situations slow violence produces in the Nature&Culture relationship.

NOTES

1. *Beyond Fordlândia: An Environmental Account of Henry Ford's Adventure in the Amazon*. Dir. Marcos Colón. Amazônia Latitude Films, 2017.

2. DeLoughrey and Handley (2011: 19) comment "Certainly postcolonial ecology must engage the complexity of global environmental knowledges, traditions, and histories in a way that moves far beyond the discourses of modernization theory on the one hand, which relegates the global south to a space of natural poverty, and the discourse of colonial exploitation on the other, which relegates the global south to a place without agency, bereft of complicity or resistance. Our conviction that dominant ecocritical methodologies that make universalist claims (from the unmarked Anglophone viewpoint of the United States) must address colonial legacies and postcolonial contexts is in concert with critics who have cautioned against turning to indigenous and postcolonial ecologies to simply provide moral, spiritual, or financial redemption for the capitalist metropole." The authors add that "Nixon has argued that the vast body of postcolonial theorists from Homi Bhabha to Glissant foreground hybridity and cross-cultural exchange as constitutive of history and literary discourse, whereas mainstream American ecocritics often emphasize a desire for a primordial natural purity in the wilderness, a retreat from the social and environmental pollution of modernity" (p. 23).

3. Nixon presents an inspiring trio and how the itinerary of their commitment to integrity of the intellectual in its reflection in the world moved and convinced him. This is an indication of how Edward Said, Rachel Carson, and Ramachandra Guha—respectively a literature professor, a science writer, and a sociologist—refine their interpretation of the nature-culture relationship. In all three, Nixon indicates that they exemplify a public intellectual ideal as someone without fear of opening channels of investigation, in an angle of mainstream thought; the three are critically positioned before conventional professional disciplinary expectations; all are ready to confront hostility; all of them have demonstrated a communicative passion that responds to

various publics. This passion has molded audiences, and, is a mobilizing passion in its pedagogy of persuasion.

4. Among the environmental concerns articulated by Nixon (2011) the categories of "slow violence," "the environmentalism of the poor," and "the environmental writer-activist" stand out. Slow violence explains a "violence that occurs gradually and out of sight, a violence of delayed destruction that is dispersed across time and space, an attritional violence that is typically not viewed as violence at all" (p. 2). In the second category Nixon defines his concern "for it is those people lacking resources who are the principal casualties of slow violence. Their unseen poverty is compounded by the invisibility of the slow violence that permeates so many of their lives" (p. 4). And finally, what Rob Nixon calls "the environmental writer-activist," according to him, informs "not just literary but more broadly rhetorical and visual challenges posed by slow violence." He himself concludes, "I place particular emphasis on combative writers who have deployed their imaginative agility and worldly ardor to help amplify the media-marginalized causes of the environmentally dispossessed" (p. 5).

5. The venture of the Ford Industrial Company in Brazil, for the extraction of latex, occupying an area of 14,568 km^2 in the municipality of Aveiro, on the banks of the Tapajós River, was conceded to Henry Ford, on September 30, 1927, by governor Dionísio Bentes, with the approval of the Legislative Assembly of Pará.

6. Rob Nixon's emphasis on activist writers can be seen in the following quote: "Alongside that activism, a diverse group of writer-activists is espousing the causes of the environmentally dispossessed. These writers are geographically wide ranging and work in a variety of forms—novels, poetry, essays, memoirs, theater, blogs. Figures like Wangari Maathai, Indra Sinha, Ken Saro-Wiwa, Abdul Rahman Munif, Njabulo S. Ndebele, Nadine Gordimer, Jamaica Kincaid, Arundhati Roy, and June Jordan have recorded the long-term inhabited impact of corrosive transnational forces, including petro-imperialism, the megadam industry, the practice of shipping rich nations' toxins (like e-waste) to poor nations' dumping grounds, tourism that threatens indigenous peoples, conservation practices that drive people off their historic lands, environmental deregulation for commercial or military demands, and much more." That is mentioned in Nixon, R., Slow Violence: Literary and Postcolonial Studies Have Ignored the Environmentalism that Often only the Poor Can See. *The Chronicle of Higher Education*. June 26, 2011, available at https://www.chronicle.com/article/Slow-Violence/127968.

7. The notion of "actance" is recovered by Latour (1993) from the ethical and religious perspective of Baruch Spinoza (1632–1677). "Actance" is the capacity of acting or actor's power of action. The terms "actance" and "agency" are synonyms.

8. Albuquerque Junior (1999: 192) defines that "the visibility and speakalibity of the [North] and the Northeast are composed principally of products of the imagination, to which reality is attributed. They are composed of facts that, once seen, heard, recounted, or read, become fixed, are repeated, are imposed as truth, become consistent, create "roots." They are facts, characters, images, texts that become archetypes, myths, which appear to float beyond or fall short of history, which, however, possess a positivity on being embodied in practices, in institutions, in social subjectivities.

They are images, pronouncements, themes and "prejudices" necessarily negotiated by the author, by the painter, by the musician or by the cineaste that wishes to make their narrative or work of art realistic. They are discursive regularities that are crystalized as expressive, typical, essential characteristics of the region." For the author "In the realist mimesis, representation of the social is bound by the limits imposed by the situation of that experienced and known by the receiver. As such, the word is blocked in its power of creation of worlds by the obedience to a visibility and speakalibity of reality that are already crystalized, under punishment of not being a work considered true to what is real" (p. 207).

9. Ethos is a set of characteristics of a people or group that differentiates it from others. The research of Gomes de Matos (2015), through the Norbeth Elias's perspective, illuminates other perceptions of ethno-knowledge and of sociocultural practices of Amazonia.

10. Baggio (2009: 169) states that "Cinéma Vérité has Jean Rouch as its founder and, more than that, its idealizer. Rouch was an engineer, a Doctor of Literature, ethnographer, and explorer. In connection with the center of studies in anthropology of the Museum of Man in Paris, he had filmed his voyages of ethnographic exploration since the 1940s. Under the influence of Direct Cinema and, mainly using the recently created Nagra apparatus for the capture of direct sound and lightweight cameras, Rouch films produced in 1960, in collaboration with sociologist Edgar Morin, *Chronique d' un Été* (Chronicle of a Summer). This film shows the behavior and opinions of residents of Paris and inaugurates Cinéma Vérité."

11. "The characteristics of the film lie not only in the manner in which man presents himself to mechanical equipment but also in the manner in which, by means of this apparatus, man can represent his environment. [. . .] As compared with painting, filmed behavior lends itself more readily to analysis because of its incomparably more precise statements of the situation. [. . .] This circumstance derives its chief importance from its tendency to promote the mutual penetration of art and science. Actually, of a screened behavior item which is neatly brought out in a certain situation, like a muscle of a body, it is difficult to say which is more fascinating, its artistic value or its value for science to demonstrate the identity of the artistic and scientific uses of photography which heretofore usually were separated will be one of the revolutionary functions of the film" (Benjamin, 1973: 235–36).

12. Nature&Culture, characterized as a social relations phenomenon at the end of the eighteenth and the nineteenth centuries, indicates a process of long-term cultural transformation within a western environmental tradition that dates back to modernity. German critic, Gadamer (1999: 455) states that "the horizon of the past, from which all human life lives and that is there under the form of tradition, puts a comprehensive horizon in motion." Such an inflection leads me to understand the Nature&Culture relationship, as a phenomenon coupled with a historic and mobile horizon. The historical perceptions of the Nature&Culture relationship that we trace have their place in the seventeenth century, connected, reimagined, and sedimented throughout modern history by the idyllic perceptions of English naturalism and the new bourgeois romantic sensibilities of the eighteenth and nineteenth centuries. However, at the end of the nineteenth century the cultural transformations of

European romanticism, started at the beginning of the eighteenth century, presented a new way of imagining the nature and culture pairing, that continues to the present. It is therefore fitting to seek alternative pathways to the modern separation between Nature&Culture.

13. The spelling adopted of Nature&Culture is a way of standardizing my interpretations of the nature and culture relationship, and to avoid the redundancy. It is also a way in which I argue that the conceptual pair of Nature&Culture, do not contemplate opposition or separation when at maximum tension. I understand that Nature is culture. Culture is nature. Therefore, we consider that "Nature and Culture," spelled separately, characterizes a position of opposition and a counterpoint of subalternity of nature as a result of the anthropic and cultural action of man.

14. Although my grandfather never physically visited the Amazon, like many other Brazilians, he knew the region through literature. This did not impede him from observing complex interactions between the environment and violent human intervention in the region. His southern sensibility experienced the consequences of climate change in south-east Brazil, which, according to him, was a result of the human violence in the Amazon rainforest.

15. Once again Nixon comes to meet my tensions as citizen and researcher when the author addresses, in the guise of activist writer, a living instrument of intervention and reflection. For Nixon, the writer who masters the comprehension of phenomena and the surrounding processes can see and read the signs of slow violence and unravel the meanings to those that are deprived of this interpretative ability. The intellectual may give voice to those who, weakened and destitute of their citizenship, cannot make themselves heard nor have the channels of expression for their suffering and anguish. As individuals or a social segment, these subjects of knowledge are leaders, disseminators that give visibility to the problems of the community on a transnational level, as translators of the language and effects of slow violence. They therefore become essential in the grasping and diffusion of invisible or hidden violence—as it is practiced under the shroud of various types of natural, cultural, or political isolation.

16. Gadelha (2002: 68), citing Capistrano de Abreu, highlights, "the importance of the *bandeiras* (ensigns) that left from various points on the coast of Brazil, conquering the hinterland territory: first the *bandeiras* from São Paulo, who, trekking from the Vicentino plateau, the way known by the Piabirú Tupi, connected Paraná to Paraguay, progressively increasing their voyages to the south of Mato Grosso, from where they reached the limits of Spanish territory of the Moxo and Chiquito (Bolivia), going up to the Guaporé River. Or, from the second half of the seventeenth century, penetrated Madeira, Tapajós, and Araguaia-Tocantins, establishing connections with the valley of Amazonas."

17. Carvalho (2019) points out that "the process of natural recomposition of the forest is unbalanced under the negative synergy between agriculture and livestock, on the one hand, and the ecosystem, on the other. If an area above 25 percent is deforested, the original forest will be transformed into savanna. The Brazilian part of the Amazon forest has a little more than 3 million km^2. Between 1991 and 2000, the area deforested for agriculture, livestock, and the extraction of timber grew from 415,000 to 587,000 km^2. From 2000 to 2017, an average of 3,000 km^2 has been deforested

each year. At the present rate, it won't take long for the Amazonian ecosystem to go from tropical forest to savanna, and then, to desert."

18. This is my characterization of the physiography of the Amazon (its climate, atmosphere, man, and environment). It is situated between the physical and human geography, as Tocantins suggests, it is related to a specific type of forest and hydrography, distinct from the semi-arid, as not all tropics are humid.

19. Mário de Andrade was a photographer, ethnographer, anthropologist, and interpreter of the Amazonian region to which he was a visitor. The author of *Macunaíma* describes his study trip to the North and North-east of Brazil avoiding talking about these places from the foreign point of view of a "southerner." He puts himself in the position of apprentice, trying to know the communities of Amazonia and its nature with the interest of recording in order to preserve it. In *O Turista Aprendiz* (The Apprentice Tourist), the author thinks on the nation as a meeting of different eras, places, images, and voices. I came to the Amazon with a similar attitude, to appreciate and see it slowly.

20. The Amazonian region has been referenced as the "lungs of the world." Although this is not strictly true, the Amazonian region has an impact as a regulator of Earth's climate. If Amazonia were to suffer an extensive process of deforestation, there would be a fall in the amount of humidity that is released into the atmosphere through evaporation. This would lead to decreased rainfall, triggering a cycle in which the surviving forest coverage would become increasingly arid. Consequently, this transformed ecology and its arid climate would lead to extinction of the remaining tropical forest. Over time, the impact would spread like a voracious disease. It would extend to the South of Brazil, bringing drought to its extensive agricultural lands and commodity frontiers. Beyond Brazil, these changes would negatively impact the entire planet. Globally, this process induced by human beings would contribute to the Earth becoming progressively hotter. Such a dramatic and prolonged increase in the temperature of the planet would have terrible effects for nature and society, including intense, concentrated rains, forest fires, stronger storms, floods, erosion, and fatal landslides. Furthermore, the soil would become drier and less fertile, culminating in prolonged desertification. Perhaps we already know this, or we have already experienced such effects personally, or we have witnessed them through local and international television news and other means of informative propagation. Depending on where one lives, some of these threatening forecasts may already be part of our catalogue of ecological "knowledge" on the Amazon.

21. Between August 2017 and July 2018, the preliminary rate of increase in deforestation in Amazônia Legal registered by the Deforestation Monitoring by Satellite Project in Amazônia Legal (PRODES), was 13.7 percent of the biome in relation to that registered in the previous period. The deforested area reached 7,900 km². The observed deforestation in 2018 corresponds to a reduction of 72% in relation to the registered area in 2004, the year in which the federal government began the Plan of Action for the Prevention and Control of Deforestation in Amazônia Legal (PPCDAm) and 60 percent of the target foreseen in the National Policy on Climate Change. The highest values of deforestation were registered in the states of Pará, Mato Grosso, Rondônia, and Amazonas. Mapping used satellite images to register

and quantify the deforested areas greater than 6.25 hectares. The complete removal of primary forest cover by clear cutting was considered deforestation, regardless of the future use of these areas; according to the Ministries of the Environment and Science, Technology, Innovations and Communications, available at: http://www.mma.gov.br/informma/item/15259-governo-federal-divulga-taxa-de-desmatamento-na-amaz%C3%B4nia.html.

22. MATOPIBA is an economic region that covers the complex of *cerrado* regions in the states of Maranhão, Tocantins, Piauí, and Bahia (North and North-east regions). Its territorial extension is 73,173,485 hectares, in 337 municipalities.

23. The slogan belongs to TV Globo, a part of the media conglomerate *Grupo Globo de Comunicação,* and has been used in national campaigns supporting agribusiness.

24. R. Nixon, "Slow Violence: Literary and Postcolonial Studies Have Ignored the Environmentalism that Often only the Poor Can See." *The Chronicle of Higher Education.* June 26, 2011, https://www.chronicle.com/article/Slow-Violence/127968.

25. Ingold (2000: 3) explains that "Perception is not the achievement of a mind in a body, but of the organism as a whole in its environment, and is tantamount to the organism's own exploratory movement through the world. If mind is anywhere, then it is not 'inside the head' rather than 'out there' in the world." At a later time Ingold clarifies that finding a way on a landscape is less a question of reading a map than of memory: "To perceive the landscape is to carry out an act of remembrance, and remembering is not so much a matter of calling up an internal image, stored in the mind, as of engaging perceptually with an environment that is itself pregnant with the past" (p. 189).

26. The prescriptive symmetrical anthropology of Latour proposes to articulate a vision of the western world and its intricate game between intellectual production and modern thought in order to reconsider the dichotomy of nature and culture. His analysis deciphers the modern axiom through the theoretical agents of nature and culture. For Latour, the term modern brings in its essence two asymmetries, one in relation to time and one in relation to others. The modern belief in marching in the direction of progress, breaking up the passage of time through revolutions and establishing specific differences between yesterday, today, and tomorrow. Besides which, they also establish a confrontation between themselves, the old and others. For the author, both asymmetries originate from a single asymmetry: nature and culture. Being modern, explains Latour, implies two sets of practices which, to be effective, should remain distinct. In the separation process of both practices: "translation"—mixtures between hybrids of nature and culture—or "purification"—two distinct anthological zones: that of humans and that of nonhumans—resides the greatest contradiction of the modern. In other words, when hybrids are taken into consideration, they are placed before mixtures of nonhumans and humans, but in the work of "purification" they oppose in two distinct categories, nature and culture. Therefore, for Latour, the modern, in order to comprehend himself, needs to take the work of these intermediaries or hybrids more seriously. In other words, it is necessary to provide a statute to the hybrids and realize how nature is intrinsically connected to the construction of society. As such, we would stop historicizing only man and we would include the history of natural things, taking into consideration both

humans and nonhumans and the way they are interconnected in the construction of the western collective, as in other collectives. Lastly, Latour advocates that nature and culture cannot be seen as explanatory terms, but as that which touts a joint explanation. His symmetrical anthropology establishes the principal that both nature and society need to be explained, and that this explanation, according to Latour, comes from the center, from the quasi-objects, or rather, from the hybrids of humans and nonhumans.

27. Maguari is one of twenty-five communities belonging to the Tapajós National Forest (Flona), a Federal Conservation Unit located on the banks of the Tapajós River. The Tapajós Flona has an area of 549,066 hectares, with more than 160 kilometers of beaches. In the entire Flona area, there is a wide diversity of landscapes: rivers, lakes, wetlands, dry land, hills, plateaus, forest, fields, açaí plantations, etc. It has around 5,000 traditional inhabitants, between all the communities and populations with their own cultural traditions.

BIBLIOGRAPHY

Albuquerque Junior, D. M. 1999. *A Invenção do Nordeste e Outras Artes*. Fundação Joaquim Nabuco/Editora Massangana: Recife.

Atiles-Osoria, J. M. 2013. Slow Violence and the Environmentalism of the Poor by Rob Nixon. *Revista Crítica de Ciências Sociais*, 100, 235–40.

Baggio, E. T. 2009. O Cinema Verdade de Jean Rouch no Filme *Di Cavalcanti di Glauber. Revista Científica/FAP, Curitiba*, 4(2), 166–79.

Benjamin, W. 1973. *Illuminations: Essays and Reflections*. Trans. H. Zohn. Fontana: London.

Benjamin, W. 2009. *One-way Street and Other Writings*. Penguin: London.

Bennett, J. 2010. *Vibrant Matter: A Political Ecology of Things*. Duke University Press: Durham.

Berger, J. 1977. *Ways of Seeing*. Penguin: London.

Bernucci, L. 2017. *Paraíso Suspeito: A Voragem Amazônica*. Trans. G. G. Souza. Edusp: São Paulo.

Bosi, A. 2002. *Literatura e Resistência*. Companhia das Letras: São Paulo.

Bourdieu, P. 2012. *Bosquejo de una Teoría de la Práctica*. Prometeo Libros: Buenos Aires.

Carvalho, J. F. 2019. O Valor da Amazônia. *Folha de São Paulo*, January 4, 2019. Available at: http://www.ilumina.org.br/o-valor-da-amazonia-artigo-folha-de-sp/.

Costa, S. V. 2017. Beyond Fordlândia: Ai de Ti, Amazônia! *Jornal da Ciência*, October 23, 2017. Available at: http://www.jornaldaciencia.org.br/edicoes/?url=http://jcnoticias.jornaldaciencia.org.br/28-beyond-fordlandia-ai-de-ti-amazonia/.

Cunha, E. 1909. *À Margem da História*. Livraria Chardron de Lello & Irmão: Porto.

Crutzen, P. J. and Stoermer, E. F. 2000. The "Anthropocene." *Global Change Newsletter*, 41, 17–18.

DeLoughrey, E. and Handley, G. B. 2011. *Postcolonial Ecologies: Literatures of the Environment*. University Press: Oxford.

Faulkner, W. 1951. *Requiem for a Nun*. Random House: New York.
Foucault, M. 1988. *The History of Sexuality*. Vintage Books: New York.
Gadamer, H. -G. 1999. *Verdade e Método*. Trans. F.P. Meurer. Vozes: Petrópolis.
Gadelha, R. M. A. F. 2002. Conquista e Ocupação da Amazônia: A Fronteira Norte do Brasil. *Estudos Avançados*, 45, 63–80.
Gomes de Matos, G. C. 2015. *Ethos e Figurações na Hinterlândia Amazônica*. Valer: Manaus.
Hardman, F. F. 1988. *Trem Fantasma: A Modernidade na Selva*. Campanhia das Letras: São Paulo.
Hardman, F. F. 2009. *A Vingança da Hiléia: Euclides da Cunha, a Amazônia e a Literatura Moderna*. Unesp: São Paulo.
Hecht, S. B. 2019. Beyond Fordlândia: An Environmental Account of Henry Ford's Adventure in the Amazon. Directed by Marcos Colón. *Environmental History*, 24(2), 376–39.
Ingold, T. 2000. *The Perception of the Environment: Essays on Livelihood*. Routledge: New York.
Latour, B. 1993. *We Have Never Been Modern*. Harvard University Press: Cambridge.
Latour, B. 2005. *Reassembling the Social: An Introduction to Actor-Network-Theory*. Oxford University Press: Oxford and New York.
Laurance, W. F., Cochrane, M. A., Bergen, S., Fearnside, P. M., Delamonica, P., Barber C., D'Angelo, S., and Fernandes, T. 2001. The Future of the Brazilian Amazon. *Science*, 291, 438–49.
Leopold, A. 2001. *A Sand County Almanac*. Oxford University Press: New York.
Moraes, P. 2011. *Os Intérpretes da Amazônia*. Valer/Governo do Estado do Amazonas: Manaus.
Morton, T. 2010. *The Ecological Thought*. Harvard University Press: Cambridge.
Nixon, R. 2011. *Slow Violence and the Environmentalism of the Poor*. Harvard University Press: Cambridge.
Nunes, B. 1998. *Amazônia: O Olhar Sem Fronteiras*. Fundação Nacional de Arte: Rio de Janeiro.
Pacheco de Oliveira, J. 2016. *O Nascimento do Brasil e Outros Ensaios*. Contracapa: Rio de Janeiro.
Paes Loureiro, J. J. 2019. Cultura Amazônica: Uma Diversidade Diversa. *Amazônia Latitude Digital Magazine*. April 10, 2019. Available at: https://amazonialatitude.com/2019/04/10/cultura-amazonica-uma-diversidade-diversa.
Rangel, A. 2001. *Inferno Verde: Cenas e Cenários do Amazonas*. Valer: Manaus.
Renan, E. 2018. *What Is a Nation? and Other Political Writings*. Columbia University Press: New York.
Skidmore, T. E. 2014. Mapping the Amazon. *Luso-Brazilian Review*, 51(2), 200–2.
Tocantins, L. 1978. *Euclides da Cunha e o Paraíso Perdido*. Civilização Brasileira: Rio de Janeiro.
Viveiros de Castro, E. B. 2010. *Metafisicas Canibales: Líneas de Antropologia Postestructural*. Katz: Buenos Aires.

Index

About the Editors and Contributors

ABOUT THE EDITORS

Antonio Augusto Rossotto Ioris is a reader in human geography and director of the graduate program on Environment and Development at the School of Geography and Planning, Cardiff University. Trained as a political geographer and agricultural engineer, Ioris is an international scholar with an extensive research portfolio of research projects and publications. Previous responsibilities included agricultural production in Brazil, project management at the Brazilian Ministry of the Environment, Senior Policy Officer at the Scottish Environment Protection Agency, and academic positions at the universities of Aberdeen and Edinburgh. His main areas of interest are the relationship between environment and development, natural resources management, indigenous studies, and controversies related to agriculture modernization. His most relevant books include *Tropical Wetland Management: The South-American Pantanal and the International Experience* (Ashgate, 2012), *The Political Ecology of the State: The Basis and the Evolution of Environmental Statehood* (Routledge, 2012), *Agriculture, Environment and Development: International Perspectives on Water, Land and Politics* (Palgrave Macmillan, 2016), and *Agribusiness and the Neoliberal Food System in Brazil: Frontiers and Fissures of Agro-neoliberalism* (Routledge, 2017). Ioris is the coordinator of the Agrocultures International Network, which provided the intellectual space for the publication of this book.

Rafael R. Ioris is an associate professor of Latin American History at the University of Denver. He is an affiliated faculty at the Latin American Center at the Josef Korbel School of International Studies. He holds a PhD in Latin American History from Emory University, and MA in International Studies from the University of Brasilia, and a BA in Social Sciences from the Federal University of Rio Grande do Sul, in Brazil. He is the author *Transforming Brazil: A History of National Development in the Cold War Era* (Routledge, 2014) and numerous articles published in different international journals on topics ranging from the history of development ideologies in the twentieth century, the history of labor activism in modern Brazil, to Brazilian foreign policy at the turn of the twenty-first century. He is currently working on the expansion of the economic frontiers in the Amazon region in Brazil, while also developing a new research on the history of US-Brazil relations in the 1960s and 1970s. He is a frequent author of op-ed pieces on Brazilian newspapers, magazines, and blogs dealing with current political and cultural issues in Brazil, Latin American, and the United States.

Sergei V. Shubin is a professor in human geography at Swansea University. His research interests are in poverty, mobility, and inequality. He is the author of several peer-reviewed publications on poverty in Europe, South America, and Asia. He is the member of the Newton-Bhabha Global Research Partnership in Aquaculture with the focus on Poverty Alleviation. In the last five years, he has been involved in several projects, including Horizon2020 EU award for research on Perceptions of Europe, UK government (ESRC) award for research on social security of migrants in Scotland, Newton Fund award for poverty alleviation in Bangladesh and India, and UK government (ESPA) funding for inter-disciplinary research for poverty alleviation in the Amazon. Shubin was the coordinator of the Newton-Fund sponsored workshop on "Supporting Sustainable Ecosystems for Poverty Alleviation in the Amazon," which brought together in 2016 in Manaus a team of inter-disciplinary Latin American and UK researchers and helped to identify the key discussion points for this book.

ABOUT THE CONTRIBUTORS

Gustavo S. Azenha is the executive director of Columbia University's Institute of Latin American Studies (ILAS). He runs the daily operations of the Institute and its affiliated centers and programs, serving as the chief budget and human resources officer, and overseeing programming, outreach, and development. He also serves as the director of Graduate Studies for the MSc program in Latin American and Caribbean Studies, as well as the director of the Lemann Center for Brazilian Studies. Azenha holds a PhD from Cornell University, where his training and research focused on environmental anthropology and political ecology. He specializes in development and socio-environmental politics in Brazil, with his primary research efforts directed toward examining the historical and contemporary interrelations between environmental, indigenous, and rural development policies. More broadly, his interests and expertise are concerned with the politics of inequality and with the role of civil society and social movements in public policy dynamics. His research and professional experiences have included engagement with the politics of sustainable development, public health, and education and technology.

Fabio de Castro is a political ecologist (PhD Indiana University, USA) lecturer at the Centre for Latin American Research and Documentation at the University of Amsterdam. His research interest is related to natural resource governance in Brazil, with particular focus on how discourses, interactions, and practices are shaped and transformed in forest-rich areas. His research projects address local sustainable production systems such as agroecology, agroforestry, and small-scale fisheries, environmental and agrarian policies, socio-environmental movements and environmental justice. His recent research program includes co-production of local solutions in agroecological systems, commoning initiatives in Latin America, and rural-urban connections in socio-environmentalism. Fabio has published over fifty articles and book chapters. Among his publications are the edited volume *Environmental Governance in Latin America*, the monograph *Fishing Accords: The Political Ecology of Fishing Intensification in the Amazon*, and the articles "Multi-scale Environmental Citizenship: Traditional Populations and Protected Areas in Brazil" and "Environmental Policies in the Lula Era: Accomplishments and Contradictions."

Marcos Colón received his PhD in Spanish and Portuguese Cultural Studies from the University of Wisconsin-Madison in 2019. His research focuses on Brazilian literary and cultural studies, with a particular emphasis on representations of the Amazon in twentieth- and twenty-first-century Brazilian literature and film. He is currently working on a book-project based on his experiences filming the Amazon. He has produced and directed two documentary films that represent diverse perspectives on humanity's complex relations with the natural world: "Beyond Fordlândia: An Environmental Account of Henry Ford's Adventure in the Amazon" (2018) and "Zo'é" (2018), based on his experiences with the Zo'é tribe, an Amazonian indigenous community that has had little to no contact with the outside world. He is particularly interested in examining a variety of perspectives on the post-rubber era in the Amazon. Colón's scholarship uses the post-rubber era as a springboard for re-envisioning the region in a "relational" way, challenging hegemonic representations of the tropics in literature and culture. He is the editor and creator of Amazonia Latitude, a digital environmental magazine.

Frederico Freitas is an assistant professor of Latin American and Digital History at North Carolina State University where he is also a core faculty at the Visual Narrative Initiative. He holds a PhD (2016) and an MA (2011) in history from Stanford University, and a BA (2008) in history from University of São Paulo. In 2018, he co-edited with Jake Blanc *Big Water: The Making of the Borderlands Between Brazil, Argentina, and Paraguay* (Arizona). He is currently working on a book manuscript titled "Boundaries of Nature: National Parks and Environmental Change at the Argentine-Brazilian Border, 1890–1990."

Maria Fernanda Gebara is an independent post-doctoral researcher working in the areas of climate change, instruments for forest conservation, reducing emissions from deforestation and degradation (REDD+), land-use and environmental governance, participatory forest management, and indigenous cultures. In the last eight years she has worked as a researcher, lecturer, and consultant in the leading centers of environmental research and governance in Brazil, Indonesia, and United Kingdom, and has vast first-hand experience of working with indigenous tribes in Brazilian Amazon. She has a number of publications in top-ranked journals in development, environmental, and educational studies. She has been a visiting researcher at the Oxford Centre for Tropical Forests, associate researcher at the Department for International Development at the London School of Economics, and consultant for various institutions such as the Nature Conservancy (TNC), the Inter-American Development Bank, the Brazilian Ministry of Environment and the Center for International Forestry Research (CIFOR).

Lynn Holland received her PhD in political science from the University of California at Los Angeles (UCLA) and has taught courses in international political economy and USA–Latin American relations at the University of Denver's Josef Korbel School of International Studies (JKSIS) for fifteen years. She has written on the adverse effects of metal mining and rise of environmental activism in Mexico and Peru, the politics of deportation in the United States, relations between drug trafficking organizations and the state in Honduras, and the origins and transnational impacts of the war on drugs. She was also a senior research associate for the Council on Hemispheric Relations (COHA) involved in editing "The Washington Report on the Hemisphere" for several years. She coordinates an annual trip to the Mexico-Guatemala border area where students, faculty, and human rights advocates are able to learn about the impacts of metal mining on the environment and local residents. She is a past recipient of the University of Denver's Ruth Murray Underhill Teaching Award.

Paul E. Little is an anthropologist and specialist in Amazonian studies, with a research focus on the ecological relations of forest peoples within a context of rapid land-use change. He is the author of "Amazonia: Territorial Disputes on Perennial Frontiers" (Johns Hopkins University Press, 2001), and the organizer and editor of "Os Novos Desafios da Política Ambiental Brasileira" (IEB Mil Folhas, 2014) and "Conhecimentos Tradicionais para o Século XXI: Etnografias da Intercientificidade" (AnnaBlume, 2010); and author of numerous peer-reviewed articles on Amazonia. He currently works as an independent international consultant in policy analysis and program evaluation on socio-environmental issues.

João Pacheco de Oliveira is a professor of anthropology at the Museu Nacional's (National Museum) post-graduation program. He is the author or editor of fifteen books and has published over 130 articles in scientific journals or book chapters. He is a researcher "1A" (top rank) of the National Research Council (CNPq). Oliveira has done field research in the Amazon with the Ticuna indigenous people and also in the northeast of the country. His most recent book, "O Nascimento do Brasil: Pacificação, Regiime Tutelar e Formação de Alteridades" (The Birth of Brazil. Pacification, Tutelary Regime, and Formation of Alterities), won the ANPOCS award for best book in social sciences published in 2017. It was translated into Spanish ("Exterminio y Tutela: Processes of Formation of Alteridades in Brazil"), available since July 2019 and is also being translated into French. Oliveira served as Visiting Professor at several Graduate Programs in Brazil (UNICAMP/SP, UFPE/PE, UFBA/BA, Joaquim Nabuco Foundation/PE, and UFAM/AM), and abroad (Universidad Nacional de La Plata/Argentina, Università di Roma La Sapienza/Italy, École des Hautes Études en Sciences So-ciales/France, Universidad Nacional de San Martin/Argentina, and Institute des Hautes Études de l`Amérique Latine/Paris-Sorbonne/France). He was president of the Brazilian Association of Anthropology (ABA) and coordinates the initiative "Os Brasis e Suas Memórias" ("The Brazis and Their Memories"), dedicated to the collection of indigenous biographies.

Aaron Schneider is a Leo Block Chaired Professor of International Studies at the Josef Korbel School of International Studies at the University of Denver. He has published over seventy peer-reviewed chapters and journal articles, and two singly authored manuscripts, "State-building and Tax Regimes in Central America" and "Renew Orleans? Globalized development and Working Class Resistance." He is currently engaged in a comparative research project on boom and bust development cycles in emerging powers, with a focus on Brazil and India.

Tatiana Schor has a degree in economics, M.Sc in Human Geography and Ph.D in Environmental Science from the University of São Paulo. She received the award Cathedra Ruth Cardoso, Lehman Center Brazil—Columbia University, Fulbright-CAPES Fellowship (2016–2017). She is professor at the Geography Department and researcher at the Program in Environmental Sciences and Sustainability in the Amazon of the Center for Environmental Sciences, both of the Federal University of Amazonas (UFAM). She coordinates the Study and Research Program of the Urban Network of the Solimões-Amazonas River by the Center of Studies and Research of the Cities in the Amazon (NEPECAB). In 2012, she coordinated the State Center for Conservation Units (CEUC) of the State Secretariat for the Environment and Sustainable Development (SDS) of Amazonas, and served as the Deputy Secretary for Planning of the Secretariat of Rural Production of the State of Amazonas (2013–2014). She is also President of the Observatory of the Metropolitan Region of Manaus (2017–2019) and Representative of the Brazilian Society for the Progress of Science in Amazonas (2018). She currently holds the position of Executive Secretary of Science, Technology and Innovation in the Secretariat of Economic Development, Planning, Science, Technology and Innovation (SEDECTI) in the state of Amazonas, Brazil.

www.ingramcontent.com/pod-product-compliance
Ingram Content Group UK Ltd.
Pitfield, Milton Keynes, MK11 3LW, UK
UKHW020838290926
15470UKWH00004B/1430

9 781498 594714